Advances in Mechanical Systems Dynamics

Advances in Mechanical Systems Dynamics

Special Issue Editors

Alberto Doria
Giovanni Boschetti
Matteo Massaro

MDPI • Basel • Beijing • Wuhan • Barcelona • Belgrade

Special Issue Editors

Alberto Doria
University of Padova
Italy

Giovanni Boschetti
University of Padova
Italy

Matteo Massaro
University of Padova
Italy

Editorial Office
MDPI
St. Alban-Anlage 66
4052 Basel, Switzerland

This is a reprint of articles from the Special Issue published online in the open access journal *Applied Sciences* (ISSN 2076-3417) from 2019 to 2020 (available at: https://www.mdpi.com/journal/applsci/special_issues/Mechanical_Systems_Dynamics).

For citation purposes, cite each article independently as indicated on the article page online and as indicated below:

LastName, A.A.; LastName, B.B.; LastName, C.C. Article Title. *Journal Name* **Year**, *Article Number*, Page Range.

ISBN 978-3-03928-188-6 (Pbk)
ISBN 978-3-03928-189-3 (PDF)

© 2020 by the authors. Articles in this book are Open Access and distributed under the Creative Commons Attribution (CC BY) license, which allows users to download, copy and build upon published articles, as long as the author and publisher are properly credited, which ensures maximum dissemination and a wider impact of our publications.

The book as a whole is distributed by MDPI under the terms and conditions of the Creative Commons license CC BY-NC-ND.

Contents

About the Special Issue Editors . . . . **vii**

Alberto Doria, Giovanni Boschetti and Matteo Massaro
Advances in Mechanical Systems Dynamics
Reprinted from: *Appl. Sci.* **2020**, *10*, 61, doi:10.3390/app10010061 1

Xuanqi Zeng, Songyuan Zhang, Hongji Zhang, Xu Li, Haitao Zhou and Yili Fu
Leg Trajectory Planning for Quadruped Robots with High-Speed Trot Gait
Reprinted from: *Appl. Sci.* **2019**, *9*, 1508, doi:10.3390/app9071508 6

Lorenzo Scalera, Ilaria Palomba, Erich Wehrle, Alessandro Gasparetto and Renato Vidoni
Natural Motion for Energy Saving in Robotic and Mechatronic Systems
Reprinted from: *Appl. Sci.* **2019**, *9*, 3516, doi:10.3390/app9173516 27

Tetsunori Haraguchi, Ichiro Kageyama and Tetsuya Kaneko
Study of Personal Mobility Vehicle (PMV) with Active Inward Tilting Mechanism on Obstacle
Avoidance and Energy Efficiency
Reprinted from: *Appl. Sci.* **2019**, *9*, 4737, doi:10.3390/app9224737 53

Sharad Singhania, Ichiro Kageyama and Venkata M Karanam
Study on Low-Speed Stability of a Motorcycle
Reprinted from: *Appl. Sci.* **2019**, *9*, 2278, doi:10.3390/app9112278 78

Xiao Ling, Jianfeng Tao, Bingchu Li, Chengjin Qin and Chengliang Liu
A Multi-Physics Modeling-Based Vibration Prediction Method for Switched Reluctance Motors
Reprinted from: *Appl. Sci.* **2019**, *9*, 4544, doi:10.3390/app9214544 93

Shangwen He, Wenzhen Jia, Zhaorui Yang, Bingbing He and Jun Zhao
Dynamics of a Turbine Blade with an Under-Platform Damper Considering the Bladed
Disc's Rotation
Reprinted from: *Appl. Sci.* **2019**, *9*, 4181, doi:10.3390/app9194181 109

Galibjon M. Sharipov, Dimitrios S. Paraforos and Hans W. Griepentrog
Validating the Model of a No-Till Coulter Assembly Equipped with a Magnetorheological
Damping System
Reprinted from: *Appl. Sci.* **2019**, *9*, 3969, doi:10.3390/app9193969 125

Jianqiang Zhang, Yongkai Liu, Shijie Gao and Chengshan Han
Control Technology of Ground-Based Laser Communication Servo Turntable via a Novel
Digital Sliding Mode Controller
Reprinted from: *Appl. Sci.* **2019**, *9*, 4051, doi:10.3390/app9194051 139

Mingming Zhang and Anping Hou
Numerical Investigation on Unsteady Separation Flow Control in an Axial Compressor Using
Detached-Eddy Simulation
Reprinted from: *Appl. Sci.* **2019**, *9*, 3298, doi:10.3390/app9163298 158

Enrico Pipitone, Christian Maria Firrone and Stefano Zucca
Application of Multiple-Scales Method for the Dynamic Modelling of a Gear Coupling
Reprinted from: *Appl. Sci.* **2019**, *9*, 1225, doi:10.3390/app9061225 172

Jiadui Chen, Bo Wang, Dan Liu and Kai Yang
Study on the Dynamic Characteristics of a Hydraulic Continuous Variable Compression Ratio System
Reprinted from: *Appl. Sci.* **2019**, *9*, 4484, doi:10.3390/app9214484 . . . . **197**

Zhengzheng Zhu, Yunwen Feng, Cheng Lu and Chengwei Fei
Efficient Driving Plan and Validation of Aircraft NLG Emergency Extension System via Mixture of Reliability Models and Test Bench
Reprinted from: *Appl. Sci.* **2019**, *9*, 3578, doi:10.3390/app9173578 . . . . **211**

About the Special Issue Editors

Alberto Doria is currently Associate Professor of Mechanisms and Machine Science with the Department of Industrial Engineering of the University of Padova, where he teaches Mechanical Vibrations and Applied Mechanics. His research interests are in vibrations, vehicle dynamics, and vibrations energy harvesting. He was among the organizers of ASME 2019 AVT Conference and of Bicycle and Motorcycle Dynamics Conference 2019. He is currently involved in a number of projects related to vibration control and harvesting. He is a member of the Editorial Board of *Applied Sciences*, Section Mechanical Engineering.

Giovanni Boschetti is currently Associate Professor of Mechanisms and Machine Science with the Department of Management and Engineering of the University of Padova, where he teaches Industrial Robotics and Mechanics of Machines. He is President of the Degree Course in Mechatronics Engineering. His main research interests are in Serial and Parallel Industrial Robots, Cable Direct Driven Robots, and Collaborative Robotics. He is an active member of the IFToMM ITALY Association; he was the chief organizer of their first conference and member of the organizing committee of the subsequent ones. He is currently Associate Editor of the *International Journal of Mechanics and Control*.

Matteo Massaro is currently Associate Professor of Applied Mechanics with the Department of Industrial Engineering of the University of Padova, where he teaches Vehicle Dynamics, Applied Mechanics, and Modelling and Simulation of Mechanical Systems. His research interests are in vehicle dynamics and control, driver–vehicle interactions, multibody systems, and driving simulators. He was among the organizers of the Bicycle and Motorcycle Dynamics Conference 2019. He is currently involved in a number of projects related to minimum time problems of race vehicles, experimental characterization of tyres, modelling of race tracks, and assessment of the performance of airbag systems for two-wheeled vehicles. He recently published the book *"Dynamics and Optimal Control of Road Vehicles"* with Oxford University Press.

Editorial

Advances in Mechanical Systems Dynamics

Alberto Doria [1],*, Giovanni Boschetti [2] and Matteo Massaro [1]

1 Department of Industrial Engineering, University of Padova, 35131 Padova, Italy; matteo.massaro@unipd.it
2 Department of Management and Engineering, University of Padova, 36100 Vicenza, Italy; giovanni.boschetti@unipd.it
* Correspondence: alberto.doria@unipd.it; Tel.: +39-049-827-6803

Received: 12 December 2019; Accepted: 15 December 2019; Published: 20 December 2019

1. Introduction

Modern dynamics was established many centuries ago by Galileo and Newton before the beginning of the industrial era. Presently, we are in the presence of the fourth industrial revolution, and mechanical systems are increasingly integrated with electronic, electrical, and fluidic systems. This trend is present not only in the industrial environment, which will soon be characterized by the cyber-physical systems of industry 4.0 [1,2], but also in other environments like mobility, health and bio-engineering, food and natural resources, safety, and sustainable living. In this context, purely mechanical systems with a quasi-static behavior will become less common and the state-of-the-art will soon be represented by integrated mechanical systems, which need accurate dynamic models to predict their behavior. Therefore, mechanical systems dynamics is going to play an increasingly central role. Significant research efforts are needed to improve the identification of the mechanical properties of systems in order to develop models which take non-linearity into account, and to develop efficient simulation tools. This Special Issue aims at disseminating the latest research achievements, findings, and ideas in mechanical systems dynamics, with particular emphasis on the applications which are strongly integrated with other systems and require a multi-physical approach.

2. Advances in Mechanical Systems Dynamics

The papers collected in this Special Issue can be grouped into some topical areas of dynamics, as follows: Trajectory and motion planning, dynamic stability, vibration control and damping, control, modelling, and simulation. Most of them deal with multi-physical systems applications, including robotics, turbomachinery, vehicles, agricultural, and industrial machinery.

2.1. Trajectory and Motion Planning

Trajectory and motion planning are increasingly relevant in robotics and in other mechanical systems [3,4]. Indeed, the goal of achieving ever higher speeds is extending into all fields of mechanics. In order to preserve accuracy and repeatability, proper strategies should be adopted in order to generate trajectories that could be executed at high speed, while avoiding excessive motor accelerations and mechanical structure vibrations.

In Reference [5], a single leg platform for quadruped robots is designed in order to achieve high-speed locomotion. For this purpose, the foot-end trajectory for quadruped robots with a high-speed trot gait is proposed. The gait trajectory is planned for swing and stance phases. These phases are separately designed with position control and impedance control, while guaranteeing continuous and smooth transitions. Such an approach allows avoiding great rigid impact and achieving stable walking or running.

In Reference [6], a classification and a discussion of several approaches that adopt the concept of natural motion to enhance the energetic performance in robotic and mechatronic systems is presented.

In the first part of the paper, the physical requirements that a system has to fulfill in order to exploit the natural motion are identified. While in the second part, the approaches related to natural motion are classified by trajectory types, as follows: Given trajectory, optimized trajectory, free-vibration response, and periodic trajectory learning. In the end, the methods which are able to reduce energy consumption while preserving task flexibility are highlighted.

2.2. Dynamic Stability

Dynamic stability is a classic topic of dynamics that, presently, has important applications in many fields of engineering, including manned and unmanned aircraft [7,8], ground vehicles [9,10], and walking robots [11,12]. In recent years, there have been important research developments in the field of light vehicles for urban mobility, for example, electric scooters, Segways, electrical bicycles, three-wheeled vehicles, and motorcycles [13]. In the Special Issue there are two papers which address vehicle stability.

In the first paper [14], a three-wheeled vehicle with double front wheels and single rear wheel and an active tilting mechanism is studied. A comprehensive analysis including stability, obstacle avoidance, and energy management is carried out considering the effect of both mechanical and control parameters. Results show that the developed vehicle has good handling and stability properties and is more efficient than a standard car.

The second paper [15] deals with the low speed stability of a scooter-type motorcycle. This problem is closely related to urban mobility, since congested traffic conditions limit vehicle speed, generating stability problems that require the continuous effort of the rider to stabilize the vehicle. A theoretical model is developed and validated by means of road tests. The validated model is able to predict regions of low speed stability and will be used for developing a controller.

2.3. Vibration Control and Damping

In recent years the interest in vibration has increased, owing to the rapid development of vibration energy harvesting technologies [16]. However, vibrations are also a potential problem for any application that includes moving components [17]. Thus, control and damping of the dynamic response is a very relevant topic. In the Special Issue there are three papers which deal with with vibration control and damping in three very different fields, as follows: Electric motors, turbines, and agricultural machines.

In Reference [18], the problem of prediction of vibrations in switched reluctance motors (SRMs) is tackled with a multi-physics approach. The comparison between the numerical and experimental data shows that the method is accurate. Therefore, it can be applied to the structural and control design optimization of SRMs.

In Reference [19], the vibrations of turbine blades are considered, while removing one of the assumptions often employed, i.e., the bladed disc's rotation. The work contributes to a better understanding of the dynamics contributions to be considered when designing under-platform dampers.

In Reference [20], the dynamic behavior of a No-Till Coulter Assembly is analyzed, with a focus on the effect of magneto-rheological (MR) dampers, aimed at giving a consistent seeding depth. The comparison between the simulated and measured vertical dynamics shows good agreement with the numerical model developed. Therefore, the model can be used for the optimization of the MR dampers.

2.4. Control

Presently, more and more mechanical systems are being controlled, with the aim of adjusting their dynamics, e.g., pantograph/catenary [21], suspension bridges [22], gas turbines [23], motorcycles [24], etc. In this issue two challenging scenarios are considered. The first is related to laser communication, while the second is related to the control of flow separation in axial compressors. In the first paper [25], the design of a sliding mode controller to solve the nonlinear disturbance problem of a ground-based laser communication turntable is discussed—indeed, the alignment of the platform is a key issue in this field. Experimental results on the pitch closed-loop behavior show a better performance of the proposed (chatter-free) controller when compared to the traditional proportional, integrative, derivative (PID) and existing sliding mode.

In the second paper [26], the flow separation in axial compressors is controlled by the unsteady (pressure) excitation, not only at the shredding vortex frequency (traditional method), but also at other frequencies, demonstrating the impact on the structure of shredding vortices.

2.5. Modelling and Simulation

Mechanisms, gears, and transmissions are still key elements of advanced industrial systems [27]. To improve the performance of the system, detailed models of machine elements, taking into account non-linearities or time-variant properties, are needed [28,29]. This Special Issue includes three papers that cover the modeling and simulation of machine elements. The first paper [30] investigates the dynamics of thin walled gears and takes into account time-variant properties due to gear meshing. The method of multiple scales [31] is adopted to solve the equations of motion in the frequency domain. This method requires shorter calculation times than direct time integration methods. The results presented in this paper are important for aeronautical applications.

The second paper [32] covers a variable compression ratio engine and presents a non-linear model which includes both mechanical and hydraulic equations, similar to the models adopted for studying vehicle suspensions and shock absorbers [33]. Results show that the proposed system can achieve a continuous variation in the compression ratio of an engine, with advantages in terms of efficiency and pollution.

The third paper [34] addresses the problem of the emergency extension of nose landing gear. An interesting combination of mechanism analysis methods and statistical methods for reliability analysis is presented. The most important failure factors of an existing mechanism for emergency extension are highlighted and a more reliable mechanism is designed.

3. Final Remarks

In summary, this Special Issue contains a series of interesting research works focused on advances in mechanical systems dynamics, covering a wide area of applications. This collection shows the actuality of this topic and sheds light on future developments.

Author Contributions: Conceptualization, A.D., G.B., M.M.; writing—original draft preparation, A.D., G.B., M.M.; writing—review and editing, A.D., G.B., M.M.; supervision, A.D. All authors have read and agreed to the published version of the manuscript.

Funding: This research received no external funding.

Conflicts of Interest: The authors declare no conflict of interest.

References

1. Kagermann, H.; Wahlster, W.; Helbig, J. *Final Report of the Industrie 4.0 Working Group, Securing the Future of German Manufacturing Industry Recommendations for Implementing the Strategic Initiative INDUSTRIE 4.0*; National Academy of Science and Engineering: Frankfurt, Germany, 2013.
2. Nolting, L.; Priesmann, J.; Kockel, C.; Rödler, G.; Brauweiler, T.; Hauer, I.; Robinius, M.; Praktiknjo, A. Generating Transparency in the Worldwide Use of the Terminology Industry 4.0. *Appl. Sci.* **2019**, *9*, 4659. [CrossRef]
3. Boschetti, G.; Trevisani, A. Cable robot performance evaluation by Wrench exertion capability. *Robotics* **2018**, *7*, 15. [CrossRef]
4. Carbone, G.; Gomez-Bravo, F. *Motion and Operation Planning of Robotic Systems*; Springer: Heidelberg, Germany, 2015.
5. Zeng, X.; Zhang, S.; Zhang, H.; Li, X.; Zhou, H.; FuLeg, Y. Trajectory Planning for Quadruped Robots with High-Speed Trot Gait. *Appl. Sci.* **2019**, *9*, 1508. [CrossRef]

6. Scalera, L.; Palomba, I.; Wehrle, E.; Gasparetto, A.; Vidoni, R. Natural Motion for Energy Saving in Robotic and Mechatronic Systems. *Appl. Sci.* **2019**, *9*, 3516. [CrossRef]
7. Pounds, E.I.P.; Bersak, D.R.; Dollar, A.M. Stability of small-scale UAV helicopters and quadrotors with added payload mass under PID control. *Auton. Robot* **2012**, *33*, 129–142. [CrossRef]
8. Sheng, S.; Sun, C. Design of a Stability Augmentation System for an Unmanned Helicopter Based on Adaptive Control Techniques. *Appl. Sci.* **2015**, *5*, 575–586. [CrossRef]
9. Bulsink, V.; Doria, A.; Van De Belt, D.; Koopman, B. The effect of tire and rider properties on the stability of a bicycle. *Adv. Mech. Eng.* **2015**, *7*, 1–19. [CrossRef]
10. Cossalter, V.; Doria, A.; Formentini, M.; Peretto, M. Experimental and numerical analysis of the influence of tyre's properties on the straight running stability of a sport touring motorcycle. *Veh. Syst. Dyn.* **2012**, *50*, 357–375. [CrossRef]
11. Aoi, S.; Tsuchiya, K. Generation of bipedal walking through interactions among the robot dynamics, the oscillator dynamics, and the environment: Stability characteristics of a five-link planar biped robot. *Auton. Robot.* **2011**, *30*, 123–141. [CrossRef]
12. Figliolini, G.; Ceccarelli, M. Walking programming for an electropneumatic biped robot. *Mechatronics* **1999**, *9*, 941–964. [CrossRef]
13. Limebeer, D.J.N.; Massaro, M. *Dynamics and Optimal Control of Road Vehicles*; Oxford University Press: Oxford, UK, 2018.
14. Haraguchi, T.; Kageyama, I.; Kaneko, T. Study of Personal Mobility Vehicle (PMV) with Active Inward Tilting Mechanism on Obstacle Avoidance and Energy Efficiency. *Appl. Sci.* **2019**, *9*, 4737. [CrossRef]
15. Singhania, S.; Kageyama, I.; Karanam, V.M. Study on Low-Speed Stability of a Motorcycle. *Appl. Sci.* **2019**, *9*, 2278. [CrossRef]
16. Tian, W.; Ling, Z.; Yu, W.; Shi, J. A Review of MEMS Scale Piezoelectric Energy Harvester. *Appl. Sci.* **2018**, *8*, 645. [CrossRef]
17. Boschetti, G.; Caracciolo, R.; Richiedei, D.; Trevisani, A. A Non-Time Based Controller for Load Swing Damping and Path-Tracking in Robotic Cranes. *J. Intell. Robot. Syst. Theory Appl.* **2014**, *76*, 201–217. [CrossRef]
18. Ling, X.; Tao, J.; Li, B.; Qin, C.; Liu, C. A Multi-Physics Modeling-Based Vibration Prediction Method for Switched Reluctance Motors. *Appl. Sci.* **2019**, *9*, 4544. [CrossRef]
19. He, S.; Jia, W.; Yang, Z.; He, B.; Zhao, J. Dynamics of a Turbine Blade with an Under-Platform Damper Considering the Bladed Disc's Rotation. *Appl. Sci.* **2019**, *9*, 4181. [CrossRef]
20. Sharipov, G.M.; Paraforos, D.S.; Griepentrog, H.W. Validating the Model of a No-Till Coulter Assembly Equipped with a Magnetorheological Damping System. *Appl. Sci.* **2019**, *9*, 3969. [CrossRef]
21. Poetsch, G.; Evans, J.; Meisinger, R.; Kortüm, W.; Baldauf, W.; Veitl, A.; Wallaschek, J. Pantograph/catenary dynamics and control. *Veh. Syst. Dyn.* **1997**, *28*, 159–195. [CrossRef]
22. Bakis, K.N.; Massaro, M.; Williams, M.S.; Limebeer, D.J.N. Aeroelastic control of long-span suspension bridges with controllable winglets. *Struct. Control Health Monit.* **2016**, *23*, 1417–1441. [CrossRef]
23. Richards, G.A.; Straub, D.L.; Robey, E.H. Passive Control of Combustion Dynamics in Stationary Gas Turbines. *J. Propul. Power* **2003**, 19. [CrossRef]
24. Savino, G.; Lot, R.; Massaro, M.; Rizzi, M.; Symeonidis, I.; Will, S.; Brown, J. Active safety systems for powered two-wheelers: A systematic review. *Traffic Inj. Prev.* **2020**, in press.
25. Zhang, J.; Liu, Y.; Gao, S.; Han, C. Control Technology of Ground-Based Laser Communication Servo Turntable via a Novel Digital Sliding Mode Controller. *Appl. Sci.* **2019**, *9*, 4051. [CrossRef]
26. Zhang, M.; Hou, A. Numerical Investigation on Unsteady Separation Flow Control in an Axial Compressor Using Detached-Eddy Simulation. *Appl. Sci.* **2019**, *9*, 3298. [CrossRef]
27. Barbazza, L.; Faccio Oscari, F.; Rosati, G. Agility in assembly systems: A comparison model. *Assem. Autom.* **2017**, *37*, 411–421. [CrossRef]
28. Palomba, I.; Richiedei, D.; Trevisani, A. Energy-Based Optimal Ranking of the Interior Modes for Reduced-Order Models under Periodic Excitation. *Shock Vib.* **2015**, art. no. 348106. [CrossRef]
29. Belotti, R.; Caracciolo, R.; Palomba, I.; Richiedei, D.; Trevisani, A. An Updating Method for Finite Element Models of Flexible-Link Mechanisms Based on an Equivalent Rigid-Link System. *Shock Vib.* **2018**, art. no. 1797506. [CrossRef]

30. Pipitone, E.; Firrone, C.M.; Zucca, S. Application of Multiple-Scales Method for the Dynamic Modelling of a Gear Coupling. *Appl. Sci.* **2019**, *9*, 1225. [CrossRef]
31. Huang, J.; Zhang, A.; Sun, H.; Shi, S.; Li, H.; Wen, B. Bifurcation and Stability Analyses on Stick-Slip Vibrations of Deep Hole Drilling with State-Dependent Delay. *Appl. Sci.* **2018**, *8*, 758. [CrossRef]
32. Chen, J.; Wang, B.; Liu, D.; Yang, K. Study on the Dynamic Characteristics of a Hydraulic Continuous Variable Compression Ratio System. *Appl. Sci.* **2019**, *9*, 4484. [CrossRef]
33. Cossalter, V.; Doria, A.; Pegoraro, R.; Trombetta, L. On the non-linear behaviour of motorcycle shock absorbers. *Proc Inst. Mech Eng Part D J. Automob. Eng.* **2010**, *224*, 15–27. [CrossRef]
34. Zhu, Z.; Feng, Y.; Lu, C.; Fei, C. Efficient Driving Plan and Validation of Aircraft NLG Emergency Extension System via Mixture of Reliability Models and Test Bench. *Appl. Sci.* **2019**, *9*, 3578. [CrossRef]

© 2019 by the authors. Licensee MDPI, Basel, Switzerland. This article is an open access article distributed under the terms and conditions of the Creative Commons Attribution (CC BY) license (http://creativecommons.org/licenses/by/4.0/).

Article

Leg Trajectory Planning for Quadruped Robots with High-Speed Trot Gait

Xuanqi Zeng, Songyuan Zhang *, Hongji Zhang, Xu Li, Haitao Zhou and Yili Fu

State Key Laboratory of Robotics and System, Harbin Institute of Technology, Harbin 150001, China; 18s008064@stu.hit.edu.cn (X.Z.); 18s008067@stu.hit.edu.cn (H.Z.); hitlx@hit.edu.cn (X.L.); htzhouhit@hit.edu.cn (H.Z.); meylfu@hit.edu.cn (Y.F.)
* Correspondence: zhangsy@hit.edu.cn; Tel.: +86-0451-8640-3679

Received: 14 March 2019; Accepted: 9 April 2019; Published: 11 April 2019

Abstract: In this paper, a single leg platform for quadruped robots is designed based on the motivation of high-speed locomotion. The leg is designed for lightweight and low inertia with a structure of three joints by imitating quadruped animals. Because high acceleration and extensive loadings will be involved on the legs during the high-speed locomotion, the trade-off between the leg mass and strength is specifically designed and evaluated with the finite element analysis. Moreover, quadruped animals usually increase stride frequency and decrease contact time as the locomotion speed increases, while maintaining the swing duration during trot gait. Inspired by this phenomenon, the foot-end trajectory for quadruped robots with a high-speed trot gait is proposed. The gait trajectory is planned for swing and stance phase; thus the robot can keep its stability with adjustable trajectories while following a specific gait pattern. Especially for the swing phase, the proposed trajectory can minimize the maximum acceleration of legs and ensure the continuity of position, speed, and acceleration. Then, based on the kinematics analysis, the proposed trajectory is compared with the trajectory of Bézier curve for the power consumption. Finally, a simulation with Webots software is carried out for verifying the motion stability with two trajectory planning schemes respectively. Moreover, a motion capture device is used for evaluating the tracking accuracy of two schemes for obtaining an optimal gait trajectory suitable for high-speed trot gait.

Keywords: quadruped robots; high-speed locomotion; leg trajectory planning; trot gait; motion capture sensor

1. Introduction

Legged robots are more suitable for applications with rough terrain and complex cluttered environments compared to wheeled robots [1–3]. With respect to the leg length, quadruped robots can freely select contact points while making contact with the environment. Therefore, it is promising that they can be used for rescuing people in forests and mountains, climbing stairs, to carry payloads in construction sites, and so on [4]. Recently, several quadruped robots are being developed which are equipped with hydraulic actuators or electric actuators. For example, Boston Dynamics developed a series of robots equipped with hydraulic actuators with high energy density. Hydraulic drive quadruped robots, such as Big Dog [5] and HyQ [6], make use of the characteristics of the high power density of hydraulic pressure to realize a stronger carrying capacity and motion ability. However, their large noise and size have limited their applications to outdoor applications. Moreover, they require additional hydraulic source and oil circuit, which often makes the structure more complex. Different from that, quadruped robots equipped with electric actuators can even be used to indoor environments, such as SpotMini [7], ANYmal [8], and MIT Cheetah [9–11]. Amongst them, MIT Cheetah realized a fast, efficient design with advanced proprioceptive actuator design [9]. ANYmal applies a bioinspired actuator design, making it robust against impact, while allowing accurate torque measurement at the joints. However, the complicated actuator design also increases cost and compromises the power output of the robot [4]. SpotMini did not publish their detailed researches; however, public media has released its capabilities, such as climbing stairs and stabilizing several gaits [7].

On the other hand, high-speed locomotion involves high acceleration and extensive loadings on the legs, which brings a trade-off between weight and strength of legs [12]. There are generally two main ways to apply the leg motors for quadruped robots: putting the motor at each joint and concentrating the motor at the shoulder. Quadruped robots, such as ANYmal, adopt the method of putting motors at each joint to achieve a fully actuated motion with high compliant series elastic actuation, suitable for highly dynamic motions. Although by using this method the quadruped robots are easier to control and can walk steadily, the swing speed of their legs is slow, which leads to low walking speed. Some quadruped robots, such as SpotMini and MIT Cheetah, adopt the method of putting motors at the shoulder. This strategy can make the leg mass and inertia of the swinging parts of the legs very low and achieve a high running locomotion. Moreover, as for the leg structure of quadruped robots, the main leg structure is a two-joint structure, which is used in most quadruped robots, such as ANYmal, Wildcat [13], SpotMini, etc. This structure is simple, intuitive, and easy to control. However, from a biological point of view, toed animals, such as tigers, leopards, lions, and wolves, have a three-joint structure in their legs, which imparts great advantages in running speed and energy efficiency. MIT Cheetah adopted the three-part structure, which achieved a 6 m/s running speed, and the ability to cross obstacles with high energy efficiency [14]. Also, Cheetah-cub designed with pantograph leg configuration to simplify the control of three links with only two joints and reached a running trot with short flight phases [15]. Besides, a real "cat-sized" robot called Pneupard also explored this same pantograph method [16]. The method of putting motors at the shoulder has been proved to be effective on the high running locomotion. However, there is still a lack of analysis with infinitesimal kinematics for a favorable leg structure of quadruped robots with high-speed locomotion.

Except for the suitable leg design, the stable gait trajectory is the basis of movement stability for quadruped robots. Quadruped animals can achieve various gait patterns such as walk, trot, pronk, bound, half-bound, rotary gallop, transverse gallop, etc. [17]. Among them, the trot gait is the most widely used gait, which is very simple and effective. The duration of leg swing is always constant for both trot and gallop gait, although quadruped animals adjust their stride frequency while trotting, and adjust stride length for gallop gait [17]. Also, the leg trajectory planning is important for robots' stability with adjustable trajectories. Saputra et al. designed a Bézier curve based passive neural control method applied in bioinspired locomotion for decreasing the computational cost [18]. Lee et al. designed a trajectory with a Nonuniform Basis Spline (NUBS) curve to effectively overcome obstacles [19]. Hyun et al. used properties of the Bézier curve for desirable swing-leg dynamics [11]. However, suitable gait trajectories considering energy efficiency are seldom researched.

Therefore, in this paper, the motivation is to design an electrically actuated leg platform for quadruped robots with high-speed locomotion. The method of putting motors at the shoulder is adopted for reducing the leg inertia during the swing. In particular, the suitable leg structure and high-efficiency gait trajectory are analyzed and verified on the leg platform. The novel designed leg trajectory combines cosine curve, quintic curve, and hexagonal curve to minimize the maximum acceleration of legs and ensure the continuity of position, speed, and acceleration during high-speed trot gait. The efficiency is evaluated by measuring the endpoint acceleration, stability, and power consumption compared with the Bézier curve [9]. Furthermore, an optotrak sensor is used for measuring the actual foot-end trajectory of the two schemes. The sensor can track and analyze kinetics and dynamic motion in real-time, which is a more direct and precise method to measure the actual endpoint position, especially when the motor encoder has low precision.

The remainder of this paper is organized as follows. Section 2 introduces the detailed design of the leg and the finite element analysis of the leg for the trade-off between the leg weight and strength. Moreover, the inverse kinematic analysis of the single leg gait and two different gait trajectory planning schemes are given. Section 3 gives the simulation result of two schemes in trajectory planning with Webots and experimental results measured with the optotrak sensor. Section 4 will discuss the simulation and experimental results. At last, the conclusion and direction of future works are given.

2. Materials and Methods

2.1. Leg Design of the Quadruped Robot

2.1.1. Leg Structure Comparison with Kinematics Analysis

The two-joint and three-joint articulated leg structures were considered during the initial leg structure design. By comparing the differences on the aspect of geometry and kinematics, the manipulability, obstacle avoiding ability, and occupied space for legs were analyzed.

Figure 1 shows the schematic for the three-joint leg structure on the left and two-joint leg structure on the right. It is assumed that the total link lengths for both two leg structures are identical and decided by the height of robots. The manipulability which measures at state θ with respect to manipulation vector r is defined as Equation (1) [20]:

$$w = \sqrt{\det J(\theta) J^T(\theta)} \qquad (1)$$

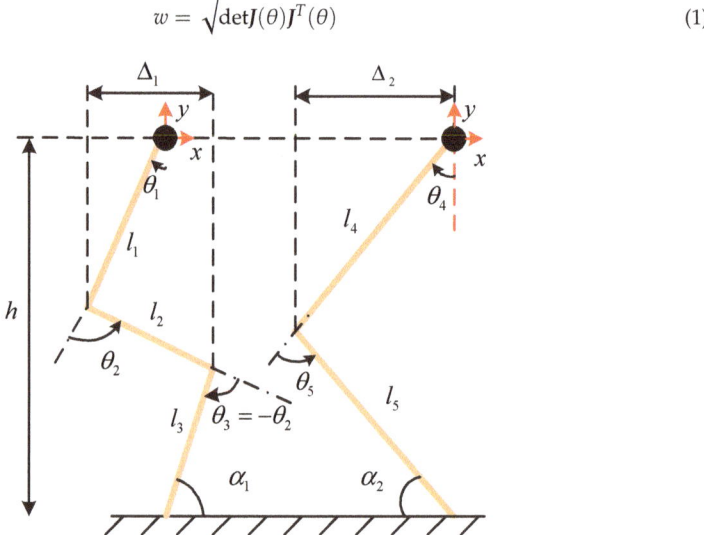

Figure 1. Schematic for the three-joint and two-joint leg structures.

For the three-joint leg structure, the Jacobian matrix can be calculated as Equation (2):

$$J(\theta_1, \theta_2) = \begin{bmatrix} (l_3 + l_1)\cos\theta_1 + l_2\cos(\theta_1 + \theta_2) & l_2\cos(\theta_1 + \theta_2) \\ (l_3 + l_1)\sin\theta_1 + l_2\sin(\theta_1 + \theta_2) & l_2\sin(\theta_1 + \theta_2) \end{bmatrix} \qquad (2)$$

Therefore, the manipulability can be calculated with $w = |\det J(\theta_1, \theta_2)| = (l_1 + l_3)l_2|\sin(\theta_2)|$. Similar, the Jacobian matrix for the two-part leg structure can be calculated as Equation (3):

$$J(\theta_4, \theta_5) = \begin{bmatrix} l_4\cos\theta_4 + l_5\cos(\theta_4 + \theta_5) & l_5\cos(\theta_4 + \theta_5) \\ l_4\sin\theta_4 + l_5\sin(\theta_4 + \theta_5) & l_5\sin(\theta_4 + \theta_5) \end{bmatrix} \qquad (3)$$

The manipulability can be calculated with $w = |\det J(\theta_4, \theta_5)| = l_4 l_5 |\sin(\theta_5)|$. Here the lengths of l_1, l_2, and l_3 are given in the Table 1. For the two-joint leg structure, the manipulability can get a maximum value when the l_4 equates to l_5. For comparing the manipulability of two leg structures during one gait cycle, the Bézier curve for the swing phase and the cosine function for the stance phase were used. The manipulability during one gait cycle is shown in Figure 2.

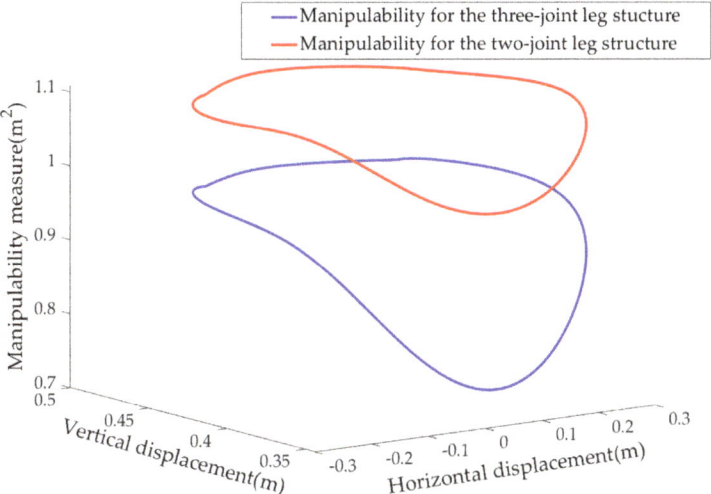

Figure 2. Manipulability for the three-joint and two-joint leg structure.

The leg structure should also protract the leg with enough ground clearance to avoid obstacles. Therefore, we compared the angles α_1 and α_2, as shown in Figure 1, which can be calculated with Equations (4) and (5):

$$\alpha_1 = 90° - \arccos\left(\frac{(l_1+l_3)^2 + l_2^2 + h^2}{2l_2(l_1+l_3)}\right) \quad (4)$$

$$\alpha_2 = 90° - \arccos\left(\frac{l_5^2 + h^2 - l_4^2}{2l_5 h}\right) \quad (5)$$

where h represents the height between the shoulder part of leg to the ground. The obstacle avoiding ability for the three-joint and two-joint leg structures is shown in Figure 3.

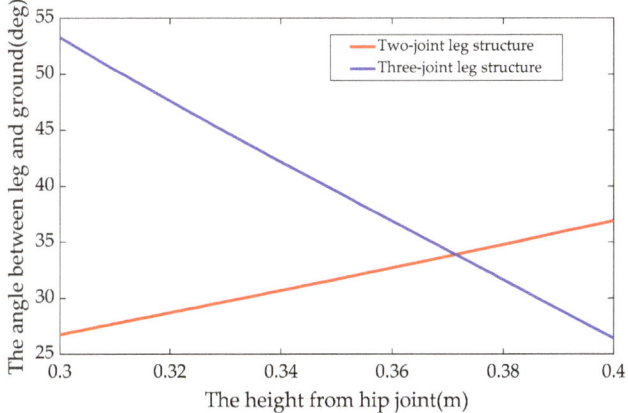

Figure 3. Obstacle avoiding ability of the three-joint and two-joint leg structures.

At last, the space occupied for two leg structures was also compared. Assuming that $p_1 = (l_4 + l_5 + h)/2$ and $p_2 = (l_1 + l_2 + l_3 + h)/2$, we can get

$$\Delta_1 = \frac{2\sqrt{p_1(p_1 - l_4)(p_1 - l_5)(p_1 - h)}}{h} \quad (6)$$

$$\Delta_2 = \frac{2\sqrt{p_2(p_2 - l_1 - l_3)(p_2 - l_2)(p_2 - h)}}{h} \quad (7)$$

Setting $h \in (0.3, 0.4)$ and considering the height of the quadruped, the result can be obtained as shown in Figure 4.

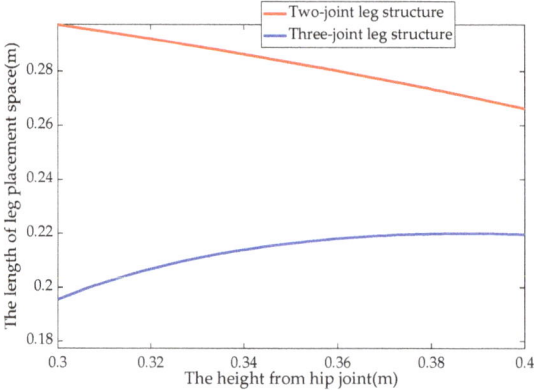

Figure 4. Occupied space after retracting the leg for the three-joint and two-joint leg structures.

The leg configuration between the three-joint and two-joint structures can be more clearly found in Figure 5. From these kinematics analyses, it can be found that, when h is selected as 0.343 in our leg design, the three-part leg structure has enough ground clearance to avoid obstacles. The occupied space after retracting the leg for the three-joint leg structure is much smaller than that of the two-joint leg structure when h is set as 0.343 m; the manipulability is similar between the two leg structures. By combining the biological point of view and these kinematics analyses, the three-joint leg structure was determined.

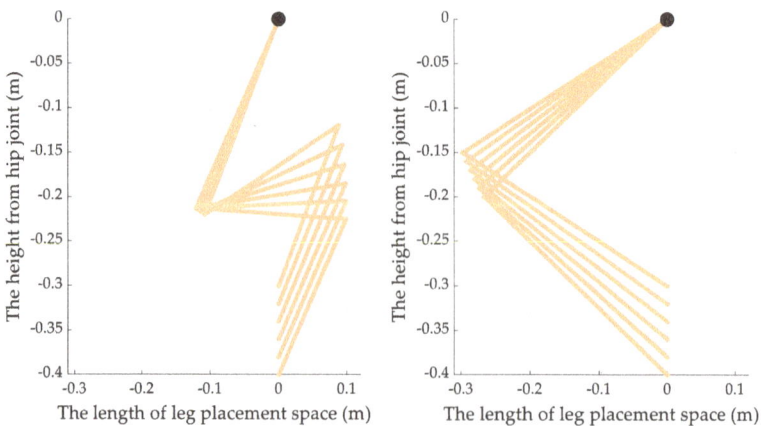

Figure 5. Leg configuration comparisons between the three-joint and two-joint leg structures.

2.1.2. Detailed Leg Design with Three-joint Leg Structure

The leg was designed for high-speed locomotion of quadrupeds and inspired by the design of MIT Cheetah [9]. The physical leg parameters are shown in Table 1. The total mass of the leg is ~5.5 kg, where the swing part (1.284 kg) only occupies 23% of total mass. Moreover, the center of leg mass is only 84 mm away from the rotation center of the shoulder module, which helps reduce leg inertia during the swing. Shoulder and knee actuators with planetary gears are coaxially located at the shoulder module part for achieving a coupled motion of two joints, as shown in Figure 6, thus that the mass and leg inertia can be decreased. In detail, the knee actuator drives the knee joint through a parallel linkage, while the shoulder actuator directly drives the shoulder joint. The leg consists of the thigh, calf, and foot, where a parallel linkage connects the calf and foot. Therefore, two parallel linkages can drive the motion of the foot and thigh in parallel without an extra actuator on foot.

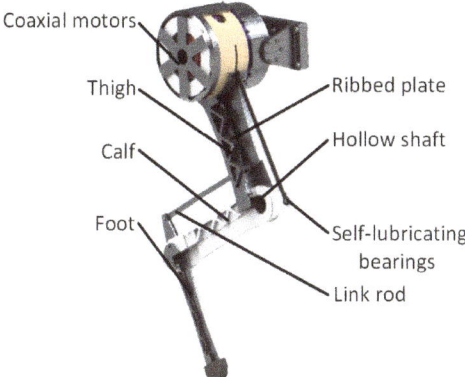

Figure 6. Structural design of one leg of the quadruped.

Table 1. Physical leg parameters.

Parameter	Symbol	Value	Units
Leg Mass	m	5.500	kg
Thigh Mass	m_1	0.670	kg
Thigh Length	l_1	0.245	m
Calf Mass	m_2	0.446	kg
Calf Length	l_2	0.220	m
Foot Mass	m_3	0.168	kg
Foot Length	l_3	0.201	m

The detailed design of the shoulder module is shown in Figure 7. There are two identical motors, including a knee motor and a shoulder motor in the shoulder module. The shoulder motor drives the knee motor part and thigh part to rotate through the hollow motor shaft and a planetary gear reducer. The shoulder module is designed as a thin-walled structure and is very compact in order to reduce its mass. The hollow motor shafts and gears are made of T4 titanium alloy to ensure their strength. On the contrary, the elements bearing less stress are made of 7075 aluminum alloy to ensure their light weight.

Actuators for quadruped robots with high-speed locomotion should provide high torque density which manages the dynamic physical interactions well. The proprioceptive actuator paradigm achieves a combination of high-bandwidth force control, high torque density, as well as impact mitigation [9]. For example, NABi-V2 utilizes six-back drivable high-power density electromagnetic actuators with a low gear ratio single-phase planetary gearbox to realize proprioceptive, force-controlled dynamic locomotion [12]. Similarly, a single-stage planetary gear reduction (main material is titanium alloy, and weight is 0.113 kg) with low gear ratio (6:1) combined with high torque density frameless motors was applied during our leg design.

Moreover, during the high-speed locomotion, when the leg touched the ground, both linkages were pulled by two opposite forces, and neither bending moment nor torque is applied on the linkages. Therefore, these two linkages can be very thin. Hollow shafts are used in the joints of the leg, and ribbed plates are applied on the legs, which also decrease the total mass. In particular, the joints of linkages are so small that standard bearings cannot be used, so wear washers and dry bushing were adopted, which can effectively reduce the friction of linkages' joint by the self-lubrication characteristics of the dry bushing.

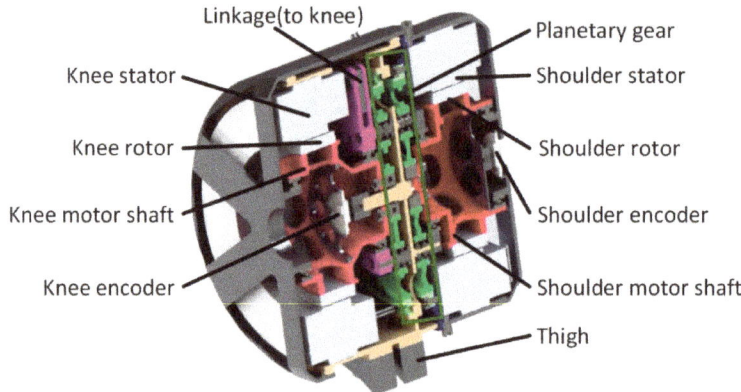

Figure 7. Detailed design of the shoulder module.

2.2. Finite Element Analysis (FEA) of the Leg

As mentioned above, there is a trade-off between the total leg weight and its strength. Therefore, in order to reduce the total weight and inertia of the leg without losing too much strength, the leg structure is mainly connected by ribbed plates with only 4-mm-thin plates in the middle, and the inclination angle of ribbed plates is 45 degrees to facilitate better performance of bearing pressure and bending moment. What is more, many thin-walled structures and lightweight materials were used in both motor and leg structure, which is likely to lead to insufficient strength and stiffness. Therefore, it is necessary to confirm the strength of the leg by Finite Element Analysis (FEA).

During the locomotion of a quadruped robot, the leg movement mainly consists of two parts: a stance phase and a swing phase. The stance phase is a moment when the legs are stressed, and the weight of the robot body is supported by only two legs in the middle of the stance phase during trot and bound gaits. Considering that the quadruped robot needs to complete tasks like jumping over obstacles, which may generate an impact 2 to 3 times the weight of body [13], it is estimated that the total weight of the designed quadruped robot is ~40 kg, so the maximum impact distributed to each leg is !600 N. At the same time, the legs need to give the body a forward motion force, which was set to 1/10 of the impact (60 N). These two forces are applied to the endpoint of the leg model imported into Adams as load forces, and the forces generated in each part of the leg can be measured as shown in Figure 8.

From the force analysis result, we know that each joint of the leg bears a large force under the ultimate load at the endpoint of the leg. In order to increase accuracy during the strength and stiffness checking, the forces obtained by the force analysis in Adams was further used as the load input for the deformation and stress analyses of each part by FEA with ANSYS software. The results are shown in Figure 9.

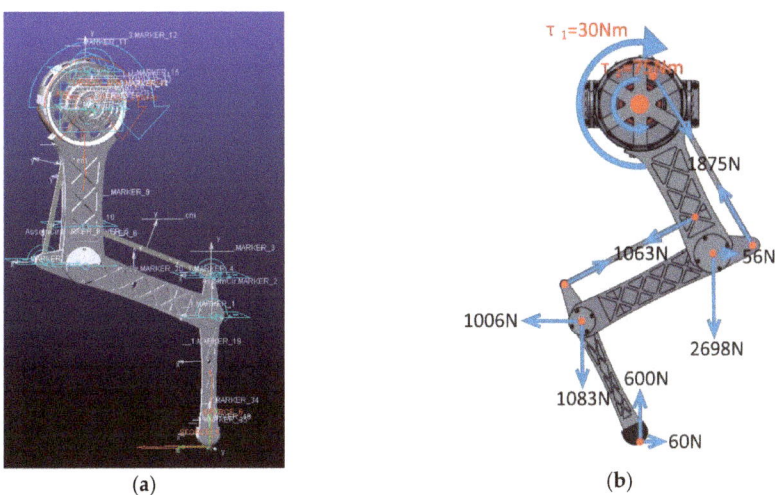

Figure 8. Force calculation by Adams. (**a**) Leg model imported in Adams. (**b**) Forces measured in each part of the leg.

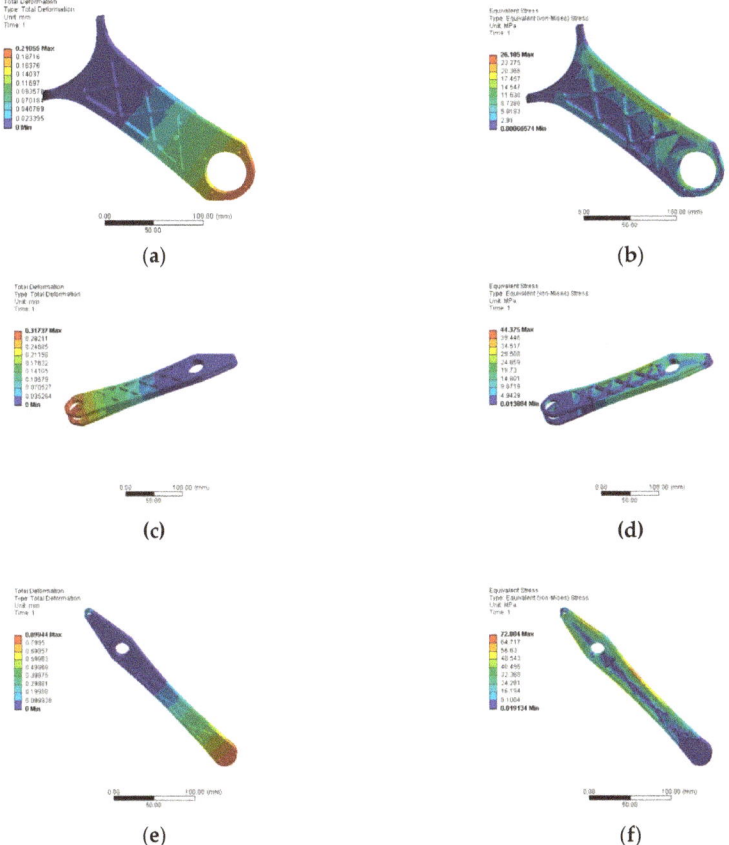

Figure 9. *Cont.*

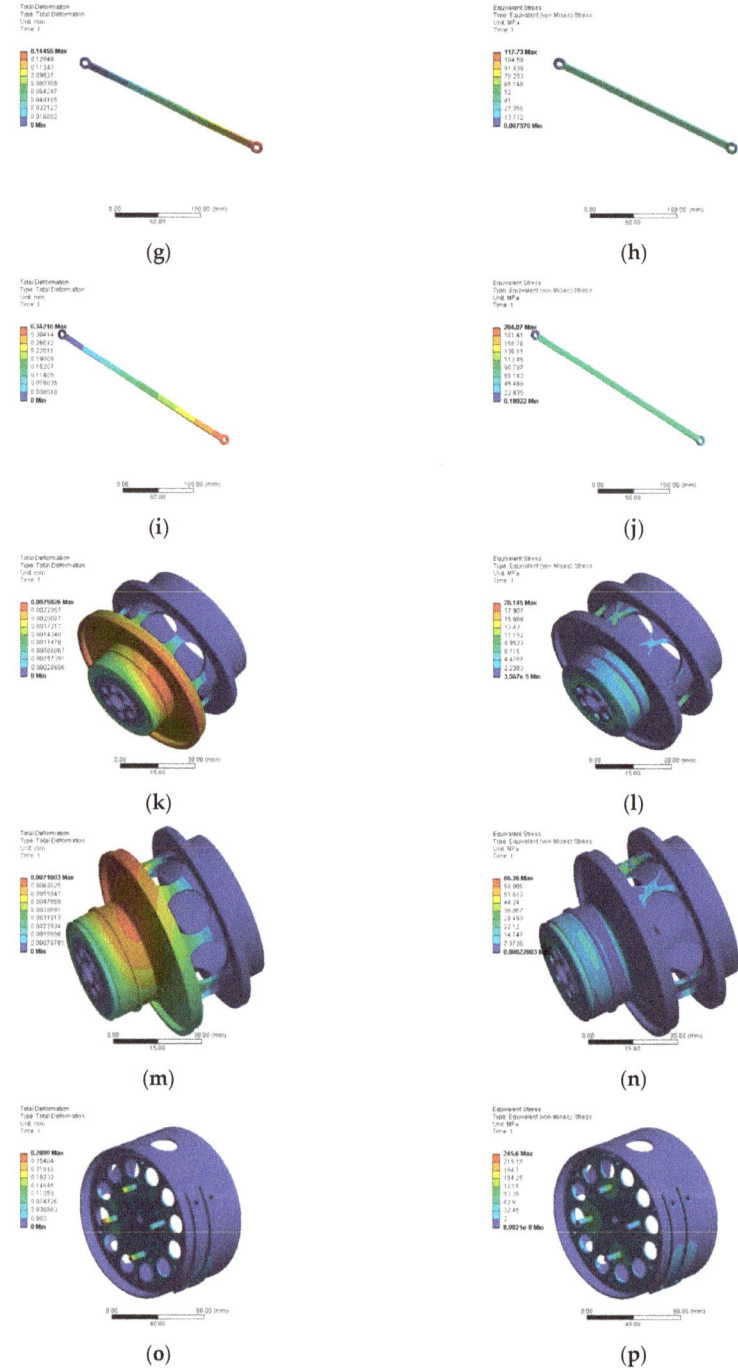

Figure 9. Finite Element Analysis (FEA): (**a**) The total deformation of the thigh; (**b**) the equivalent stress of the thigh; (**c**) the total deformation of the calf; (**d**) the equivalent stress of the calf; (**e**) the total deformation of the foot; (**f**) the equivalent stress of the foot; (**g**) the total deformation of the linkage

connecting the knee motor and knee joint; (h) the equivalent stress of the linkage connecting the knee motor and knee joint; (i) the total deformation of the linkage connecting the thigh and foot; (j) the equivalent stress of the linkage connecting the thigh and foot; (k) the total deformation of the shoulder motor shaft; (l) the equivalent stress of the shoulder motor shaft; (m) the total deformation of the knee motor shaft; (n) the equivalent stress of the knee motor shaft; (o) the total deformation of the component connecting the thigh and the shoulder motor; and (p) the equivalent stress of the component connecting the thigh and the shoulder motor.

It can be found that the maximum stress on each part of the leg is 245.6 Mpa, that is, less than 455 Mpa, which is the material strength of the 7075 aluminum alloy, and the structural strength of the leg meets the requirements with a high safety factor of 1.85. Also, there is a certain amount of deformation as a result of using a thin-walled structure and light aluminum alloy material resulting in lower stiffness brought about by the relatively lower weight of the leg. Moreover, from the view of natural animal running, the contact between their legs and the ground is a flexible contact rather than a rigid contact, which can decrease the peak force between feet and the ground and make the legs bear larger force. Therefore, the structure of the leg having a certain amount of deformation during the movement can be beneficial for buffering.

2.3. Inverse Kinematic Analysis

In order to accomplish the foot-point trajectory control of the leg, it is necessary to derive the inverse kinematics formula for the leg. The leg model is shown in Figure 10, where q_1 represents the rotation angle of the shoulder part. Because the knee motor rotates by the shoulder motor, the actual rotation angle of the knee part will be $(q_1 + q_2)$.

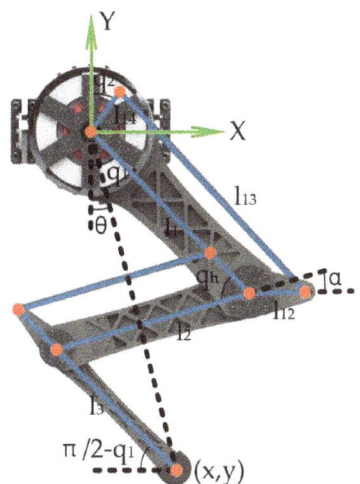

Figure 10. Inverse kinematics model of the leg.

The final expression formulas are shown as Equations (8)–(10):

$$q_1 = \arccos\left(\frac{l_1^2 + 2l_1 l_3 + l_3^2 - l_2^2 - x^2 - y^2}{2l_2 \sqrt{x^2 + y^2}}\right) - \arccos\left(\frac{l_1^2 + 2l_1 l_3 + l_2^2 + l_3^2 - x^2 - y^2}{2l_2(l_1 + l_3)}\right) - \arctan\left(\frac{x}{y}\right) \quad (8)$$

$$q_h = \arccos\left(\frac{l_1^2 + 2l_1 l_3 + l_2^2 + l_3^2 - x^2 - y^2}{2l_2(l_1 + l_3)}\right) \quad (9)$$

$$q_2 = \pi - q_1 - \arccos\frac{l_1 + l_{12}\cos(\alpha - qh)}{\sqrt{l_1^2 + 2l_1 l_{12}\cos(\alpha - qh) + l_{12}^2}} - \arccos\frac{l_1^2 + 2l_1 l_{12}\cos(\alpha - qh) + l_{12}^2 + l_{14}^2 - l_{13}^2}{2l_{14}\sqrt{l_1^2 + 2l_1 l_{12}\cos(\alpha - qh) + l_{12}^2}} \quad (10)$$

where, q_h is the angle between the calf and the thigh; θ is the angle of leg's endpoint with the opposite direction of Y; l_1, l_2, l_3, l_{12}, l_{13}, and l_{14} are the length of the linkages in the four-bar structure; α is the angle between the calf and the linkage l_{12}; and (x,y) is the coordinate of the endpoint of the leg.

2.4. Leg Trajectory Planning

2.4.1. Stance Phase Trajectory Design

The leg movement of a quadruped robot can be divided into two phases: the stance phase and the swing phase. Consider that high acceleration and extensive loadings will be exerted on the legs during the stance phase. If only the position control is used, it will easily bring a greater rigid impact to the quadruped robot's body, which does not benefit stable walking or running and will also cause damage to the structure. Therefore, the stance and swing phases are separately designed with position control and impedance control. When the leg contacts the ground, an impedance control method will be applied to ensure the flexible contact between the leg and ground, which can reduce the impact effectively [21,22].

In order to reduce the impact between the leg and ground when the leg leaves or touch the ground, the X-direction motion of the stance phase is designed as a uniform velocity. The formulas of the X-direction of the stance phase are shown as Equations (11) and (12):

$$S_{ST,x} = V_{desire} t_{ST} \quad (11)$$

$$T_{ST} = \frac{L}{V_{desire}} \quad (12)$$

where L is the leg stride in a gait cycle, T_{st} is the time of the stance phase, and V_{desire} is the desired speed of the robot.

Consider that the cosine function has a good performance in smoothness, so the cosine function, which was also used in other robots [19,23], was used in the trajectory of the stance phase. Considering that the leg will bear an impact and generate deformation when it touches the ground and the impedance control is used to resist impact in the stance phase, it needs a virtual displacement Δ to ensure the body stability of quadruped robot. The Y-direction formula of support phase is shown as Equation (13):

$$S_{ST,y} = -\Delta \times \cos(\pi(\frac{1}{2} - \frac{V_{desire} t_{ST}}{L})) - P_0 \quad (13)$$

2.4.2. Swing Phase Trajectory Design

The objective of the swing phase trajectory design is to protract the leg with enough ground clearance for avoiding obstacles and to have desirable swing leg retraction rate for reducing energy losses of running during touch down motion [24]. Consider that the leg is not affected by environmental forces so it can achieve higher speed and acceleration. Therefore, the period of the swing phase T_{sw} is defined as 0.25 s, calculated according to the maximum rotation speed of motors. Two schemes of the trajectory in the swing phase are proposed based on the spline curve and Bézier curve respectively.

Scheme I: Trajectory design with spline curve

The basic principle of this scheme is to minimize the maximum acceleration of legs as much as possible during the swing phase and ensure the continuity of position, speed, and acceleration. The planning trajectory is shown in Figure 11 with four increased velocities.

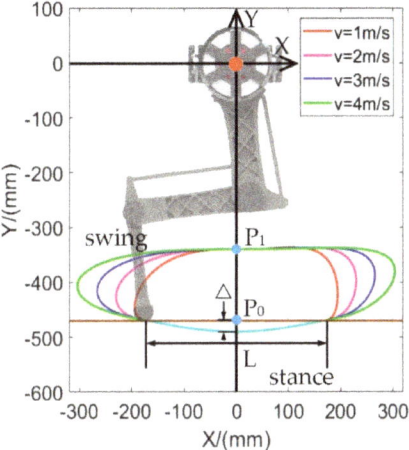

Figure 11. The spline curve trajectory.

In order to avoid the high order polynomial trajectory planning, the trajectory in the swing phase is divided into 0 to $0.5T_{SW}$ and $0.5T_{SW}$ to T_{SW} in the X direction. During 0 to $0.5T_{SW}$, for guaranteeing the touch-down and leave-off points between the swing phase and stance phase have better smoothness, it is necessary to ensure the position, speed, and acceleration continuously, which are given by Equations (14), (15), and (16):

$$S_{SW,x}|_{t_{SW}=0} = S_{ST,x}|_{t_{ST}=T_{ST}} \tag{14}$$

$$\frac{dS_{SW,x}}{dt_{SW}}\bigg|_{t_{SW}=0} = \frac{dS_{ST,x}}{dt_{ST}}\bigg|_{t_{ST}=T_{ST}} \tag{15}$$

$$\frac{d^2S_{SW,x}}{dt_{SW}^2}\bigg|_{t_{SW}=0} = \frac{d^2S_{ST,x}}{dt_{ST}^2}\bigg|_{t_{ST}=T_{ST}} \tag{16}$$

Similarly, for ensuring the acceleration changes slightly and smoothly in the X direction and limiting the trajectory length of the swing phase to be appropriate, the time reaching zero velocity, $T_{SW,V_{SW,x}=0}$, which is a variable parameter with V_{desire}, should be limited by Equation (17):

$$\frac{dS_{SW,x}}{dt_{SW}}\bigg|_{t_{SW}=T_{SW},V_{SW,x}=0} = 0 \tag{17}$$

In the X direction, the trajectory is in the acceleration stage before $0.5T_{SW}$, and then in the deceleration stage. At the time of $0.5T_{SW}$, in order to ensure that the acceleration is small and continuous between the two stages, the acceleration and position should be zero, which are given by Equations (18) and (19):

$$S_{SW,x}|_{t_{SW}=\frac{1}{2}T_{SW}} = 0 \tag{18}$$

$$\frac{d^2S_{SW,x}}{dt_{SW}^2}\bigg|_{t_{SW}=\frac{1}{2}T_{SW}} = 0 \tag{19}$$

From Equations (14) to (19) above, the five-order spline curves in the direction of X can be fitted from 0 to $0.5T_{SW}$, and the curves between $0.5T_{SW}$ and T_{SW} are obtained symmetrically.

The trajectory in the Y direction is also divided into two parts: 0 to $0.5T_{SW}$ and $0.5T_{SW}$ to T_{SW}. It is also necessary to ensure that the position, velocity, and acceleration are continuous at the moment connecting the stance phase and swing phase in the part of 0 to $0.5T_{SW}$, which are given by Equations (20)–(22):

$$S_{SW,y}|_{t_{SW}=0} = S_{ST,y}|_{t_{ST}=T_{ST}} \tag{20}$$

$$\frac{dS_{SW,y}}{dt_{SW}}\bigg|_{t_{SW}=0} = \frac{dS_{ST,y}}{dt_{ST}}\bigg|_{t_{ST}=T_{ST}} \tag{21}$$

$$\frac{d^2S_{SW,y}}{dt_{SW}^2}\bigg|_{t_{SW}=0} = \frac{d^2S_{ST,y}}{dt_{ST}^2}\bigg|_{t_{ST}=T_{ST}} \tag{22}$$

The highest position of the swing phase is limited at P_1, in which the velocity and acceleration are zero, which are given by Equations (23)–(25):

$$S_{SW,y}|_{t_{SW}=\frac{1}{2}T_{SW}} = P_1 \tag{23}$$

$$\frac{dS_{SW,y}}{dt_{SW}}\bigg|_{t_{SW}=\frac{1}{2}T_{SW}} = 0 \tag{24}$$

$$\frac{d^2S_{SW,y}}{dt_{SW}^2}\bigg|_{t_{SW}=\frac{1}{2}T_{SW}} = 0 \tag{25}$$

Therefore, the five-order spline curve in the Y direction during the period from 0 to $0.5T_{SW}$ can be fitted by Equations (20) to (25). For a smooth change between 0 to $0.5T_{SW}$ and $0.5T_{SW}$ to T_{SW}, one extra condition is added for ensuring the jerk continuously. Thus, a six-order spline curve can be fitted.

Scheme II: Bézier curve

In this scheme, the trajectory of the swing phase can be obtained by Bézier curve, which is a smooth curve controlled by several points. MIT Cheetahs also used this curve to obtain a smooth swing phase trajectory [9], and a similar method was used in a quadruped robot, called AiDIN-IV [19]. The formula is written as Equation (26):

$$B(t) = \sum_{i=0}^{n} \binom{n}{i} P_i (1-t)^{n-i} t^i \tag{26}$$

where P_i is the fitting point. The Bézier curve has several properties: (1) double coincidence fitting points determine a zero-velocity point; (2) triple coincidence fitting points determine a zero acceleration point; (3) and $V_{SW,i}|_{t_{SW}=0} = (n+1)(P_1 - P_0)/T_{sw}$, $V_{SW,i}|_{t_{SW}=T_{SW}} = (n+1)(P_n - P_{n-1})/T_{sw}$. According to these properties, a smooth curve can be obtained by twelve control points as shown in Table 2. The shape of the trajectory is shown in Figure 12.

Table 2. The 12-control point of Bézier curve.

P_n	X (mm)	Y (mm)
P_0	−170	−470
P_1	$-(170 + V_{desire}/((n+1)T_{sw}))$	−470
P_2	−300	−360
P_3	−300	−360
P_4	−300	−360
P_5	0	−360
P_6	0	−360
P_7	0	−320
P_8	300	−320
P_9	300	−320
P_{10}	$170 + V_{desire}/((n+1)T_{sw})$	−470
P_{11}	170	−470

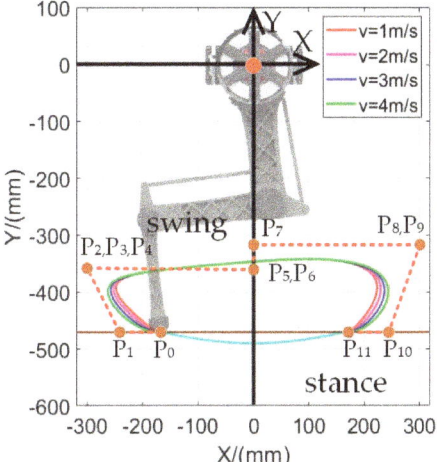

Figure 12. The Bézier curve trajectory.

3. Results

3.1. The Features of Two Schemes

Comparing the swing phase trajectory acceleration of spline curve with that of the Bézier curve as shown in Figure 13, a curve with continuous acceleration cannot be obtained. Moreover, the acceleration of the Bézier curve at a contact point with the stance phase cannot reach 0, which means that there will be an impact force on the ground, and the maximum value of its acceleration curve is also larger than that in the spline curve. As to the spline curve trajectory, it will be more difficult to obtain the trajectory. Although there are many fitting conditions, it can obtain continuous curves with the stance phase, velocity, and acceleration, because the impact on the ground will be relatively small. However, from the analysis of Figure 14, in both schemes, the rotation angle for the shoulder and knee motors are changed continuously.

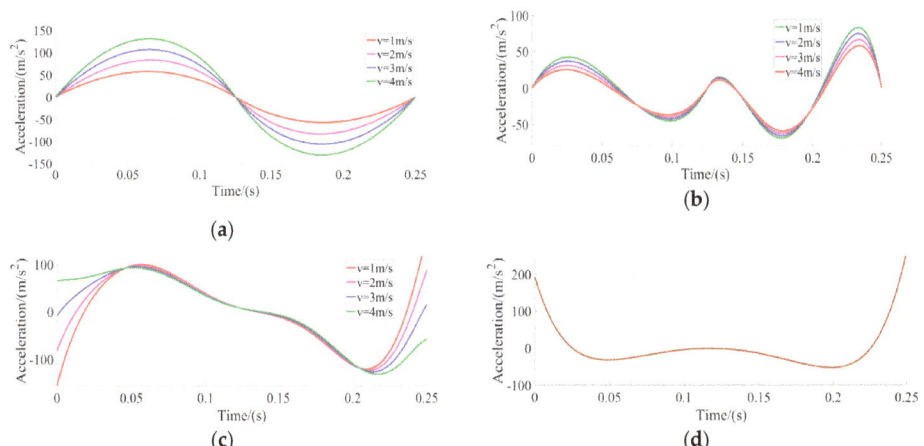

Figure 13. The endpoint acceleration of the leg calculated by MATLAB during the swing phase: (**a**) The acceleration in the X direction for the spline curve trajectory; (**b**) the acceleration in the Y direction for the spline curve trajectory; (**c**) the acceleration in the X direction for the Bézier curve trajectory; and (**d**) the acceleration in the Y direction for the Bézier curve trajectory.

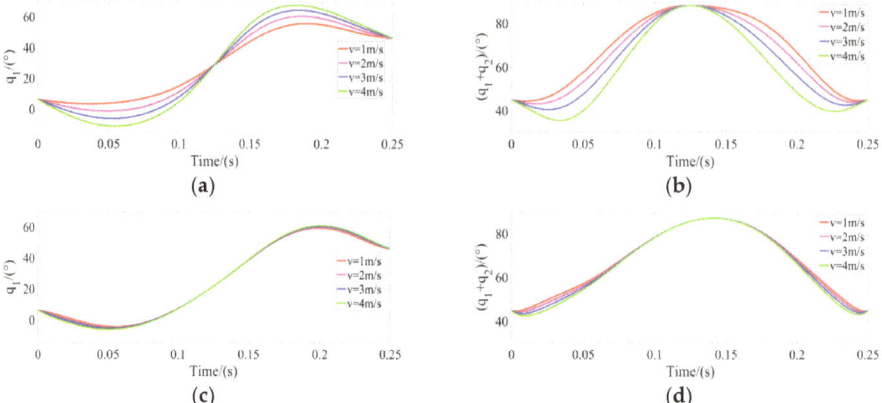

Figure 14. The motor rotation angle calculated by MATLAB during the swing phase: (**a**) The shoulder motor rotation angle in the spline curve trajectory, (**b**) the knee motor rotation angle in the spline curve trajectory, (**c**) the shoulder motor rotation angle in the Bézier curve trajectory, and (**d**) the knee motor rotation angle in the Bézier curve trajectory.

3.2. Simulation

From the above results, it can be seen that the two leg trajectories have different properties, but their movement stability performance on a quadruped robot is still unknown. Therefore, in this section, the Webots software (Cyberbotics Ltd.) is introduced to simulate the motion stability of a quadruped robot under two high-speed gait trajectories. A model, which has the same physical parameters of the designed leg, was built, and two different gait trajectories were applied. The simulation results are shown in Figure 15. The desired speed was set as 2 m/s, which is a relatively high speed for quadruped robots. During the stance phase, an impedance controller was added for both of two schemes with appropriately set parameters of K = 80 Nm/rad and C = 5 Nms/rad. A distance sensor was placed in the center of the robot to measure the height change of the center of gravity of the quadruped robot from the ground for testing the new proposed gait trajectory and comparing the stability between two gait trajectories as shown in Figure 16.

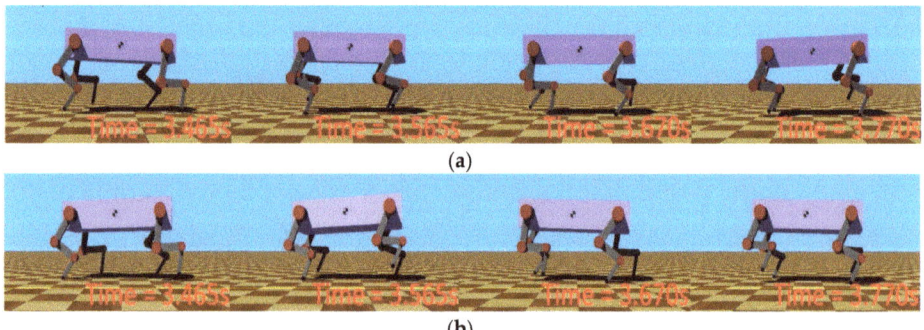

Figure 15. The simulation of the movement running at $V_{desire} = 2$ m/s with trot gait: (**a**) The quadruped robot runs by the spline curve trajectories and (**b**) the quadruped robot runs by the Bézier curve trajectories.

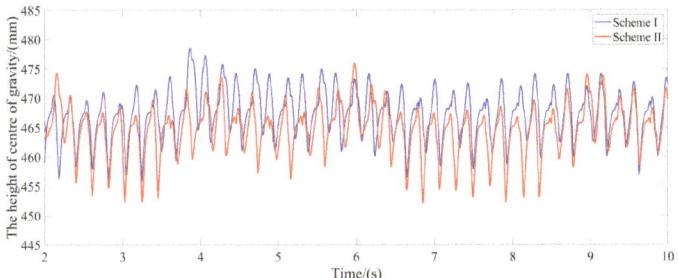

Figure 16. Simulation results with Webots for the height of the center of mass (CoM).

3.3. Experiment

The hardware configuration of the single leg platform main consists of a PC, a controller (C6015, Beckhoff), two motor drives (G-SOLTWI10/200EE1, Elmo), two frameless motors (127P1, 380W, Allied Motion), and rotary magnetic encoders (RMB20SC13BC10, Renishaw), as shown in Figure 17. The controller has a dual-core CPU of 1.92 GHz with a rapid data computing capability. Moreover, the size of the controller is only 96 mm x 91 mm x 41 mm, which can greatly save space. The custom driven unit with a frameless motor with high torque density was specific designed, which significantly reduces the weight of the leg structure. Additionally, this motor can provide a sufficient torque of 21 Nm for high-speed running and jumping. Considering that the legs will bear periodic impact load during movement, magnetic encoders were used in the leg that can separate the encoder from the rotating shaft and avoid damage caused by the impact between the encoder and the shaft.

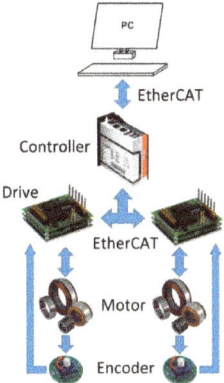

Figure 17. The hardware configuration.

During the high-speed locomotion of the quadruped robot, its legs need to swing periodically and quickly, so the legs need to complete a swing trajectory in a very short time, which requires that the control cycle time be short enough to follow the trajectory [24]. Therefore, both the controller and drives need to have high real-time computing ability and high signal transmission speed. Therefore, an EtherCAT bus was chosen from PC to driver. The EtherCAT bus has a high transmission rate, and the control cycle time can reach the level of a microsecond, which can meet the high real-time requirement of the movement of the leg.

The control program was designed by TwinCAT 3 which can provide abundant modules to realize different kinds of motion control in Visual Studio. The Beckhoff C6015 controller has a dual-core structure, which can process different data in parallel, and the communication between the two cores is simple and efficient. Therefore, one of the cores can be used for trajectory selection and trajectory planning, and the other core can be used to send command signals to the motor to ensure real-time movement of the legs.

The experimental platform was constructed as shown in Figure 18. The NDI Optotrak Certus was used to trace the endpoint of the leg to measure the actual trajectories of the two schemes ($V_{desire} = 2$ m/s). The marker was fixed on the endpoint of the foot. The experimental results are shown in Figure 19, from which we can see that the leg can trace the desired trajectory.

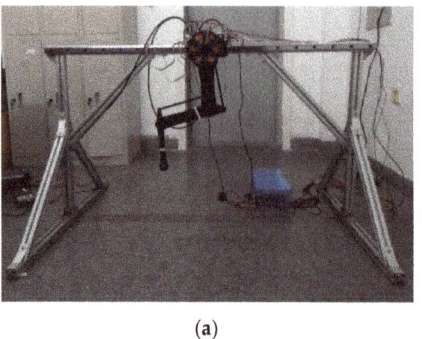 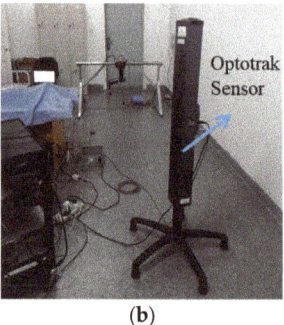

Figure 18. Experimental platform. (a) One leg experimental platform. (b) Trajectory measurement with Optotrak sensor.

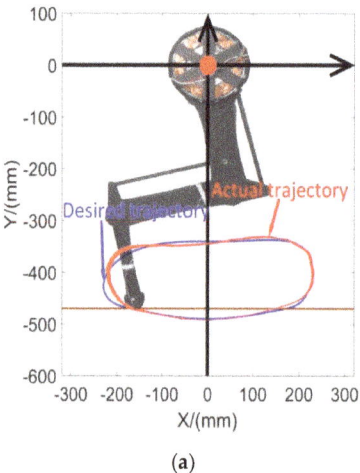 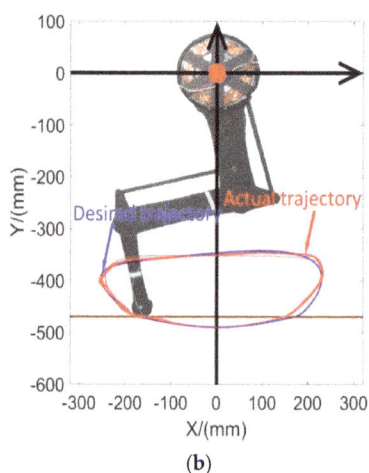

Figure 19. The desired and actual trajectory with $V_{desired} = 2$ m/s. (a) The trajectory of the scheme I. (b) The trajectory of the scheme II.

Whether the movement is stable and the power consumption is low are directly related to the efficiency of the trajectory design. For evaluating the efficacy of two trajectory planning methods, during the test, the voltage and current of the battery, the angular velocity of the motors, and the torque output of the motors in two schemes were sampled simultaneously. Also, the following Formulas (27) and (28) were used to calculate the consumption power. Finally, the power consumption can be obtained as shown in Figure 20, and the energy consumption is compared in Table 3. In particular, the time traces of the energetics of legs swing with $V_{desired} = 2$ m/s for two schemes can be found in Figure 21.

$$Battery\ power = \sum\nolimits_{2\ motors} U \times I \tag{27}$$

$$\text{Mechanical power} = \sum\nolimits_{2\ motors} \tau \times \omega \tag{28}$$

Table 3. Energy consumption of two schemes in $V_{desired} = 2$ m/s within 17 s.

	Mechanical Energy Consumption (J)	Battery Energy Consumption (J)	Efficiency	RMS Shoulder Current (A)	RMS Knee Current (A)
Scheme I	71.60	104.28	68.66%	0.7735	0.2885
Scheme II	64.89	108.24	59.95%	0.8163	0.2733

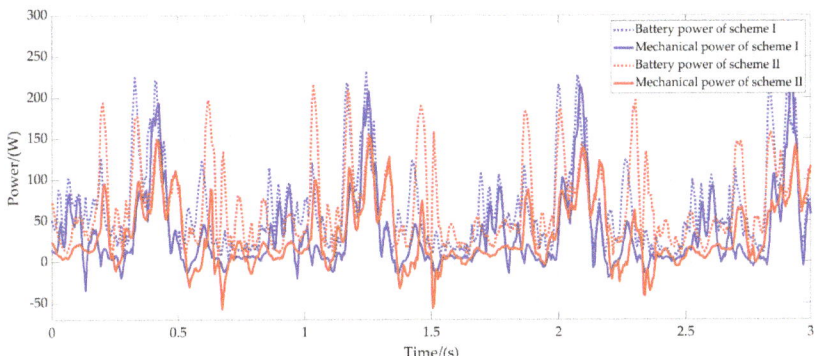

Figure 20. Power consumption of two schemes in $V_{desired} = 2$ m/s.

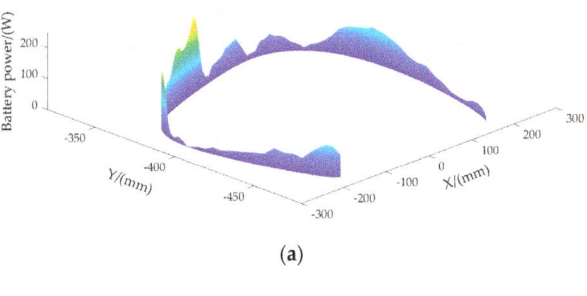

(a)

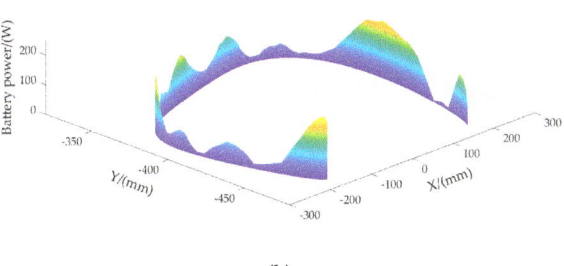

(b)

Figure 21. Time traces of the energetics of legs swing with two schemes. (**a**) The battery power consumption of the scheme I with spline curve in swing phase. (**b**) The battery power consumption of the scheme 2 with Bézier curve in swing phase.

4. Discussion

Our leg design concentrates all the motors on the shoulder joint. Hence the center of mass is near to the center of rotation, which makes the inertia of the swing leg change smaller so that the leg can achieve higher acceleration and a faster response. This leg configuration is also adopted by several quadrupeds, such as the MIT cheetah [25], Laikago robot [26], and SpotMini [7]. For the manufactured leg, it can be found that the swing inertia of the knee joint, which mainly provides the movement in the vertical direction, is very low and its movement is very flexible. Also, the force the quadruped robots bear most is also in the vertical direction. Hence such a leg structure is suitable to the high-speed movement and jumping for quadruped robots. Differing from the leg of MIT cheetahs, which was made with a bioinspired fabrication method [14], our leg structure is mainly connected by ribbed plates with only 4-mm-thin plates in the middle. The structure has a better performance of bearing pressure and bending moment and is simpler for manufacturing. In particular, wear washers and dry bushings were adopted in our joints design to reduce the friction of linkages' joint by the self-lubrication characteristics of the dry bushing.

Moreover, two high torque density motors are used in the single leg, which can provide a strong power regardless of a low gear ratio. The coaxial motor configuration can achieve a coupled motion of two shoulder and knee joints, which has also been adopted by several quadrupeds, such as MIT cheetah [25] and Laikago robot [26]. The used single-stage gear can greatly decrease the size and mass of the reducer as well as reducing the energy loss in the process of mechanical transmission, making a significant contribution to energy efficiency.

From the acceleration curves of the two schemes, it is obvious that the gait trajectory of spline curve in scheme 1 can ensure the continuity of acceleration, and the acceleration of the contact point between swing phase and stance phase is almost zero in scheme 1, which means no additional impact is generated from the ground. Because of the characteristics of the Bézier curve, the gait trajectory planned finds it difficult to achieve continuity of acceleration, and the acceleration of the legs once in contact with the ground is not zero. Therefore, there will be an impact on the legs, which will affect the walking stability of the quadruped robot. Besides, the shape of the curve in Figure 13c resembles human gait ground reaction forces [27]. According to Marc Raibert's virtual leg principle [28], a quadruped gait that uses four legs in two pairs, such as trot and pace, could be viewed as an equivalent biped, which means that human gait somewhat resembles quadruped gait. That may be the reason why the curves are similar, and it also proves that the gait trajectory is close to a biological gait. The characteristics of these two schemes were also evaluated with the simulation in Webots. During the simulation, the robot used the gait of scheme 1 and can walk more steadily with less body shaking; however, the quadruped robot, while using the gait of scheme 2, suffers from minor shaking, although it can walk normally. According to the change of the height of gravity center, the maximum difference of gravity center height in the scheme I is 22.62 mm, while that of scheme II is 23.79 mm. Overall, it can be seen that the gait stability of scheme I is better than that of scheme II especially for the change of pitch angle.

Furthermore, the power consumption for the two schemes indicates that scheme 1 consumes less battery energy and has higher efficiency than that of scheme 2, as shown in Figure 20 and Table 3 in the results section, although scheme 1 consumes more mechanical energy. Besides, the RMS knee current, which corresponds to heating in scheme 1, is similar to that of scheme 2. While the RMS shoulder current in scheme 1 is much smaller than that in scheme 2. Hence, scheme 1 is superior to scheme 2 in energy consumption. In particular, from Figure 21, which shows the time traces of the energetics of legs swing with two schemes, it can be found that, due to the acceleration discontinuity of scheme 2, there is greater power consumption for the leave-off motion than the touch-down motion between the swing leg and ground.

5. Conclusions

In this paper, a leg structure for the quadruped robot was designed, which has low inertia, lightweight, and is suitable for high-speed locomotion. Moreover, it was verified to be robust enough by stress and deformation analysis. After that, two schemes based on the spline curve and Bézier curve were proposed respectively to plan the leg trajectories. Swing phase and stance phase trajectories were designed individually in different perspectives while guaranteeing continuous and smooth transitions. By comparing their acceleration curves, Webots simulation results, actual trajectory, and energy consumption, we conclude that the spline curve trajectory has better stability, lower energy consumption, and higher energy efficiency for the movement of quadruped robots. This paper also provides an available method to design the leg trajectory according to the stability and efficiency of robots. In future, the planned gait trajectory following a specific gait pattern will be tested on the aspect of stability and high-speed locomotion. The adaptability to terrain obstacles/slopes will also be evaluated.

Author Contributions: Conceptualization, S.Z., X.Z., X.L. and H.Z. (Haitao Zhou); methodology, X.Z., S.Z. and H.Z. (Hongji Zhang); formal analysis, X.Z. and H.Z. (Hongji Zhang); writing—original draft preparation, X.Z.; funding acquisition, S.Z. and Y.F.; writing—review and editing, S.Z., X.Z., X.L., H.Z. (Haitao Zhou) and Y.F.

Funding: This research was funded by the National Natural Science Foundation of China (Grant No. 61703124), the Self-Planned Task of State Key Laboratory of Robotics and System (HIT) (SKLRS201801A02), and the Innovative Research Groups of the National Natural Science Foundation of China (51521003).

Conflicts of Interest: The authors declare no conflicts of interest.

References

1. Li, X.; Zhou, H.; Feng, H.; Zhang, S.; Fu, Y. Design and experiments of a novel hydraulic wheel-legged robot (WLR). In Proceedings of the 2018 IEEE/RSJ International Conference on Intelligent Robots and Systems (IROS), Madrid, Spain, 1–5 October 2018; pp. 3292–3297.
2. Fu, Y.; Luo, J.; Ren, D.; Zhou, H.; Li, X.; Zhang, S. Research on impedance control based on force servo for single leg of hydraulic legged robot. In Proceedings of the 2017 IEEE International Conference on Mechatronics and Automation (ICMA), Takamatsu, Japan, 6–9 August 2017; pp. 1591–1596.
3. Bellicoso, C.D.; Jenelten, F.; Gehring, C.; Hutter, M. Dynamic locomotion through online nonlinear motion optimization for quadrupedal robots. *IEEE Robot. Autom. Lett.* **2018**, *3*, 2261–2268. [CrossRef]
4. Hwangbo, J.; Lee, J.; Dosovitskiy, A.; Bellicoso, D.; Tsounis, V.; Koltun, V.; Hutter, M. Learning agile and dynamic motor skills for legged robot. *Sci. Robot.* **2019**, *4*, 1–13. [CrossRef]
5. Raibert, M.; Blankespoor, K.; Nelson, G.; Playter, R. BigDog, the rough-terrain quadruped robot. *Int. Fed. Autom. Control* **2008**, *41*, 10822–10825. [CrossRef]
6. Semini, C.; Tsagarakis, N.G.; Guglielmino, E.; Focchi, M.; Cannella, F.; Caldwell, D.G. Design of HyQ—A hydraulically and electrically actuated quadruped robot. *J. Syst. Control Eng.* **2011**, *225*, 831–849. [CrossRef]
7. Spotmini Autonomous Navigation. Available online: https://youtu.be/Ve9kWX_KXus (accessed on 11 August 2018).
8. Hutter, M.; Gehring, C.; Lauber, A.; Gunther, F.; Bellicoso, C.D.; Tsounis, V.; Fankhauser, P.; Diethelm, R.; Bachmann, S.; Bloesch, M.; et al. ANYmal—Toward legged robots for harsh environments. *Adv. Robot.* **2017**, *31*, 918–931. [CrossRef]
9. Hyun, D.J.; Seok, S.; Lee, J.; Kim, S. High speed trot-running: Implementation of a hierarchical controller using proprioceptive impedance control on the MIT Cheetah. *Int. J. Robot. Res.* **2014**, *33*, 1417–1445. [CrossRef]
10. Bledt, G.; Powell, M.J.; Katz, B.; Carlo, J.D.; Wensing, P.M.; Kim, S. MIT Cheetah 3: Design and control of a robust, dynamic quadruped robot. In Proceedings of the 2018 IEEE/RSJ International Conference on Intelligent Robots and Systems (IROS), Madrid, Spain, 1–5 October 2018; pp. 2245–2252.
11. Wensing, P.M.; Wang, A.; Seok, S.; Otten, D.; Lang, J.; Kim, S. Proprioceptive actuator design in the MIT cheetah: Impact mitigation and high-bandwidth physical interaction for dynamic legged robots. *IEEE Trans. Robot.* **2017**, *33*, 509–522. [CrossRef]

12. Yu, J.; Hooks, J.; Zhang, X.; Ahn, M.S.; Hong, D. A proprioceptive, force-controlled, non-anthropomorphic biped for dynamic locomotion. In Proceedings of the 2018 IEEE-RAS 18th International Conference on Humanoid Robots (Humanoids), Beijing, China, 6–9 November 2018; pp. 489–496.
13. WildCat-The World's Fastest Quadruped Robot. Available online: https://www.bostondynamics.com/wildcat (accessed on 14 March 2019).
14. Ananthanarayanan, A.; Azadi, M.; Kim, S. Towards a bio-inspired leg design for high-speed running. *Bioinspir. Biomin.* **2012**, *7*, 046005. [CrossRef]
15. Spröwitz, A.; Tuleu, A.; Vespignani, M.; Ajallooeian, M.; Badri, E.; Ijspeert, A.J. Towards dynamic trot gait locomotion-design, control, and experiments with Cheetah-cub, a compliant quadruped robot. *Int. J. Robot. Res.* **2013**, *32*, 932–950. [CrossRef]
16. Rosendo, A.; Liu, X.; Nakatsu, S.; Shimizu, M.; Hosoda, K. A combined cpg-stretch reflex study on a musculoskeletal pneumatic quadruped. Biomimetic and Biohybrid Systems. *Living Mach.* **2014**, 417–419. [CrossRef]
17. Walter, R.M.; Carrier, D.R. Ground forces applied by galloping dogs. *J. Exp. Biol.* **2007**, *210*, 208–216. [CrossRef] [PubMed]
18. Saputra, A.A.; Tay, N.N.W.; Toda, Y.; Botzheim, J.; Kubota, N. Bézier curve model for efficient bio-inspired locomotion of low cost four legged robot. In Proceedings of the 2016 IEEE/RSJ International Conference on Intelligent Robots and Systems (IROS), Daejeon, South Korea, 9–14 October 2016.
19. Lee, Y.L.; Lee, Y.H.; Lee, H.; Phan, L.T.; Kang, H.; Kim, U.; Jeon, J.; Choi, H.R. Trajectory design and control of quadruped robot for trotting over obstacles. In Proceedings of the 2017 IEEE/RSJ International Conference on Intelligent Robots and Systems (IROS), Vancouver, BC, Canada, 24–28 September 2017; pp. 4897–4902.
20. Yoshikawa, T. Manipulability of Robotics Mechanism. *Inter. J. Robot. Res.* **1985**, *4*, 3–9. [CrossRef]
21. Hogan, N. Impedance control—An approach to manipulation. I. Theory. II. Implementation. III. Applications. *J. Dyn. Syst. Meas. Control* **1985**, *107*, 1–24. [CrossRef]
22. Jung, S.; Hsis, T.C.; Bonitz, R.G. Force tracking impedance control of robot manipulators under unknown environment. *IEEE Trans. Control Syst. Technol.* **2004**, *12*, 474–483. [CrossRef]
23. Liu, M.; Xu, F.; Jia, K.; Yang, Q.; Tang, C. A stable walking strategy of quadruped robot based on foot trajectory planning. In Proceedings of the 2016 3rd International Conference on Information Science and Control Engineering (ICISCE), Beijing, China, 8–10 July 2016; pp. 799–803.
24. Haberland, M.; Karssen, J.G.D.; Kim, S.; Wisse, M. The effect of swing leg retraction on running energy efficiency. In Proceedings of the 2011 IEEE/RSJ International Conference on Intelligent Robots and Systems, San Francisco, CA, USA, 25–30 September 2011; pp. 3957–3962.
25. Lee, J. Hierarchical Controller for Highly Dynamic Locomotion Utilizing Pattern Modulation and Impedance Control: Implementation on the MIT Cheetah Robot. Master's Thesis, MIT, Cambridge, MA, USA, 2013.
26. Laikago: A Four Leg Robot Is Coming to You. Available online: https://www.youtube.com/watch?v=d6Ja643GqL8 (accessed on 14 March 2019).
27. Handzic, I.; Reed, K.B. Validation of a passive dynamic walker model for human gait analysis. In Proceedings of the 2013 35th Annual International Conference of the IEEE Engineering in Medicine and Biology Society (EMBC), Osaka, Japan, 3–7 July 2013; pp. 6945–6948.
28. Raibert, M.; Chepponis, M.; Brown, H.B., Jr. Running on four legs as though they were one. *IEEE J. Robot. Autom.* **1986**, *2*, 70–82. [CrossRef]

© 2019 by the authors. Licensee MDPI, Basel, Switzerland. This article is an open access article distributed under the terms and conditions of the Creative Commons Attribution (CC BY) license (http://creativecommons.org/licenses/by/4.0/).

Review

Natural Motion for Energy Saving in Robotic and Mechatronic Systems

Lorenzo Scalera [1], Ilaria Palomba [1], Erich Wehrle [1,*], Alessandro Gasparetto [2] and Renato Vidoni [1]

1. Faculty of Science and Technology, Free University of Bozen-Bolzano, 39100 Bolzano, Italy
2. Polytechnic Department of Engineering and Architecture, University of Udine, 33100 Udine, Italy
* Correspondence: erich.wehrle@unibz.it

Received: 3 August 2019; Accepted: 22 August 2019; Published: 27 August 2019

Abstract: Energy saving in robotic and mechatronic systems is becoming an evermore important topic in both industry and academia. One strategy to reduce the energy consumption, especially for cyclic tasks, is exploiting natural motion. We define natural motion as the system response caused by the conversion of potential elastic energy into kinetic energy. This motion can be both a forced response assisted by a motor or a free response. The application of the natural motion concepts allows for energy saving in tasks characterized by repetitive or cyclic motion. This review paper proposes a classification of several approaches to natural motion, starting from the compliant elements and the actuators needed for its implementation. Then several approaches to natural motion are discussed based on the trajectory followed by the system, providing useful information to the researchers dealing with natural motion.

Keywords: natural motion; natural dynamics; energy saving; robotic system; trajectory planning; optimization

1. Introduction

Industrial robotic and mechatronic systems are required to have high energy efficiency, especially when high-speed continuous operations and high-volume production are needed [1]. Operating a robot or a mechatronic system at high speed produces significant losses at high velocities, as well as energy surpluses in the deceleration phases. These losses have repercussions on the amount of electric energy that is needed to operate the manufacturing system. Furthermore, increasing energy prices and environmental awareness encourage to reduce the power consumption. This is highlighted by the policy applied by the European Union, which aims to reduce the whole energy consumption up to 30% by 2030 [2]. Moreover, in the last years, the demand for industrial robots has accelerated due to the ongoing trend toward automation [3]. Therefore, their energy efficiency is becoming crucial since manufacturers are incentivized to install eco-friendly solutions for plants and production systems. For these reasons, engineers and researchers have been motivated to investigate and develop novel strategies to increase energy efficiency in industrial robots and mechatronic systems.

Several energy-saving methods for robotic and automatic systems can be found in the literature. In Reference [4], G. Carabin et al. present a classification of these methods, drawing a distinction between *hardware*, *software* and *mixed* approaches. In particular, hardware solutions include the implementation of new kinds of actuating systems, regenerative drives [5] and the design of lightweight manipulators [6–8]. The software approach is focused on time minimization, operations scheduling and trajectory optimization [9–13]. Mixed approaches rely upon the concurrent improvement of both hardware and software components of the automatic system. Among those, a particular method for enhancing energy efficiency is based on the concept of *natural motion*.

The definition of natural motion used here is as follows: a system response caused by the conversion of potential elastic energy into kinetic energy. Natural motion can be both a forced response assisted by a motor or a free response.

In the early 1980s, D. Koditschek defined the natural motion as the hope that the concurrent use of dynamic models and suitable control strategies might match the internal dynamics of the robot, affording more accurate performance with less effort [14,15]. The natural motion of a system is the one that occurs thanks to the transformation of its potential elastic energy into kinetic energy. The exploitation of this kind of motion allows to save energy in performing repetitive or cyclic motions, such as pick-and-place tasks [16,17] and legged locomotion (in walking and running robots) [18,19]. The more energy effective natural motion is the free response of the system: the unforced motion of a perturbed system that moves back to its equilibrium position. Indeed, in the ideal case of a conservative system, it keeps oscillate requiring only the initial energy to perturb its equilibrium. However, due to damping, friction forces and disturbances, the free response does not always coincide with a continuous oscillation that can be used in practice. Nevertheless, it is still possible to reduce energy consumption by exploiting the system natural motion by exciting it at resonance.

Systems working excited at resonance are quite common in several industrial applications. Examples are resonant conveyors, employed for material moving [20,21] and ultrasonic horns, used for cleaning, welding, cutting, machining and drilling [22,23]. To obtain a perfect match between the desired motion and the resonator natural dynamics, the resonant system is typically modified through changes in mass and/or stiffness parameters [24,25]. Modifications of system parameters are generally needed also to make a robot resonant [26], as well as to exploit its free response. Indeed, the vibratory behavior of traditional robotic systems can be neglected, since they are composed of links that are assumed to be rigid, which are connected through stiff joints and stiff actuators. Modifications are generally needed also dealing with lightweight robots that have elastic links, since link vibration could be too small compared with the displacements required by the task [27]. Large oscillations are typically obtained by installing lumped springs at the actuated joints. By properly tuning the compliance of the added springs, the natural dynamics of the robot can match the desired motion.

The aim of this review paper is to provide a classification and a discussion of existing technologies and methods to exploit natural motion for energy efficiency in robotic and mechatronic systems.

The remaining of this paper is organized as follows: Section 2 describes how a robotic system can be modified to exploit the natural motion by adding elasticity. Section 3 provides a classification and a description of the different methods developed on the basis of the trajectory followed by the robotic system. In Section 4 the optimal design problems in natural motion are analyzed, since the implementation of natural motion is often related to the definition of an optimization problem. Then, Section 5 analyzes some open issues and critical points in the field of natural motion. Finally, Section 6 summarizes the conclusions of this survey together with possible future developments.

2. Mechanical Design for Natural Motion

The natural motion can only be applied to vibratory dynamic systems. This means that the system has to be capable of conveying kinetic energy and of storing potential energy. Every robotic system is capable of conveying kinetic energy thanks to its mass and inertia but it is not always equipped with compliance elements to store elastic energy, that hence have to be added. The concept of adding springs or additional elements to a robotic system in order to reduce the efforts of the actuators is first adopted in gravity-balancing techniques [28,29]. Examples of gravity balancing techniques can be found in References [30–35]. These methods are proposed to compensate the input effort required to move the links of a pick-and-place robot and to reduce energy consumption in slow movements. However, static-balancing techniques cannot be implemented in high-speed manipulators in which the dynamic effects are not negligible. To exploit the natural motion, the elastic elements are installed at robot joints. The possible connections between elastic elements and joints are three: *serial, parallel* and their combination *serial and parallel*. These three configurations are analyzed in the following. Furthermore, a comparison of these configurations is discussed at the end of this section.

2.1. Serial Compliance Elements

Serial elastic elements can be added to a robotic system by connecting an electric motor to a joint by means of an elastic element (see Figure 1a). Such an arrangement is not simply obtained by adding a spring between them but instead by substituting stiff actuators with compliant ones. Series elastic actuators (SEA), first introduced by G.A. Pratt in Reference [36], are constituted of a linear compliant element between a high impedance actuator and the load. Their use is demonstrated to be beneficial in compliant actuation, energy storage, interaction safety, in the absorption of contact shocks and in the reduction of the peak forces due to the impacts in bipedal walking robots [36,37]. Examples of SEA for minimizing energy consumption can be found in References [38,39]. In these works, a convex optimization problem is presented for the design of a SEA capable of energy regeneration such that energy consumption and peak power are minimized for arbitrary periodic reference trajectories.

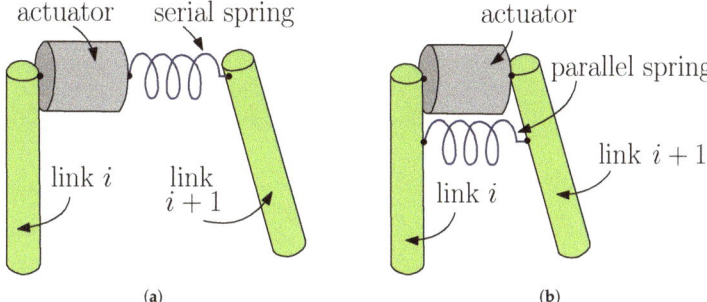

Figure 1. Compliance configurations: (**a**) spring in series and (**b**) spring in parallel.

The main drawback of SEA is that the stiffness is fixed and cannot be changed during motion, thus limiting the level of compliance to adapt for several tasks. To overcome this problem, a novel class of actuators was proposed: the variable stiffness actuators (VSA). These actuators consist of a motor connected to the output link by a spring in series, whose stiffness is variable. Therefore, the stiffness can be properly controlled to reduce energy requirements during the execution of repetitive tasks, as explained in Reference [40]. Reviews on variable stiffness actuators can be found in References [41–44], whereas a comparison of several design of VSA based on spring pretension is illustrated in Reference [45].

VSA were first introduced to decrease contact shocks, to enhance soft collisions in human-robot interaction [46,47] and to efficiently actuate legged locomotion systems [48,49]. Furthermore, they were employed to decrease energy consumption in cyclic operations of robotic systems. An example is given by the actuator with adjustable stiffness proposed by A. Jafari et al. [50–52].

2.2. Parallel Compliance Elements

Compliant elements can be connected in parallel to the main actuators as shown in Figure 1b. Although some parallel elastic actuators can be found in the literature [53–55], it is not necessary to replace the original actuators to install parallel springs. For example, M. Iwamura et al. equipped a serial robot with two linear springs placed at joints, between neighboring links and a special spring holder [56]. In this way, they overcome the difficulties in adjusting the stiffness value and the mounting positions of torsional springs.

The research in this field mainly addresses the development of mechanisms to realize variable stiffness springs [57–59] and non-linear off-the-shelf springs [60–63]. Indeed, non-linear stiffness allows a larger energy saving in robots, since actuators torques and end-effector trajectory are always related by non-linear behavior, as demonstrated in References [58,64,65].

A mechanism to change spring stiffness is proposed by M. Uemura et al. in Reference [57]. It consists of a sliding screw with a self-lock function and a linear spring. The self-lock function guarantees that a constant elastic value is kept when the motor of the variable elastic mechanism does not exert a torque. The linear spring is attached by one end to the actuated link of the system and by the other to a point on the lead screw mechanism. By changing the position of the lead screw the length of moment arm of the elastic force exerted by the linear spring changes and hence the equivalent torsional stiffness.

R. Nasiri et al. realize a parallel variable torsional spring by means of two linear springs and two worm-gear motors [58]. One end of the two linear springs is connected to a point of the actuated link, whereas the other end to the worm-gear motor. By independently controlling the strain-length of the springs with the two motors, an arbitrary compliance profile at the actuated joint can be obtained.

A mechanism capable of realizing a constant non-linear torsional springs is presented by N. Schmit and M. Okada [60,61]. The mechanism consists of a linear spring connected to a cable wound around a non-circular spool. A non-circular spool (or variable radius drum) is characterized by the variation of the spool radius along its profile [66]. In References [60,61], the spring mechanism is attached to each actuated joint, with respect to which the linear spring behaves as a non-linear rotational spring with a described torque profile given by the shape of the spool (Figure 2). A similar design is proposed also by B. Kim and A.D. Deshpande [67], who additionally design an antagonistic spring configuration for bilateral torque generation.

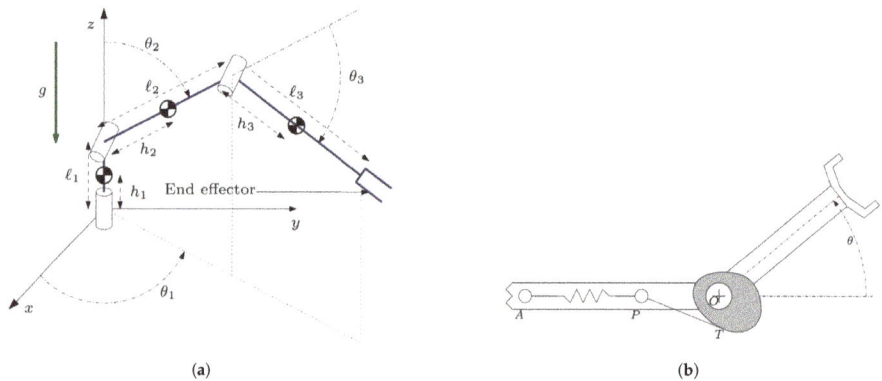

Figure 2. (a) The 3-DOF serial manipulator employed in the simulations by N. Schmit and M. Okada and (b) the design of the proposed non-linear spring mechanism [61].

H.J. Bidgoly et al. [63] realize a non-linear torsional spring with an arbitrary stiffness profile by combining a linear spring and a non-linear transmission mechanism. The last consists of a non-circular cam connected to the actuated joint and a roller, which moves along the outer circumference of the cam. The stretched linear spring is hinged to the centers of the cam and the roller. The desired torque-angle profile is obtained by properly designing the shape of the cam.

Another system to attach springs in parallel with the actuator is proposed by M. Plooij and M. Wisse [62]. A parallel spring mechanism converts the linear stiffness of a linear spring in an equivalent torsional non-linear stiffness. The mechanism consists of two pulleys of different size connected through a timing belt. The larger pulley is attached to the actuated link. The two ends of the spring are connected to two points placed on the outer circumference of the smaller pulley and of the larger one, respectively. In this way, the spring is non-linearly stretched with respect to the rotation of the link. One peculiarity of this mechanism is that it has two different configurations in which the elastic energy of the spring is null. This means that if the mechanism is properly designed for a specific pick-and-place operation, the actuators does not have to counteract the spring during the task.

It is worth noting that there are two mounting possibilities for parallel springs, that is, they can be connected to just one joint, or they can span over two joints. In the first arrangement, largely adopted, springs are called mono-articular, whereas in the second, bi-articular (Figure 3). Bi-articular springs take inspiration from biological studies [68]—bi-articular muscles actuate human limbs as well as paws of birds, reptile and insects. Non-linear bi-articular springs are designed for example by B. Kim and A.D. Deshpande [67], by means of pulley-cable mechanisms and antagonist linear springs. Bi-articular springs are mostly employed in the realization of walking robot [69–71]. The effectiveness of such springs in serial manipulators to perform pick-and-place operations are investigated by G. Lu et al. [72] and H.J. Bidgoly et al. [73], with different conclusions. In the case of a variable linear bi-articular spring added on a SCARA robot, G. Lu et al. [72] conclude that such a spring alone cannot effectively save energy. Conversely, H.J. Bidgoly et al. [73] assert that bi-articular spring contribute more in the actuation cost minimization with respect to mono-articular springs, in the case of a redundant 4-DOF serial manipulator with non-linear mono- and bi-articular springs.

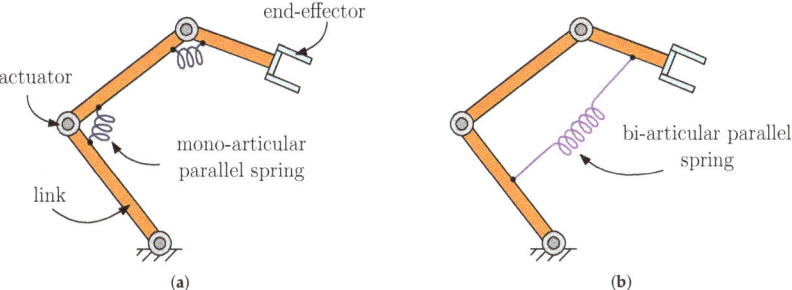

Figure 3. (a) Mono-articular and (b) bi-articular parallel springs mounted on a robotic manipulator.

2.3. Serial and Parallel Compliance Elements

In the context of natural motion, some actuators are developed combining both serial and parallel compliant elements. In these systems, the serial compliance is employed for impact mitigation, whereas the parallel one is adopted for efficient energy storage.

In Reference [74], a mixed series-parallel approach is presented by G. Mathijssen et al. Multiple series elastic actuation branches are placed in parallel, each engaged depending on torque requirements. This design leads to a significant torque effort reduction, as well as an increased output torque range, compared to traditional stiff or SEA configurations.

This actuation design is adopted by N.G. Tsagarakis et al. [75] and by W. Roozing et al. [76]. A novel asymmetric actuation scheme that consists of two actuation branches that transfer their power to a single joint through two compliant elements with diverse stiffness and storage capacity properties is presented. The serial branch is used for impact absorption and it is connected to the main actuator, whereas the parallel branch is adopted for its large potential energy storage capability in cyclic tasks.

2.4. Discussion of Elastic Elements Configurations

Both serial and parallel configurations can be implemented with springs of *variable* or *non-variable* stiffness, as well as *linear* or *non-linear* behavior. Variable compliance allows the *online* tuning of spring parameters. Therefore, control systems that act on the stiffness to perform varying operations and to change system dynamics can be adopted, as in References [16,58,77,78]. As a drawback, variable compliance springs need a more complex actuator design and sensory feedback. On the contrary, configurations based on non-variable stiffness are easier to implement, the springs can be designed and tuned *offline* but are suitable only for non-varying tasks and operations.

The serial configuration (adopted for example in References [42,50,79]) provides the actuator with a compliance that can be employed to decrease contact shocks and to reduce force peaks due to impacts, for example, In human-robot interaction [46]. On the other hand, the parallel configuration [65,80–82] results in simpler mechanism and mathematics formulation [78]. Parallel compliance does not enlarge the configuration space of the robotic system and has a well-posed quadratic cost function for energy consumption minimization with respect to the compliance coefficients [83]. Moreover, the serial arrangement of springs and motors limits the operational speed due to uncontrolled robot deflection when performing high-speed tasks.

Comparisons between serial and parallel configurations can be found in References [84–88]. T. Verstraten et al. in Reference [87], provide a comparison between stiff actuators, parallel elastic actuators and serial elastic actuators in terms of power as well as mechanical and electrical energy consumption. As test case, a sinusoidal motion is imposed to a pendulum load. By means of simulations they demonstrate that, if the stiffness of the elastic actuators is properly tuned, it is possible to reduce energy up to 78% compared to rigid actuator using serial elastic actuators and up to 20% using the parallel ones.

Although the results of this study demonstrate that a serial spring arrangement outperforms the parallel one from an energetic point of view, the majority of the works dealing with natural motion adopt parallel springs because of their easiness of installation and control. An evidence of this is provided by Table 1, where the spring installation configuration adopted in the reviewed papers are highlighted.

Table 1. Literature on natural motion in robotic and mechatronic systems.

Ref.	1st Author	Year	Path and Trajectory Planning	Compliance of Elastic Elements
Defined trajectory				
[89]	Boscariol	2017	cycloidal, 5-degree polynomial trajectory	linear parallel
[90]	Carabin	2019	double-S speed profile	linear parallel
[40]	Fabian	2018	harmonic trajectory	variable parallel
[16]	Goya	2012	harmonic trajectory	variable parallel (mono- and bi-articular)
[91]	Hill	2019	adjustment of limit cycle	variable parallel
[65]	Khoramshahi	2014	circular path in the operational space	non-linear variable parallel
[92]	Lu	2012	desired motion given	linear parallel
[72]	Lu	2012	harmonic trajectory	variable parallel
[93]	Matsusaka	2016	desired motion given	variable parallel
[94]	Matsusaka	2016	harmonic trajectory	variable parallel
[58]	Nasiri	2017	harmonic trajectory	linear and non-linear variable parallel
[64]	Scheilen	2005	harmonic trajectory	linear and non-linear parallel
[95]	Scheilen	2005	harmonic trajectory	linear parallel
[96]	Shushtari	2017	sinusoidal path given	non-linear parallel
[77]	Uemura	2007	harmonic trajectory	variable parallel
[97]	Uemura	2008	harmonic trajectory	variable parallel
[98]	Uemura	2008	harmonic trajectory	variable parallel
[99]	Uemura	2009	harmonic trajectory	variable parallel
[78]	Uemura	2014	harmonic trajectory	variable parallel
[86]	Velasco	2013	harmonic trajectory	linear serial/parallel
[79]	Velasco	2015	squared trajectory	linear and variable serial
Optimized trajectory				
[73]	Bidgoly	2017	cyclic end-effector path given	non-linear parallel (mono- and bi-articular)
[80]	Hill	2018	polynomial trajectories for robot and variable stiffness springs	variable parallel
[56]	Iwamura	2011	trajectory optimization	linear parallel
[100]	Iwamura	2016	trajectory optimization	linear parallel
[82]	Mirz	2018	optimal control theory path planning	linear parallel
[101]	Scheilen	2009	trajectory optimization	linear parallel
[102]	Schmit	2012	trajectory discretized in the joint space	non-linear parallel
[61]	Schmit	2013	trajectory discretized in the joint space	non-linear parallel

Table 1. *Cont.*

Ref.	1st Author	Year	Path and Trajectory Planning	Compliance of Elastic Elements
Free-vibration response				
[103]	Barreto	2017	defined by the multibody simulator	linear parallel
[81]	Barreto	2018	defined by the multibody simulator	linear parallel
[104]	Kashiri	2017	periodic motion defined by the natural frequency	variable serial
Periodic trajectory learning				
[105]	Khoramshahi	2017	harmonic trajectory	linear/variable parallel
[83]	Nasiri	2016	sinusoidal, polynomial and sign basis functions	linear parallel
[106]	Noorani	2013	motion defined by the pendulum	linear, applied at pendulum tip

3. Design of Desired Natural Dynamics

The concept of natural motion can be exploited to perform tasks that are compatible with the natural dynamics of a mechanical systems, that is, cycle tasks. For this reason, such a concept is mainly adopted for locomotion of legged robots or, in the industrial field, for pick-and-place and palletizing tasks.

A pick-and-place (or palletizing) task has strict requirements on the positions where objects are located and the corresponding velocities. Specifications on the task time and the path could potentially be relaxed constraints.

Conversely, walking robots do not need strict requirements for the gait. In this case, if the robotic system is already equipped with elastic elements, the desired task (the robot gait) can be adapted to the system characteristics, that is, The task can be performed by adopting the trajectory naturally generated by the system. Such an approach can be referred to as *natural dynamic exploitation* [96]. Generally speaking, when there are requirements on the task to execute, it is always necessary to modify the system to fulfill them. The adaptation of the system to perform a given periodic task can be indicated as *natural dynamic modification* [73].

As discussed in Section 1, the modifications of a robotic system to exploit the natural motion rely on spring additions. Therefore, the problem can be reformulated in this way: how should the spring parameters be determined in order to fulfill the task requirements exploiting system dynamics? Such a question has not a unique and straightforward answer. Requirements on task are typically given in terms of positions, velocities and time. Dynamic models of robotic systems depend non-linearly on the first two. Additionally, to find an analytical solution for the motion equations is very difficult, if not impossible, and hence also the setting on the time requirement is impossible, unless simplifications (such as linearization) are adopted or assumptions on the system response law are made. Therefore, since the optimal spring depends on the trajectory and *vice versa*, we classify the methods concerning with natural motion according to the trajectory followed by the system:

- *Defined trajectory.* A feasible trajectory for the desired task (typically harmonic) is imposed and the spring parameters are optimized to minimize a given objective function related to energy consumption;
- *Optimized trajectory.* The spring parameters and the system trajectories are concurrently optimized thanks to a parametric representation of both or by adopting the optimal control theory;
- *Free-vibration response.* The trajectory is not imposed. The optimal spring parameters are identified so that the free response of the system fulfills task requirements. Such a result can be obtained by means of linearized dynamic models or multibody simulators;
- *Periodic trajectory learning.* The robotic system is not modified. The forced response at resonance of the system is learned by means of proper tools and used as reference trajectory.

These four categories are analyzed and discussed in the following.

3.1. Defined Trajectory

Most of the works dealing with natural motion adopt a fixed pre-defined trajectory, which usually consists of an harmonic motion law. In the case of defined trajectory, the correct springs parameters to be added in the system can be determined with different strategies: control-based methods, graphical approaches or optimization strategies. These three approaches are discussed in the following.

3.1.1. Control-Based Methods

M. Uemura et al. in Reference [77], employ a control method based on linear stiffness adjustment to reduce the actuator torque needed to follow a desired harmonic trajectory with a serial link system. The proposed adjustment law for the parallel spring stiffness depends on the angle and angular velocity tracking errors. Such a law is added to the feedback control system: when the tracking errors become smaller by the stiffness adjustment, the feedback terms become smaller as well. In this manner, the actuator torques are reduced. The variability of the linear stiffness is exploited just to tune the system at different desired frequencies of the harmonic trajectory. In particular, the stiffness varies just at the beginning of the motion, during the tuning phase, before converging to a constant value.

H. Goya et al. in Reference [16], experimentally validated the stiffness adaptation control on a 3-DOF SCARA robot equipped with two adjustable parallel springs on the two revolute joints. The desired start and end points are determined in the task-oriented coordinates and the corresponding points in joint angle coordinates through inverse kinematics. The start and end points in the joint space are connected through an harmonic trajectory with a given period and amplitude equal to the distance between the start (end) point and the equilibrium point. The equilibrium position of the elastic element is set at the middle point between the start point and the end one in the joint coordinates. The springs are tuned to perform the given motion between two points, as discussed in Reference [77].

An alternative method to perform multi-point trajectories and concurrently exploit the adaptive stiffness control to minimize the actuator efforts is proposed by K. Matsusaka et al. in Reference [93]. In this work, it is assumed that the elasticity can be instantaneously changed when the spring is in equilibrium. In this way, the amplitude of the oscillation can be modified without affecting the potential energy stored by the elastic element. With respect to the strategy proposed by H. Goya et al. based on the feedback control method [16], this approach allows to increase the number of pick-and-place points and to reduced the energy consumption by 39%, as proved experimentally. Furthermore, M. Uemura et al. prove in Reference [98] the effectiveness of the stiffness adaptation controller [77] on a 1-DOF system. Multi-frequency harmonic trajectories are tracked, while minimizing the norm of the required torque.

Variable elastic elements and the control proposed in Reference [77] are used by K. Matsusaka et al. in Reference [94] to improve the energy-efficiency of a 2-DOF robot in palletizing task. For such a task an obstacle-avoidance trajectory is proposed consisting in moving both the joint with harmonic trajectories having an angular frequency ratio 2:1. Additionally, to increase the energy savings, the authors suggest to provide the system not only with variable stiffness springs but also with linear constant springs. In this way, the gravitational force can be counteracted.

An improvement of the adaptation stiffness controller in terms of energy saving is obtained in Reference [97], by combining such a controller with a delay feedback control. However, although the new approach allows to move the system almost without any actuation force, the delay feedback control modifies the desired trajectory. Therefore, the resulting motion can be significantly different from the desired one.

A solution to improve the energy savings of the adaptive stiffness control and achieving a good trajectory tracking control is proposed in Reference [99] and experimentally validated in Reference [78]. The new control method combines the stiffness adaptation and the iterative learning control and can be applied to multi-joint robots. The proposed control optimizes the stiffness of elastic elements installed in parallel at each joint in order to save energy and to track a desired multi-frequency harmonic trajectory.

An online adaptation method suitable for non-linear compliance acting in parallel with actuators is presented in Reference [58]. The method aims at minimizing actuation forces of multi-joint robots performing given cyclic tasks. Such a method adds an adaptation rule in parallel to the closed-loop control in order to minimize the squared actuation forces. The parameters that allow to minimize the force are the adaptable coefficients used in the compliance definition. The compliance of each actuated joint is defined as a multi-basis non-linear compliance, that is, as the sum of the products between a coefficient and a smooth basis function defined over the joint position. In particular, the compliance structure is defined by the basis functions that are decided a priori and fixed, whereas the coefficients are adaptable and used to minimize the cost function. By choosing a proper set of basis functions (e.g., polynomials), the compliance force acts as a general function approximator. Hence the elastic force has more flexibility to compensate the actuation torques, which are typically in a non-linear relationship with the joint angles. Similar to the other methods discussed up to now and based on control system, this method does not require any knowledge of the controlled system or of the dynamical equations of the robot.

A recursive algorithm to adjust the configuration of a variable spring actuator for a given trajectory is proposed in Reference [40]. Unlike the previous methods to tune variable stiffness, such an algorithm does not optimize the mechanical output. It directly reduces the input electrical energy requirements of the system during the execution of repetitive tasks without requiring a precise knowledge of the controlled system. It is based on the gradient descent optimization algorithm [107]. Basically, it expresses the objective function (the total electrical input energy of the actuator) as a convex function in the design parameters (i.e., The spring stiffness). An iterative process finds the values of the configuration parameters that minimize the objective function in a repetitive task performed by the actuator. The inputs of the algorithm are real-time measurements or estimations of the objective function, the physical limits of the design parameters and the periodicity of the task.

All the control-based approaches discussed up to now do not require the knowledge of the system dynamic model. Such an aspect not only simplifies their implementation but makes these approaches more robust to typically unmodeled physical phenomena as, for example, friction or noise, as experimentally demonstrated in References [40,99]. In Reference [99], the validity of the method is proved also in the presence of static friction, Coulomb friction and backlash. In Reference [40], the robustness of the algorithm to the variation of the objective function due to the changes in the operation condition, perturbations and signal noise is demonstrated. On the other hand, all these methods, although exploit variable stiffness springs, are intended for repetitive tasks with a relatively low rate of change of the stiffness. Therefore, the convergence to the optimal stiffness value is not instantaneous but can last several cycles.

A method proposed by W. Schiehlen and N. Guse [64] takes advantage of the inverse dynamic model and considers constant linear springs mounted in parallel with the actuators. A control based on limit cycle to reduce energy consumption in robotic system performing periodic tasks is proposed. Given the period of the task and the desired trajectory of the system, this is then adapted to match the limit cycle of the mechanical system at best.

The definition of the desired trajectory includes the identification of the boundary conditions for the state of the motion as well (i.e., position and velocity of the start and end points). By adopting a modified shooting method [108], the values of the stiffness and the neutral position of the parallel spring are adjusted to meet the boundary conditions of the system featuring a limit cycle close to the desired trajectory. A low energy control is sufficient to force the system to follow the desired trajectory. With reference to a 2-DOF assembly robot performing an harmonic motion law, the limit cycle of the system can be correctly adjusted by properly choosing the coefficient of linear springs. In the case of an arbitrary, non-harmonic motion law, such as a piece-wise constant acceleration trajectory, non-linear springs are needed to correctly adjust the limit cycle to the desired trajectory. In Reference [91], R.B. Hill et al. extend the method proposed by W. Schiehlen and N. Guse to multiple-point trajectory. In this case, variable stiffness springs are employed. The spring parameters can be tuned to force the limit cycle to converge to the desired pseudo-periodic trajectory between every two consecutive points of the multiple-point motion.

G. Lu et al. propose a control method starting from a linearized model of a 2-DOF planar manipulator, lying in the horizontal plane [72]. The purpose is to save energy in performing a given harmonic trajectory. The controller adaptively tunes the stiffness of parallel springs, while compensating for viscosity effect. The adaptive control law for the stiffness is proportional to the errors between the actual and the desired joint velocities and the difference between the actual joint position and its equilibrium one. The same authors propose an inertia adaptive control for energy savings in Reference [92]. This work starts from a 1-DOF system already equipped with a constant linear spring and add a movable mass to tune and adapt the eigenfrequency of the system. In this way, the frequency of the desired harmonic trajectory can be reached.

Another control-based approach to employ stiffness variability for energy efficiency during a task is proposed by A. Velasco et al. in Reference [79]. The method aims at determining the optimal stiffness profile of serial springs that minimizes the energy consumption of the mechanical system, performing a given task. Since changing the actuator stiffness has an energetic cost, the authors add such a cost to the objective function (squared mechanical torque). By exploiting the cost function and the motion equations of the actuated system, an analytical solution for optimal stiffness as function of the desired trajectory is found. In particular, the total time is subdivided in intervals and an optimal stiffness profile, that can be constant or variable, is determined for each sub-interval. Simulations with squared trajectories with different amplitudes and frequencies show that the optimal spring profile (i.e., constant or variable) depends on the reference trajectory. Constant stiffness is preferred for low amplitudes and low frequencies trajectories, whereas variable stiffness is suggested for high frequency motion laws.

3.1.2. Methods Based on Force-Displacement Graphs

W. Schiehlen and N. Guse propose in Reference [95] an alternative method to that in Reference [64] for determining the spring design parameters that minimize the actuators work. The required control forces or torques are computed by means of the inverse dynamics and the desired trajectory. Such forces are then expressed as function of the positions of all the bodies and fitted by means of polynomials. If a linear function is used as a curve fit, which represents a linear spring, the spring coefficient is the slope of this curve fit and the axis intercept is the spring fastening. The use of curve fitting is equivalent to a minimization task with inequality constraints, since the stiffness coefficient is always positive.

The exploitation of the force-displacement graph is suggested also by M. Khoramshi et al. [65]. In particular, they compute the absolute work along the force-displacement graph and derive the value of the parallel spring stiffness that minimize it. Such a strategy results in a non-linear stiffness profile capable of improving the energy efficiency of the system. Indeed, with the adding of parallel non-linear compliance, the resulting force-displacement graph is lined up around the horizontal axis (i.e., null force).

3.1.3. Methods Based on Offline Optimization

An analytical method for the offline computation of the optimal parallel compliant elements and the frequency of the reference trajectories for serial manipulators performing cyclic tasks is presented by M. Shushtari et al. [96]. A representation of the compliant elements similar to those used by R. Nasiri et al. [58] is adopted. In particular, they take into account a multi-basis representation consisting in the product between a compliance coefficient and a basis function dependent on the joint coordinates. Such a representation leads to a cost function (squared mechanical torques multiplied by a weighting matrix) that is a quadratic function with respect to the compliance coefficients and quartic with respect to the task frequency. This function can be analytically solved. In order to address the multi-task case, M. Shushtari et al. propose a weighted sum of the original cost functions defined over the tasks.

P. Boscariol et al. [89] set up a constrained optimization problem to find the optimal stiffness value and placement of a linear spring to be added in parallel with the joint actuator of a 1-DOF system for reducing the peak torque requirement (Figure 4). The authors also shown the effects of the joint trajectory (cycloidal motion trajectory and 5th order polynomial with null initial and final acceleration) on the optimization results.

In Reference [90], G. Carabin et al. propose a methodology to reduce the electrical energy consumption in a Delta-2 robot by concurrently exploiting energy recuperation drive axles and torsional springs mounted in parallel to the actuators. Figure 5 reports the kinematic diagram and the electrical schematic of the Delta-2 robot. An optimization-based design method determines the stiffness of the two torsional springs and their equilibrium positions. The energy consumption performing cyclic pick-and-place operations, with a predefined trajectory (double-S speed profile in the workspace) is minimized.

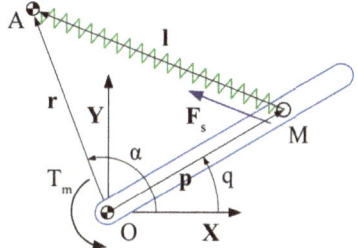

Figure 4. The 1-DOF mechanism with linear spring used by P. Boscariol et al. [89].

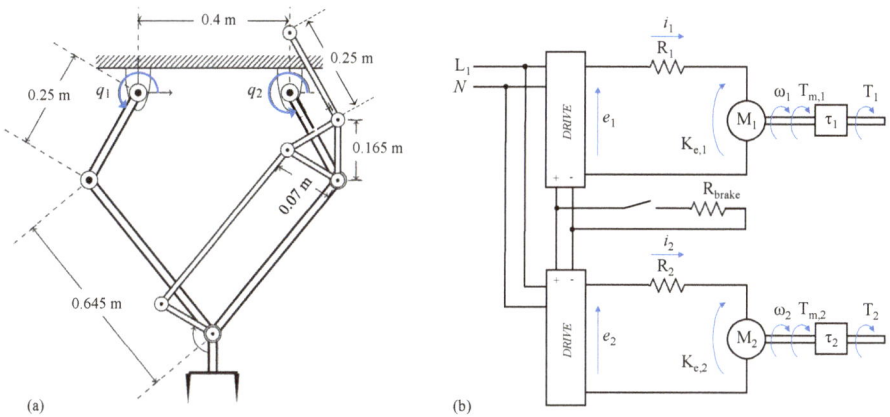

Figure 5. (a) The Delta-2 robot and (b) the electrical schematic of its energy recuperation drive axles employed by G. Carabin et al. [90].

A. Velasco et al. in Reference [86], determine the optimal stiffness value and spring preload such that a given cost functional is minimized. In particular, they consider the influence on the optimal values of different aspects: spring placements (i.e., serial or parallel actuator); parameters of a given harmonic trajectory (amplitude and frequency); cost function (squared mechanical torque or squared mechanical power). From such a study, it results that, if parallel spring are employed, the optimal values for spring stiffness and preload can be analytically found regardless the cost function employed. Conversely, for serial springs an analytical solution exists only if the cost function is the squared torque. As far as the effects of the trajectory parameters on the energy savings are concerned, the authors show that, if serial springs are employed, the savings depend also on the cost function adopted. On the other hand, parallel springs have the same trends independently from the cost function. In particular, it results in the parallel springs being more convenient for small amplitudes at low frequency or large amplitudes at high frequency. The use of serial springs is more convenient for small amplitudes at high frequency or for large amplitudes at low frequency, if mechanical torque is considered or for high frequencies independently from the amplitude, if mechanical power is considered.

3.2. Optimized Trajectory

Another strategy to improve the energetic efficiency of the system consists of the concurrent optimization of trajectory and spring stiffness. Two are the most common methodologies to reach such a goal—methods based on the optimal control theory, which find an optimal control law that minimizes a given cost function, and methods that parametrize the trajectory by means of basis functions.

3.2.1. Methods Based on the Optimal Control Theory

W. Schiehlen and M. Iwamura [101] simultaneously optimize the constant linear spring stiffness (mounted in parallel with the actuator) and the joint trajectories with respect to the energy consumption, taking advantage of the optimal control theory. Following their approach, the time to execute the task is not given but it is optimized as well. Indeed, they define a relationship between the consumed energy and the operating time and the optimal trajectory, finding a condition for the operating time to be optimal. Then, the optimal design for the springs is derived accordingly to such a time. The aforementioned relationship as well as all the optimal solutions are derived starting from linearized equations of motion. An analytical solution is provided for the optimal trajectory (which results in harmonic motions in modal coordinates), the operating time and the spring equilibrium position (corresponding to the middle point between the initial and final desired points). Conversely, the optimal stiffness is numerically found by means of an iterative method. Although the authors verify the correctness of the analytical solution proposed by comparing it with the numerical one based on the non-linear dynamics, their method is valid as long as the linearization assumptions are respected, that is, The centrifugal and Coriolis forces are negligible. This means that such a method is suitable if fairly strong springs or long operating time are adopted. A manipulator prototype, respecting the validity assumptions, is used by M. Iwamura et al. [100] for the experimental validation of this approach. The method proposed by W. Schiehlen and M. Iwamura [101] is extended by the same authors to systems working under gravity in [56].

The optimal control theory is exploited by C. Mirz et al. [82] to reduce energy consumption in parallel kinematic manipulators by means of linear torsional springs mounted in parallel to the motors. The power consumption of the drives is selected as cost function to calculate the optimal trajectory to travel between given initial and final positions in a fixed cycle time with minimum energy. By applying the Pontryagin's minimum principle, the cost function is transformed in a two points boundary value problem solved numerically by means of finite difference method. Although the characteristics of the elastic elements are determined for one specific initial and final position and a given cycle time, simulations showed satisfactory results in terms of energy savings also performing pick-and-place operations between 200 different positions arranged in a square about the initial and final points.

3.2.2. Methods Based on Parametrization through Basis Functions

N. Schmit and M. Okada et al. [102] minimize the actuator torques by simultaneously designing the robot trajectory and the torque profiles of the non-linear parallel springs located at each joint. The desired time interval to perform the task is divided into a certain number of equal sub-intervals, based on which a third-degree Hermite interpolation is used to parameterize both the joint trajectory and the spring torque profiles. By expressing the spring torques as function of the trajectory, the position and velocity of each joint at the nodes become the only design parameters. The optimal joint trajectory is found numerically minimizing a cost function composed of three terms—a term evaluating the actuator torques, a term evaluating the improvement due to the contribution of the non-linear springs and a term that weights the non-linearity of the profiles to guarantee their technical feasibility. Once the optimal joint trajectory is found, the optimum spring is designed as function of the optimal joint trajectory with a closed-form solution. However, the resulting springs may exhibit negative stiffness. Such an issue is overcome in Reference [61], where constraints to impose positive stiffness to given joint coordinates are added to the optimization problem.

The use of polynomials to parameterize the joint trajectory and spring torque profiles is suggested by H.J. Bidgoly et al. in Reference [73]. In contrast to Reference [102], the authors adopt one polynomial (with a degree to be determined by the designer) for all the time interval. Also in this case, the optimal spring are computed analytically, whereas the joint trajectories are numerically optimized: the spring torque profiles are polynomial functions of the joint trajectories, which, in turn, are polynomial functions of time. The optimal trajectories are found by solving a constrained optimization problem whose cost function takes into account actuator torques and realization complexity of the non-linear springs. The possibility to obtain an optimal trajectory, in the meaning that it is very close to the robot natural behavior, is increased by the assumption that the system is redundant with respect to the task, so that infinite solutions of inverse kinematics are possible. However, although the authors show that very good results can be obtained in terms of actuation minimization with their methodology, they do not provide an evidence that kinematic redundancy has advantages over the non-redundant case.

R.B. Hill et al. in References [80,91], propose a method to increase the energy efficiency of parallel robots, by concurrently optimizing the joint trajectory and the control law of variable springs mounted in parallel with the system actuators. Figure 6 shows the adopted five-bar mechanism performing a pick-and-place operation and the power transmission system of variable stiffness springs. The authors do not focus on the spring profiles but on the joint trajectories of the spring actuators so that a classical position controller can be used without a feedback measure of the stiffness. Similar to Reference [102], the total time to perform the task is subdivided in a certain number of equal subinterval, defined by the number of intermediate points (via points) chosen between the initial and the final positions for both the robot joints and spring motors. The via points are interconnected thanks to an expression defined by means of four polynomials, each one function of position, velocity and acceleration of the considered link and of the time, respectively. The explicit form of each polynomial is obtained by means of a heuristic approach: different functions are experimentally tested until the best polynomial form in terms of consumed energy during the motion is obtained. The optimal trajectory is defined looking for the position, velocity and acceleration of the intermediate points of both the robot joints and spring motors that minimize the energetic losses of the entire actuation chain (robot joint motors and variable stiffness motors). The larger advantage of the method proposed by R.B. Hill et al. in References [80,91] over the other discussed in this section, is the possibility of planning and optimizing multiple-point trajectories.

3.3. Free-Vibration Response

In the context of natural motion, few methods exploit the free-vibration response of the robotic system for performing a desired task. N. Kashiri et al. [104] propose a method to exploit the natural dynamics of serial robots driven by adjustable compliant actuators mounted in series. A feedback linearization control is used to linearize the system dynamics and to generate modally decoupled limit cycles. Having decoupled the system dynamics, it is possible to analytically write the response of the closed-loop system and to set its natural frequencies to the target values by properly tuning the compliant elements. The control scheme generates a smooth reference link position to excite an individual mode, as well as a combinations of modes allowing for the desired periodic motion with a minimal energy expenditure.

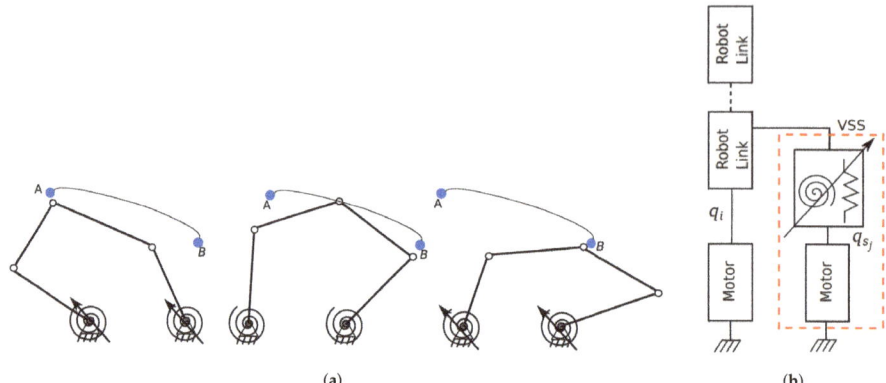

Figure 6. (**a**) Pick-and-place operation performed by a five-bar mechanism equipped with parallel variable stiffness springs (VSS). (**b**) Scheme of the VSS power transmission system employed by R.B. Hill et al. [91], where q_i represents the i-th robot joint coordinate, whereas q_{sj} the j-th variable stiffness spring coordinate.

J.P. Barreto et al. exploit the free response of a five bar linkage mechanism [103] and of a Delta robot (Figure 7) [81] to move the system between two given points for a pick-and-place operation in a prescribed time. To reach such a result, a set of springs with optimized parameters (stiffness and equilibrium positions) are added in parallel to the motors. The optimal spring parameters are found by solving the system direct dynamics with the aid of a multibody simulator. By imposing the initial desired position of the system (pick position) and taking a first guess for the spring parameters, the simulator computes the system positions and velocities after the task time, which are compared to the desired ones (place positions and velocities). Iterative simulations are carried out until small enough errors are obtained. Ideally, the resulting system does not require any actuation torque to perform the motion. However, to stop the system in the pick-and-place positions, actuators counteracting the springs or mechanical breaks are needed.

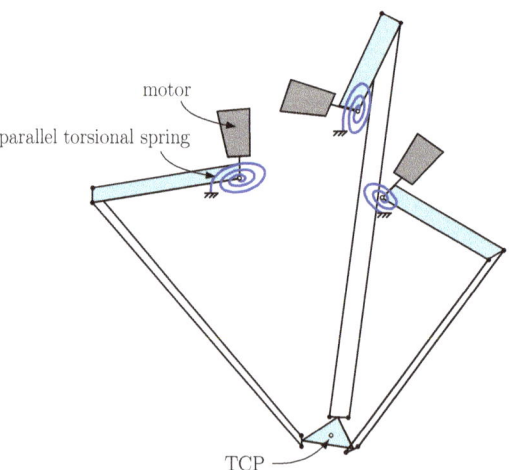

Figure 7. Delta robot equipped with parallel torsional springs.

3.4. Periodic Trajectory Learning

A mechanical system can move using, as a reference trajectory, the one that it naturally generates. Proper tools, such as adaptive oscillators, can be used to learn the periodic trajectories and adopt them as reference, especially for robot locomotion and dynamic walking [109–111]. In this manner, the reference trajectory is synchronized with the resonant frequency of the system, which results in energy saving. A tool capable of obtaining smooth and lag-free estimates of the frequency and phase of an external quasi-periodic signal is given by the adaptive frequency oscillators (AFO), introduced by L. Righetti et al. [112,113]. AFO are adopted in the literature for the estimation of cyclical movements, especially for rehabilitation purposes and in robots performing quasi-periodic motions. However, the convergence and optimality of these methods are in general not ensured [110,111]. Adaptive oscillators are used in central pattern generator to learn a specific rhythmic pattern, by synchronizing the reference trajectory with the resonant frequency of the robotic system and earning energy efficiency. For example, in Reference [109], Buchli et al. present a 4-DOF spring-mass hopper with a controller based on adaptive frequency Hopf oscillators, which adapts to general, non-harmonic signals. The adaptive oscillator adapts to the properties of the mechanical system, in particular to its resonant frequency.

In order to tune the frequency and shape of the cyclic natural motion for energy efficient, novel oscillators, that is, the adaptive natural oscillators, are introduced. M.R.S. Noorani et al. present in Reference [106] an adaptive frequency non-linear oscillator for energy efficiency that exploits the resonant mode in a leg-like mechanical system called stretchable pendulum. The system is a simple oscillating mass-spring mechanism that interacts with the ground during its oscillation. A Hopf non-linear oscillator is placed in the feedback loop and its frequency tracks the resonance frequency of the mechanical system. The system not only earns energy efficiency but has also the ability to adapt with a changing environment.

M. Khoramshahi et al. and R. Nasiri et al. present in References [83,105] a linear and a non-linear adaptive natural oscillator, ANO and NANO, respectively. These tools are capable of tuning the frequency and the shape of cyclic motions for energy efficiency and ensure optimality and convergence. Moreover, they are built upon the adaptive frequency oscillators but, in contrast to AFO that adapts to the frequency of an external signal, ANO adapts the frequency of reference trajectory to the natural dynamics of the system (Figure 8). In Reference [105], the efficiency of ANO is shown in the simulations of a hopper leg and of a compliant robotic manipulator performing a cyclic task. Furthermore, experimental results of a 1-DOF joint with variable compliance (Figure 9) show the feasibility of the approach, exploiting the natural dynamics and reducing the consumed energy. In Reference [83], the non-linear adaptive natural oscillator (NANO), adopted to optimize the shape of reference trajectories, is presented. The oscillator ensures stability, convergence and optimal solution. Three robotic models are investigated in simulation: the pendulum, the adaptive toy and the hopper-leg, showing the feasibility of the proposed approach to achieve energy efficiency.

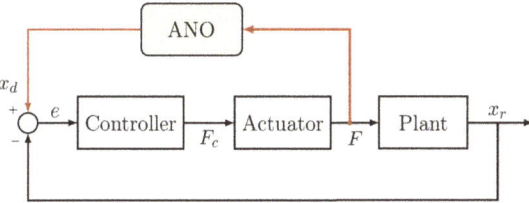

Figure 8. The adaptive natural oscillator proposed by M. Khoramshahi et al. [105] to exploit natural dynamics. The oscillator is employed as a pattern generator and the applied force is used as feedback.

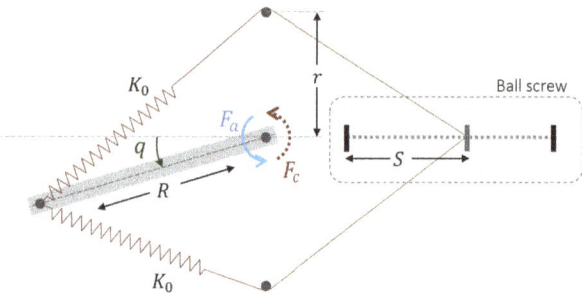

Figure 9. The 1-DOF mechanism with variable compliance presented by M. Khoramshahi et al. [105]. The linear spring acts as a parallel rotational spring at the joint and its stiffness can be tuned by controlling the ball screw mechanism.

4. Design Optimization with Natural Motion

In the previous sections, we have analyzed the mechanical design of both robotic and mechatronic systems for natural motion and of desired natural dynamics to ensure energy efficiency. It should be noted that, in the majority of the works, the exploitation of natural motion is strictly related to the definition of an optimal design problem. Indeed, spring parameters (with variable or non-variable compliance and linear or non-linear behavior) as well as trajectory parameters are designed using an optimization strategy. In this section, we provide an overview on the optimization problems adopted for the design of trajectory and spring parameters. Table 2 reports a summary on the optimal design problems, showing design variables, objective and constraint functions and algorithms adopted in the reviewed works. Furthermore, in the last column of the table the results obtained in terms of energy (or torque) efficiency are reported.

Before analyzing the optimization problems in detail, it is necessary to make a distinction between works in which the design formulation includes spring parameters, works in which it considers trajectory parameters or both. In problem formulations in which the spring parameters are to be determined, both stiffness and equilibrium position are typically design variables. The second design formulation finds the optimal values for the trajectory and task time, frequency or coefficients of the motion profile are design variables. Another formulation is the concurrent design optimization of spring parameters and trajectory properties. This is the most complex optimization formulation to be implemented, since the properties of elastic elements usually influence the trajectory and *vice versa*. Despite the complexity, this formulation allows for a high degree of design flexibility and, therefore, potential to further reduce energy consumption.

Several methods are used to solve the optimal design problem with natural motion. Due to their efficiency with convex problems, the majority of the works about natural motion utilize gradient-based optimization algorithms. Due to their use of gradients and also use of Hessians, these algorithms are referred to as first and second order algorithms. J.P. Barreto et al. [81] adopt the gradient-based trust-region-dogleg algorithm on an unconstrained problem, whereas H.J. Bidgoly et al. [73], P. Boscariol et al. [89] and R.B. Hill et al. [80] use MATLAB's gradient-based *fmincon* function. R. Fabian et al. [40], as well as M. Khoramshahi et al. [65] and R. Nasiri et al. [83] adopt a gradient descent algorithm. The unconstrained gradient-based *shooting method* was used by R.B. Hill et al. in Reference [91]. N. Schmit et al. [61,102] adopts the sequential quadratic programming (SQP) algorithm. A gradient-free method is implemented in Reference [103], where a downhill simplex algorithm is considered. In some cases, an analytical solution is found to the optimal design problem, avoiding the cost of setting up and solving numerical optimization problems. These papers include References [79,86,96,98,99].

Table 2. Optimal design formulation with natural motion.

Ref.	1st Author	Year	Design Variables	Objective Function	Optimization Constraints	Algorithm	Results
Spring parameters							
[51]	Barreto	2018	stiffness and equilibrium position	position and velocity errors	bounded unconstrained	trust-region-dogleg	100% of torque reduction (numerical) 47% (numerical) and 32% (experimental) of peak torque reduction
[89]	Boscariol	2017	stiffness and location of spring connecting points	torque norm	bounded unconstrained	MATLAB fmincon	up to 70% of energy reduction (numerical)
[90]	Carabin	2019	stiffness and equilibrium position	electrical energy	bounded unconstrained	second-order NLPQLP	53% of input torques reduction (numerical)
[91]	Hill	2019	coordinates of variable stiffness spring	position and velocity errors	bounded unconstrained	shooting method	-
[40]	Fabian	2018	configuration of variable stiffness actuator	maximum absolute energy per cycle	bounded unconstrained	gradient descent	75% of energy reduction (numerical)
[65]	Khoramshahi	2014	stiffness and equilibrium position	absolute mechanical work	unbounded	analytical solution	up to 55% of torque reduction (numerical)
[98]	Uemura	2008	stiffness	torque norm	unbounded	control method	90% of torque reduction (numerical)
[99]	Uemura	2009	stiffness	torque	unbounded	control method	up to 90% of cost function reduction (numerical)
[86]	Velasco	2013	stiffness and equilibrium position	squared-power; squared-torque	task position, velocity and acceleration	analytical solution	17% of cost function reduction (experimental)
[79]	Velasco	2015	stiffness profile	squared-torque and force needed for stiffness variation	task position, velocity and acceleration	analytical solution	
Trajectory parameters							
[73]	Bidgoly	2017	coefficients of torque profile	weighted sum of actuator torque and complexity of torque profiles	end-effector path, physical limitations on joints and motion periodicity	MATLAB interior-point	up to 99.9% of actuator cost reduction (numerical)
[82]	Mirz	2018	motion profile	energy	bounded unconstrained	optimal control theory	up to 84% of energy reduction (numerical)
[58]	Nasiri	2017	non-linear compliance parameters	actuation force	bounded unconstrained	Newton gradient descent	100% (numerical) and 24% (experimental) of energy reduction
Spring and trajectory parameters							
[103]	Barreto	2017	stiffness, equilibrium position and operating time	task time and position errors	bounded unconstrained	Nelder-Mead direct search	up to 68% of energy reduction (numerical)
[80]	Hill	2018	variable stiffness spring and trajectory parameters	power losses of joints and variable stiffness spring motors	bounded unconstrained	MATLAB fmincon	48% of energy reduction (numerical)
[56]	Iwamura	2011	stiffness, equilibrium position and operating time	energy	bounded unconstrained	optimal control theory	up to 100% of energy reduction (numerical)
[100]	Iwamura	2016	stiffness, equilibrium position and operating time	energy	bounded unconstrained	optimal control theory	94% of energy reduction (experimental)
[101]	Scheilen	2009	stiffness, equilibrium position and operating time	energy	bounded unconstrained	optimal control theory	up to 100% of energy reduction (numerical)
[102]	Schmit	2012	trajectory and torque profiles of non-linear springs	torque	function trajectory parameters at starting and ending time	SQP	up to 99% of actuators torque reduction (numerical)
[61]	Schmit	2013	trajectory and torque profiles of non-linear springs	torque	stiffness characteristic	SQP	up to 95% of actuators torque reduction (numerical)
[96]	Shushtari	2017	stiffness and task frequency	total energy	bounded unconstrained	analytical solution	-

An alternative approach is to tackle the problem with an optimal control theory approach. This is used for the optimal design problems by M. Iwamura et al. [56,100], by C. Mirz et al. [82], who along with W. Scheilen et al. [101] implement the Pontryagin's minimum principle (see Section 3.2.1).

Most of the reviewed works adopt objective functions based on actuators torque, on consumed energy or power. A few works use objective functions based on task time and positions [103], position and velocity of the end-effector [81] or position and velocity errors [91]. Results shown in the last column of Table 2 indicate that all the reviewed approaches for the optimal design problems in natural motion consistently reduce energy consumption.

The vast majority of papers utilize gradient-based optimization algorithms. Despite the great efficiency of these algorithms, the validation is typically carried out on simple test cases (as reported in Table 3) with low number of design variables and a convex objective functions. This review did not result in any research using genetic algorithms or evolutionary strategies employed in the field of natural motion. Although, these algorithms are less efficient, they can often handle non-convex problems, discrete design variables and noisy system functions.

Table 3. Mechanical systems employed for the numerical and experimental validations, grouped together with respect to their kinematic structure. The check-marks on the second and third columns indicate whether gravity is considered and whether experimental tests are performed, respectively.

Mechanical System	Gravity	Experimental Tests	Ref.	1st Author	Year
1-DOF revolute joint			[7]	Boscariol	2017
		✓	[40]	Fabian	2018
		✓	[105]	Khoramshahi	2017
	✓	✓	[58]	Nasiri	2017
	✓		[56]	Iwamura	2011
	✓		[77]	Uemura	2007
	✓		[98]	Uemura	2008
	✓		[86]	Velasco	2013
	✓	✓	[79]	Velasco	2015
2-DOF planar robot (RR)	✓		[40]	Fabian	2018
	✓		[56]	Iwamura	2011
		✓	[100]	Iwamura	2016
	✓		[104]	Kashiri	2017
	✓	✓	[94]	Matsusaka	2016
		✓	[93]	Matsusaka	2016
	✓		[58]	Nasiri	2017
			[64]	Schiehlen	2005
			[95]	Schiehlen	2005
			[101]	Schiehlen	2009
	✓		[96]	Shushtari	2017
	✓		[77]	Uemura	2007
	✓		[97]	Uemura	2008
	✓	✓	[99]	Uemura	2009
			[78]	Uemura	2014
		✓	[86]	Velasco	2013
			[79]	Velasco	2015
2-DOF planar robot (PP)	✓		[96]	Shushtari	2017
3-DOF SCARA		✓	[16]	Goya	2012
3-DOF planar robot (RRR)	✓		[73]	Bidgoly	2017
3-DOF spatial robot (RRR)	✓		[102]	Schmit	2012
	✓		[61]	Schmit	2013
3-DOF spherical robot (RRP)	✓		[96]	Shushtari	2017
4-DOF spatial robot (RRRR)	✓		[73]	Bidgoly	2017
Slider-crank mechanism			[95]	Schielen	2005
Five-bar linkages			[103]	Barretto	2017
	✓		[80]	Hill	2018
	✓		[91]	Hill	2019
			[82]	Mirz	2018
Delta-2 robot	✓		[90]	Carabin	2019
Delta robot	✓		[81]	Barretto	2018

As it is the goal to consider complex mechanical systems and expanded design problems, the optimization formulation will require a higher number design variables including the parameters of the motion law. This may lead to non-convex optimization problems for which gradient-based algorithms are not suitable. Thus, the proper optimization formulation, the system equations and corresponding choice of optimization algorithm will play a critical role moving forward with the optimal design with natural motion.

5. Discussion

In Section 2, we analyzed several approaches to perform a cyclic task with a robotic system by exploiting its the natural motion, that is, a motion mainly due to the transformation of the potential energy stored by elastic elements into kinetic energy. The application of the natural motion is demonstrated to be beneficial in terms of energy consumption on different mechanical systems (see Table 3) by means of simulation or experiments. From the table, it is evident that most of the methods are validated on very simple test cases, that is, planar systems with one or two revolute joints. In all the cases, systems with rigid-links are considered, so there are no examples of robotic systems with flexible links or fully compliant mechanisms on which the natural motion is applied. Considering the increasing interest in these kind of systems, the effect of link flexibility for the exploitation of the natural motion should be investigated, since link flexibility affects the system motion [114]. This could be taken into account in two ways—controlling the resulting unwanted vibrations or making them part of the natural motion trajectory.

For the practical implementation of the natural motion in industrial application, it is important that the robot can perform multi-point trajectories. Indeed, pick-and-place operations are typically carried out between points belonging to two areas, that is, The pick and the place positions are not always the same but can be any position in these area. The majority of the works discussed in Section 2 neglect such an aspect and only consider point-to-point trajectories. Multi-point trajectories are taken into account in References [16,80,82,91,93,96]. Among these, H. Goya et al. [16] and C. Mirz et al. [82] use spring parameters that are optimized for the center of the interested areas to move the system following a multi-point trajectory. M. Shushtari et al. [96] also consider one constant stiffness, whose value results from an optimization problem that takes into account the whole multi-point trajectory. Conversely, K. Matsusaka et al. [93] and R.B. Hill et al. [80,91] optimize the elastic elements for each segment of the trajectory by changing the stiffness values when the robot is stopped in a pick or in a place position. If small pick-and-place areas are considered, the energetic savings using constant stiffness springs is advantageous with respect to the stiff case and even comparable to the varying stiffness (considering the energy cost to change stiffness) [93], despite being a simpler approach. However, when the points of the multi-point trajectory are far from the averaged values, based on which the spring is optimized, such an approach does not work anymore. It becomes difficult to accurately follow the trajectory (because of the elastic forces to be counteracted) and the energy saving are minimal or even null. Therefore, variable stiffness springs become necessary.

The main goal of the application of the natural motion approach is to save energy. All the reviewed methods allow to improve the energy efficiency of a robotic systems, however, it is almost impossible to assert which one outperforms the others to reach such a goal. A benchmark case, with which to perform a comparative analysis between the different methods, is lacking. The benchmark is meant not only in terms of test case but above all in terms of performance evaluation. All the authors present a percentage of energy savings with respect to the stiff case (see the column of results of Table 2) but someone refers to electrical energy others to mechanical one. Most of the contributions neglect an estimate of the losses or the energy consumed, for example, to change stiffness, to activate the breaks. Additionally, the use of different objective functions (e.g., mechanical torque, mechanical energy, electrical energy) does not allow for a proper comparison, since the use of different objective function can lead to very different results as demonstrated by A. Velasco et al. [86].

Generally speaking, the use of optimized trajectories and variable stiffness springs is the more promising approach, since it leads to a trajectory that better fits the system dynamics and allows for flexibility in task execution (point-to-point or multi-point trajectories, different task frequency). On the other hand, methods based on fixed trajectories and constant springs are typically easier to set-up and implement.

6. Conclusions

In this paper we presented a review, a classification and a discussion of several approaches that adopt the concept of natural motion to enhance the energetic performance in robotic and mechatronic systems. In the first part of the paper, we identified the physical requirements that a system has to fulfill to exploit the natural motion and we discussed the technical possibilities to modify it, if necessary. To this end, the configurations in which compliant elements are installed at the joints of the mechanisms (i.e., serial and parallel, with variable and non-variable stiffness) were introduced and compared. Although a serial arrangement seems to be more convenient in terms of energy savings, most of the works deal with compliance mounted in parallel with the main actuators, because of the ease of installation and control.

In the second part of the paper, we classified the approaches related to natural motion on the basis of the trajectory followed by the system: given trajectory, optimized trajectory, free-vibration response and period trajectory learning. Moving from the first category towards the last, the resulting motion gets closer to the system dynamics. Indeed, in the last case, also known as natural dynamic exploitation, the periodic or cyclic task is designed to match the response naturally generated by a given system. In all the other cases, a modification of the system is typically required (natural dynamic modification) so that its dynamic behavior matches a desired task. Regardless the trajectory followed, the results are that, in general, the added compliance parameters can be optimized offline (in the case of non-variable compliance) or can be tuned online (if the compliance is variable). To conclude, methods based on optimized trajectory and variable stiffness springs are most promising. In fact, they allow to approximate quite well the system dynamics (and hence to reduce energy consumption), while preserving task flexibility.

Furthermore, we presented an analysis of optimal design problems in natural motion, since the implementation of natural motion is often related to the definition of an optimization problem. Numerical and experimental results, expressed as percentage reduction of required energy or torque, show that the adoption of natural motion techniques allows to highly increase energy efficiency.

Future developments in the field of natural motion for energy saving in robotic and mechatronic systems should be toward the creation of a benchmark case to better understand the improvements and the potentialities of the different strategies. Further investigations would be necessary in regards to applicability to more complex systems, for example, spatial mechanisms and robots with different kinds of joints. Furthermore, the concept of natural motion could also be extended to robots and mechanisms with flexible links or fully compliant joint-less systems. Finally, the natural motion approach is expected to be applied in more scenarios in both industrial and academic research, where novel challenges could be addressed to maximize its potentialities.

Author Contributions: All authors discussed and commented on the manuscript at all stages. More specifically, L.S. and I.P. collected the related literature, conducted the analysis and completed the draft writing under the supervision of A.G. and R.V.; E.W., A.G. and R.V. contributed to the revision of the paper structure and the presentation style, as well as the proofreading of the paper.

Acknowledgments: This work was partially supported by the Free University of Bozen-Bolzano funds within the project TN2803: "Mech4SME3—Mechatronics for Smart Maintenance and Energy Efficiency Enhancement". This work was supported by the Open Access Publishing Fund provided by the Free University of Bozen-Bolzano.

Conflicts of Interest: The authors declare no conflict of interest.

References

1. Brossog, M.; Bornschlegl, M.; Franke, J. Reducing the energy consumption of industrial robots in manufacturing systems. *Int. J. Adv. Manuf. Technol.* **2015**, *78*, 1315–1328.

2. European Comission. Proposal for a Directive of the European Parliament and of the Council Amending Directive 2012/27/EU on Energy Efficiency (52016PC0761). 2012. Available online: https://eur-lex.europa.eu/homepage.html (accessed on 19 August 2019).
3. International Federation of Robotics. Executive Summary World Robotics 2018 Industrial Robots. 2018. Available online: https://ifr.org/free-downloads/ (accessed on 19 August 2019).
4. Carabin, G.; Wehrle, E.; Vidoni, R. A review on energy-saving optimization methods for robotic and automatic systems. *Robotics* **2017**, *6*, 39. [CrossRef]
5. Inoue, K.; Ogata, K.; Kato, T. An effcient induction motor drive method with a regenerative power storage system driven by an optimal torque. In Proceedings of the 2008 IEEE Power Electronics Specialists Conference, Rhodes, Greece, 15–19 June 2008; pp. 359–364.
6. Albu-Schäffer, A.; Haddadin, S.; Ott, C.; Stemmer, A.; Wimböck, T.; Hirzinger, G. The DLR lightweight robot: Design and control concepts for robots in human environments. *Ind. Robot Int. J.* **2007**, *34*, 376–385. [CrossRef]
7. Boscariol, P.; Gallina, P.; Gasparetto, A.; Giovagnoni, M.; Scalera, L.; Vidoni, R. Evolution of a dynamic model for flexible multibody systems. In *Advances in Italian Mechanism Science*; Springer: Berlin, Germany, 2017; pp. 533–541.
8. Vidoni, R.; Scalera, L.; Gasparetto, A.; Giovagnoni, M. Comparison of Model Order Reduction Techniques for Flexible Multibody Dynamics using an Equivalent Rigid-Link System Approach. In Proceedings of the 8th ECCOMAS Thematic Conference on Multibody Dynamics, Prague, Czech Republic, 19–22 June 2017; National Technical University of Athens: Athens, Greece, 2017.
9. Lorenz, M.; Paris, J.; Schöler, F.; Barreto, J.P.; Mannheim, T.; Hüsing, M.; Corves, B. Energy-efficient trajectory planning for robot manipulators. In Proceedings of the ASME 2017 International Design Engineering Technical Conference and Computers and Information in Engineering Conference American Society of Mechanical Engineers, Cleveland, OH, USA , 6–9 August 2017.
10. Carabin, G.; Vidoni, R.; Wehrle, E. Energy Saving in Mechatronic Systems Through Optimal Point-to-Point Trajectory Generation via Standard Primitives. In Proceedings of the International Conference of IFToMM ITALY, Cassino, Italy, 29–30 November 2018; Springer: Berlin, Germany, 2018; pp. 20–28.
11. Boscariol, P.; Richiedei, D. Energy Saving in Redundant Robotic Cells: Optimal Trajectory Planning. In *IFToMM Symposium on Mechanism Design for Robotics*; Springer: Berlin, Germany, 2018; pp. 268–275.
12. Trigatti, G.; Boscariol, P.; Scalera, L.; Pillan, D.; Gasparetto, A. A new path-constrained trajectory planning strategy for spray painting robots-rev. 1. *Int. J. Adv. Manuf. Technol.* **2018**, *98*, 2287–2296. [CrossRef]
13. Boscariol, P.; Scalera, L.; Gasparetto, A. Task-Dependent Energetic Analysis of a 3 d.o.f. Industrial Manipulator. In Proceedings of the Advances in Service and Industrial Robotics, 28th Conference on Robotics in Alpe-Adria-Danube Region, RAAD, Kaiserslautern, Germany, 19–21 June 2019; Springer: Berlin, Germany, 2020; pp. 162–169.
14. Koditschek, D. Natural motion for robot arms. In Proceedings of the The 23rd IEEE Conference on Decision and Control, Las Vegas, NV, USA, 12–14 December 1984; pp. 733–735.
15. Koditschek, D.E. Adaptive strategies for the control of natural motion. In Proceedings of the 24th IEEE Conference on Decision and Control, Fort Lauderdale, FL, USA, 11–13 Decmber 1985; pp. 1405–1409.
16. Goya, H.; Matsusaka, K.; Uemura, M.; Nishioka, Y.; Kawamura, S. Realization of high-energy efficient pick-and-place tasks of scara robots by resonance. In Proceedings of the IEEE/RSJ International Conference on Intelligent Robots and Systems (IROS), Vilamoura, Portugal, 7–12 October 2012; pp. 2730–2735.
17. Boschetti, G. A picking strategy for circular conveyor tracking. *J. Intell. Robot. Syst.* **2016**, *81*, 241–255. [CrossRef]
18. Bauer, F.; Fidlin, A.; Seemann, W. Energy efficient bipedal robots walking in resonance. *ZAMM J. Appl. Math. Mech. Z. Für Angew. Math. Und Mech.* **2014**, *94*, 968–973. [CrossRef]
19. Della Santina, C.; Lakatos, D.; Bicchi, A.; Albu-Schäffer, A. Using nonlinear normal modes for execution of efficient cyclic motions in soft robots. *arXiv* **2018**, arXiv:1806.08389.
20. Despotovic, Z.V.; Urukalo, D.; Lecic, M.R.; Cosic, A. Mathematical modeling of resonant linear vibratory conveyor with electromagnetic excitation: simulations and experimental results. *Appl. Math. Model.* **2017**, *41*, 1–24. [CrossRef]

21. Comand, N.; Boschetti, G.; Rosati, G. Vibratory Feeding of Cylindrical Parts: A Dynamic Model. In Proceedings of the International Conference of IFToMM ITALY, Cassino, Italy, 29–30 November 2018; Springer: Berlin, Germany, 2018; pp. 203–210.
22. Kim, S.; Lee, J.; Yoo, C.; Song, J.; Lee, S. Design of highly uniform spool and bar horns for ultrasonic bonding. *IEEE Trans. Ultrason., Ferroelectr. Freq. Control* **2011**, *58*, 2194–2201.
23. Palomba, I.; Richiedei, D.; Trevisani, A. Mode selection for reduced order modeling of mechanical systems excited at resonance. *Int. J. Mech. Sci.* **2016**, *114*, 268–276. [CrossRef]
24. Belotti, R.; Richiedei, D.; Trevisani, A. Optimal Design of Vibrating Systems Through Partial Eigenstructure Assignment. *J. Mech. Des. Trans. ASME* **2016**, *138*, 071402. [CrossRef]
25. Wehrle, E.; Palomba, I.; Vidoni, R. In-operation structural modification of planetary gear sets using design optimization methods. *Mech. Mach. Sci.* **2019**, *66*, 395–405.
26. Babitsky, V.; Chitayev, M. Adaptive high-speed resonant robot. *Mechatronics* **1996**, *6*, 897–913. [CrossRef]
27. Vidoni, R.; Scalera, L.; Gasparetto, A. 3-D ERLS based dynamic formulation for flexible-link robots: Theoretical and numerical comparison between the finite element method and the component mode synthesis approaches. *Int. J. Mech. Control* **2018**, *19*, 39–50.
28. Wang, J.; Gosselin, C.M. Static balancing of spatial three-degree-of-freedom parallel mechanisms. *Mech. Mach. Theory* **1999**, *34*, 437–452. [CrossRef]
29. Herder, J.L. *Energy-Free Systems. Theory, Conception and Design of Statically Balanced Spring Mechanisms*; Mechanical Maritime and Materials Engineering: Delft, The Netherlands, 2001; Volume 2.
30. Agrawal, S.K.; Fattah, A. Reactionless space and ground robots: Novel designs and concept studies. *Mech. Mach. Theory* **2004**, *39*, 25–40. [CrossRef]
31. Agrawal, A.; Agrawal, S.K. Design of gravity balancing leg orthosis using non-zero free length springs. *Mech. Mach. Theory* **2005**, *40*, 693–709. [CrossRef]
32. Briot, S.; Baradat, C.; Guégan, S.; Arakelian, V. Contribution to the mechanical behavior improvement of the robotic navigation device Surgiscope. In Proceedings of the ASME 2007 International Design Engineering Technical Conference and Computers and Information in Engineering Conference American Society of Mechanical Engineers, Las Vegas, NV, USA, 4–7 September 2007; pp. 653–661.
33. Baradat, C.; Arakelian, V.; Briot, S.; Guegan, S. Design and prototyping of a new balancing mechanism for spatial parallel manipulators. *J. Mech. Des.* **2008**, *130*, 072305. [CrossRef]
34. Briot, S.; Arakelian, V. A new energy-free gravity-compensation adaptive system for balancing of 4-DOF robot manipulators with variable payloads. In Proceedings of the 14th International Federation for the Promotion of Mechanism and Machine Science World Congress (2015 IFToMM World Congress), Taipei, Taiwan, 25–30 October 2015.
35. Veer, S.; Sujatha, S. Approximate spring balancing of linkages to reduce actuator requirements. *Mech. Mach. Theory* **2015**, *86*, 108–124. [CrossRef]
36. Pratt, G.A.; Williamson, M.M. Series elastic actuators. In Proceedings of the IEEE/RSJ International Conference on Intelligent Robots and Systems, Pittsburgh, PA, USA, 5–9 August 1995; Volume 1, pp. 399–406.
37. Pratt, J.; Chew, C.M.; Torres, A.; Dilworth, P.; Pratt, G. Virtual model control: An intuitive approach for bipedal locomotion. *Int. J. Robot. Res.* **2001**, *20*, 129–143. [CrossRef]
38. Bolívar, E.; Rezazadeh, S.; Gregg, R. A General Framework for Minimizing Energy Consumption of Series Elastic Actuators With Regeneration. In Proceedings of the ASME Dynamic Systems and Control Conference American Society of Mechanical Engineers, Tysons, VA, USA, 11–13 October 2017.
39. Bolívar, E.; Rezazadeh, S.; Gregg, R. Minimizing Energy Consumption and Peak Power of Series Elastic Actuators: A Convex Optimization Framework for Elastic Element Design. *IEEE/ASME Trans. Mechatron.* **2019**, *23*, 1334–1345. [CrossRef]
40. Jimenez-Fabian, R.; Weckx, M.; Rodriguez-Cianca, D.; Lefeber, D.; Vanderborght, B. Online Reconfiguration of a Variable-Stiffness Actuator. *IEEE/ASME Trans. Mechatron.* **2018**, *23*, 1866–1876. [CrossRef]
41. Ham, R.V.; Sugar, T.G.; Vanderborght, B.; Hollander, K.W.; Lefeber, D. Compliant actuator designs. *IEEE Robot. Autom. Mag.* **2009**, *16*, 81–94. [CrossRef]
42. Vanderborght, B.; Albu-Schäffer, A.; Bicchi, A.; Burdet, E.; Caldwell, D.G.; Carloni, R.; Catalano, M.; Eiberger, O.; Friedl, W.; Ganesh, G.; et al. Variable impedance actuators: A review. *Robot. Auton. Syst.* **2013**, *61*, 1601–1614. [CrossRef]

43. Beckerle, P.; Wojtusch, J.; Rinderknecht, S.; von Stryk, O. Analysis of system dynamic influences in robotic actuators with variable stiffness. *Smart Struct. Syst.* **2014**, *13*, 711–730. [CrossRef]
44. Wolf, S.; Grioli, G.; Eiberger, O.; Friedl, W.; Grebenstein, M.; Höppner, H.; Burdet, E.; Caldwell, D.G.; Carloni, R.; Catalano, M.G.; et al. Variable stiffness actuators: Review on design and components. *IEEE/ASME Trans. Mechatron.* **2016**, *21*, 2418–2430. [CrossRef]
45. Guo, J.; Tian, G. Conceptual design and analysis of four types of variable stiffness actuators based on spring pretension. *Int. J. Adv. Robot. Syst.* **2015**, *12*, 62. [CrossRef]
46. Tsagarakis, N.G.; Laffranchi, M.; Vanderborght, B.; Caldwell, D.G. A compact soft actuator unit for small scale human friendly robots. In Proceedings of the IEEE International Conference on Robotics and Automation (ICRA), Kobe, Japan, 12–17 May 2009; pp. 4356–4362.
47. Beckerle, P.; Stuhlenmiller, F.; Rinderknecht, S. Stiffness Control of Variable Serial Elastic Actuators: Energy Efficiency through Exploitation of Natural Dynamics. *Actuators* **2017**, *6*, 282017.
48. Van Ham, R.; Vanderborght, B.; Van Damme, M.; Verrelst, B.; Lefeber, D. MACCEPA, the mechanically adjustable compliance and controllable equilibrium position actuator: Design and implementation in a biped robot. *Robot. Auton. Syst.* **2007**, *55*, 761–768. [CrossRef]
49. Vanderborght, B.; Van Ham, R.; Verrelst, B.; Van Damme, M.; Lefeber, D. Overview of the lucy project: Dynamic stabilization of a biped powered by pneumatic artificial muscles. *Adv. Robot.* **2008**, *22*, 1027–1051. [CrossRef]
50. Jafari, A.; Tsagarakis, N.G.; Caldwell, D.G. Exploiting natural dynamics for energy minimization using an Actuator with Adjustable Stiffness (AwAS). In Proceedings of the IEEE International Conference on Robotics and Automation (ICRA), Shanghai, China, 9–13 May 2011; pp. 4632–4637.
51. Jafari, A.; Tsagarakis, N.G.; Caldwell, D.G. AwAS-II: A new actuator with adjustable stiffness based on the novel principle of adaptable pivot point and variable lever ratio. In Proceedings of the IEEE International Conference on Robotics and Automation (ICRA), Shanghai, China, 9–13 May 2011; pp. 4638–4643.
52. Jafari, A.; Tsagarakis, N.G.; Caldwell, D.G. A novel intrinsically energy efficient actuator with adjustable stiffness (AwAS). *IEEE/ASME Trans. Mechatron.* **2013**, *18*, 355–365. [CrossRef]
53. Haeufle, D.F.B.; Taylor, M.D.; Schmitt, S.; Geyer, H. A clutched parallel elastic actuator concept: Towards energy efficient powered legs in prosthetics and robotics. In Proceedings of the 4th IEEE RAS EMBS International Conference on Biomedical Robotics and Biomechatronics (BioRob), Rome, Italy, 24–27 June 2012; pp. 1614–1619.
54. Toxiri, S.; Calanca, A.; Ortiz, J.; Fiorini, P.; Caldwell, D.G. A Parallel-Elastic Actuator for a Torque-Controlled Back-Support Exoskeleton. *IEEE Robot. Autom. Lett.* **2018**, *3*, 492–499. [CrossRef]
55. Batts, Z.; Kim, J.; Yamane, K. Design of a hopping mechanism using a voice coil actuator: Linear elastic actuator in parallel (LEAP). In Proceedings of the 2016 IEEE International Conference on Robotics and Automation (ICRA), Stockholm, Sweden, 16–21 May 2016; pp. 655–660.
56. Iwamura, M.; Schiehlen, W. Minimum control energy in multibody systems using gravity and springs. *J. Syst. Des. Dyn.* **2011**, *5*, 474–485. [CrossRef]
57. Uemura, M.; Matsusaka, K.; Takagi, Y.; Kawamura, S. A stiffness adjustment mechanism maximally utilizing elastic energy of a linear spring for a robot joint. *Adv. Robot.* **2015**, *29*, 1331–1337. [CrossRef]
58. Nasiri, R.; Khoramshahi, M.; Shushtari, M.; Ahmadabadi, M.N. Adaptation in variable parallel compliance: Towards energy efficiency in cyclic tasks. *IEEE/ASME Trans. Mechatron.* **2017**, *22*, 1059–1070. [CrossRef]
59. Wolf, S.; Hirzinger, G. A new variable stiffness design: Matching requirements of the next robot generation. In Proceedings of the IEEE International Conference on Robotics and Automation (ICRA), Pasadena, CA, USA, 19–23 May 2008; pp. 1741–1746.
60. Schmit, N.; Okada, M. Synthesis of a non-circular cable spool to realize a nonlinear rotational spring. In Proceedings of the IEEE/RSJ International Conference on Intelligent Robots and Systems (IROS), San Francisco, CA, USA, 25–30 September 2011; pp. 762–767.
61. Schmit, N.; Okada, M. Optimal design of nonlinear springs in robot mechanism: Simultaneous design of trajectory and spring force profiles. *Adv. Robot.* **2013**, *27*, 33–46. [CrossRef]
62. Plooij, M.; Wisse, M. A novel spring mechanism to reduce energy consumption of robotic arms. In Proceedings of the IEEE/RSJ International Conference on Intelligent Robots and Systems (IROS), Vilamoura, Portugal, 7–12 October 2012; pp. 2901–2908.

63. Jalaly Bidgoly, H.; Nili Ahmadabadi, M.; Zakerzadeh, M.R. Design and modeling of a compact rotational nonlinear spring. In Proceedings of the IEEE/RSJ International Conference on Intelligent Robots and Systems (IROS), Daejeon, South Korea, 9–14 October 2016; pp. 4356–4361.
64. Schiehlen, W.; Guse, N. Control of Limit Cycle Oscillations. In *IUTAM Symposium on Chaotic Dynamics and Control of Systems and Processes in Mechanics*; Springer: Berlin, Germany, 2005; pp. 429–439.
65. Khoramshahi, M.; Parsa, A.; Ijspeert, A.; Ahmadabadi, M.N. Natural dynamics modification for energy efficiency: A data-driven parallel compliance design method. In Proceedings of the IEEE International Conference on Robotics and Automation (ICRA), Hong Kong, China, 31 May–7 June 2014; pp. 2412–2417.
66. Scalera, L.; Gallina, P.; Seriani, S.; Gasparetto, A. Cable-Based Robotic Crane (CBRC): Design and Implementation of Overhead Traveling Cranes Based on Variable Radius Drums. *IEEE Trans. Robot.* **2018**, *34*, 474–485. [CrossRef]
67. Kim, B.; Deshpande, A.D. Design of Nonlinear Rotational Stiffness Using a Noncircular Pulley-Spring Mechanism. *J. Mech. Robot.* **2014**, *6*. [CrossRef]
68. Kumamoto, M.; Oshima, T.; Fujikawa, T. Bi-articular muscle as a principle keyword for biomimetic motor link system. In Proceedings of the 2nd Annual International IEEE-EMBS Special Topic Conference on Microtechnologies in Medicine and Biology, Madison, WI, USA, 2–4 May 2002; pp. 346–351.
69. Bipedal walking and running with spring-like biarticular muscles. *J. Biomech.* **2008**, *41*, 656–667. [CrossRef]
70. Hosoda, K.; Takayama, H.; Takuma, T. Bouncing monopod with bio-mimetic muscular-skeleton system. In Proceedings of the IEEE/RSJ International Conference on Intelligent Robots and Systems (IROS), Nice, France, 22–26 September 2008; pp. 3083–3088.
71. Eslamy, M.; Grimmer, M.; Seyfarth, A. Adding passive biarticular spring to active mono-articular foot prosthesis: Effects on power and energy requirement. In Proceedings of the IEEE-RAS International Conference on Humanoid Robots, Madrid, Spain, 18–20 November 2014; pp. 677–684.
72. Lu, G.; Kawamura, S.; Uemura, M. Proposal of an energy saving control method for SCARA robots. *J. Robot. Mechatron.* **2012**, *24*, 115–122. [CrossRef]
73. Bidgoly, H.J.; Parsa, A.; Yazdanpanah, M.J.; Ahmadabadi, M.N. Benefiting From Kinematic Redundancy Alongside Mono-and Biarticular Parallel Compliances for Energy Efficiency in Cyclic Tasks. *IEEE Trans. Robot.* **2017**, *33*, 1088–1102. [CrossRef]
74. Mathijssen, G.; Lefeber, D.; Vanderborght, B. Variable recruitment of parallel elastic elements: Series–parallel elastic actuators (SPEA) with dephased mutilated gears. *IEEE/ASME Trans. Mechatron.* **2015**, *20*, 594–602. [CrossRef]
75. Tsagarakis, N.G.; Dallali, H.; Negrello, F.; Roozing, W.; Medrano-Cerda, G.A.; Caldwell, D.G. Compliant antagonistic joint tuning for gravitational load cancellation and improved efficient mobility. In Proceedings of the 14th IEEE-RAS International Conference on Humanoid Robots (Humanoids), Madrid, Spain, 18–20 November 2014; pp. 924–929.
76. Roozing, W.; Li, Z.; Medrano-Cerda, G.A.; Caldwell, D.G.; Tsagarakis, N.G. Development and control of a compliant asymmetric antagonistic actuator for energy efficient mobility. *IEEE/ASME Trans. Mechatron.* **2016**, *21*, 1080–1091. [CrossRef]
77. Uemura, M.; Kanaoka, K.; Kawamura, S. A new control method utilizing stiffness adjustment of mechanical elastic elements for serial link systems. In Proceedings of the IEEE International Conference on Robotics and Automation (ICRA), Roma, Italy, 10–14 April 2007; pp. 1437–1442.
78. Uemura, M.; Goya, H.; Kawamura, S. Motion control with stiffness adaptation for torque minimization in multijoint robots. *IEEE Trans. Robot.* **2014**, *30*, 352–364. [CrossRef]
79. Velasco, A.; Garabini, M.; Catalano, M.G.; Bicchi, A. Soft actuation in cyclic motions: Stiffness profile optimization for energy efficiency. In Proceedings of the IEEE-RAS 15th International Conference on Humanoid Robots (Humanoids), Seoul, Korea, 3–5 November 2015; pp. 107–113.
80. Hill, R.B.; Briot, S.; Chriette, A.; Martinet, P. Increasing Energy Efficiency of High-Speed Parallel Robots by Using Variable Stiffness Springs and Optimal Motion Generation. In Proceedings of the ASME 2018 International Design Engineering Technical Conference and Computers and Information in Engineering Conference American Society of Mechanical Engineers, Quebec City, Canada, 26–29 August 2018.
81. Barreto, J.P.; Corves, B. Matching the Free-Vibration Response of a Delta Robot with Pick-and-Place Tasks Using Multi-Body Simulation. In Proceedings of the 14th International Conference on Automation Science and Engineering (CASE), Munich, Germany, 20–24 August 2018; pp. 1487–1492.

82. Mirz, C.; Schöler, F.; Barreto, J.P.; Corves, B. Optimal Control Based Path Planning for Parallel Kinematic Manipulators Utilising Natural Motion. In Proceedings of the IEEE 14th International Conference on Automation Science and Engineering (CASE), Munich, Germany, 20–24 August 2018; pp. 223–228.
83. Nasiri, R.; Khoramshahi, M.; Ahmadabadi, M.N. Design of a nonlinear adaptive natural oscillator: Towards natural dynamics exploitation in cyclic tasks. In Proceedings of the IEEE/RSJ International Conference on Intelligent Robots and Systems (IROS), Daejeon, Korea, 9–14 October 2016; pp. 3653–3658.
84. Grimmer, M.; Eslamy, M.; Gliech, S.; Seyfarth, A. A comparison of parallel-and series elastic elements in an actuator for mimicking human ankle joint in walking and running. In Proceedings of the IEEE International Conference on Robotics and Automation (ICRA), Saint Paul, MN, USA, 14–18 May 2012; pp. 2463–2470.
85. Tagliamonte, N.L.; Sergi, F.; Accoto, D.; Carpino, G.; Guglielmelli, E. Double actuation architectures for rendering variable impedance in compliant robots: A review. *Mechatronics* **2012**, *22*, 1187–1203. [CrossRef]
86. Velasco, A.; Gasparri, G.M.; Garabini, M.; Malagia, L.; Salaris, P.; Bicchi, A. Soft-actuators in cyclic motion: Analytical optimization of stiffness and pre-load. In Proceedings of the 13th IEEE-RAS International Conference on Humanoid Robots (Humanoids), Atlanta, GA, USA, 15–17 October 2013; pp. 354–361.
87. Verstraten, T.; Beckerle, P.; Furnémont, R.; Mathijssen, G.; Vanderborght, B.; Lefeber, D. Series and parallel elastic actuation: Impact of natural dynamics on power and energy consumption. *Mech. Mach. Theory* **2016**, *102*, 232–246. [CrossRef]
88. Beckerle, P.; Verstraten, T.; Mathijssen, G.; Furnémont, R.; Vanderborght, B.; Lefeber, D. Series and parallel elastic actuation: Influence of operating positions on design and control. *IEEE/ASME Trans. Mechatron.* **2017**, *22*, 521–529. [CrossRef]
89. Boscariol, P.; Boschetti, G.; Gallina, P.; Passarini, C. Spring Design for Motor Torque Reduction in Articulated Mechanisms. In Proceedings of the International Conference on Robotics in Alpe-Adria Danube Region, Turin, Italy, 21–23 June 2017; Springer: Berlin, Germany, 2017; pp. 557–564.
90. Carabin, G.; Palomba, I.; Wehrle, E.; Vidoni, R. Energy Expenditure Minimization for a Delta-2 Robot Through a Mixed Approach. *Comput. Methods Appl. Sci.* **2020**, *53*, 383–390. [CrossRef]
91. Hill, R.B.; Briot, S.; Chriette, A.; Martinet, P. Minimizing Input Torques of a High-Speed Five-Bar Mechanism by Using Variable Stiffness Springs. In *ROMANSY 22, Robot Design, Dynamics and Control*; Springer: Berlin, Germany, 2019; pp. 61–68.
92. Lu, G.; Kawamura, S.; Uemura, M. Inertia Adaptive Control Based on Resonance for Energy Saving of Mechanical Systems. *SICE J. Control Meas. Syst. Integr.* **2012**, *5*, 109–114. [CrossRef]
93. Matsusaka, K.; Uemura, M.; Kawamura, S. Realization of highly energy efficient pick-and-place tasks using resonance-based robot motion control. *Adv. Robot.* **2016**, *30*, 608–620. [CrossRef]
94. Matsusaka, K.; Uemura, M.; Kawamura, S. Highly Energy-Efficient Palletizing Tasks Using Resonance-Based Robot Motion Control. *J. Mech. Eng. Autom.* **2016**, *6*, 8–17.
95. Schiehlen, W.; Guse, N. Powersaving Control of Mechanisms. In Proceedings of the IUTAM Symposium on Vibration Control of Nonlinear Mechanisms and Structures, Munich, Germany, 18–22 July 2005; Springer: Berlin, Germany, 2005; pp. 277–286.
96. Shushtari, M.; Nasiri, R.; Yazdanpanah, M.J.; Ahmadabadi, M.N. Compliance and frequency optimization for energy efficiency in cyclic tasks. *Robotica* **2017**, *35*, 2363–2380. [CrossRef]
97. Uemura, M.; Lu, G.; Kawamura, S.; Ma, S. Passive periodic motions of multi-joint robots by stiffness adaptation and DFC for energy saving. In Proceedings of the SICE Annual Conference, Tokyo, Japan, 20–22 August 2008; pp. 2853–2858.
98. Uemura, M.; Kawamura, S. An energy saving control method of robot motions based on adaptive stiffness optimization-cases of multi-frequency components. In Proceedings of the IEEE/RSJ International Conference on Intelligent Robots and Systems (IROS), Nice, France, 22–26 September 2008; pp. 551–557.
99. Uemura, M.; Kawamura, S. Resonance-based motion control method for multi-joint robot through combining stiffness adaptation and iterative learning control. In Proceedings of the IEEE International Conference on Robotics and Automation (ICRA), Kobe, Japan, 12–17 May 2009; pp. 1543–1548.
100. Iwamura, M.; Imafuku, S.; Kawamoto, T.; Schiehlen, W. Design and Control of an Energy-Saving Robot Using Storage Elements and Reaction Wheels. In *Multibody Dynamics*; Springer: Berlin, Germany, 2016; pp. 277–297.

101. Schiehlen, W.; Iwamura, M. Minimum energy control of multibody systems utilizing storage elements. In Proceedings of the ASME International Design Engineering Technical Conference and Computers and Information in Engineering Conference American Society of Mechanical Engineers, San Diego, CA, USA, 30 August–2 September 2009; pp. 1919–1926.
102. Schmit, N.; Okada, M. Simultaneous optimization of robot trajectory and nonlinear springs to minimize actuator torque. In Proceedings of the IEEE International Conference on Robotics and Automation (ICRA), Saint Paul, MN, USA, 14–18 May 2012; pp. 2806–2811.
103. Barreto, J.; Schöler, F.F.; Corves, B. The concept of natural motion for pick and place operations. In *New Advances in Mechanisms, Mechanical Transmissions and Robotics*; Springer: Berlin, Germany, 2017; pp. 89–98.
104. Kashiri, N.; Spyrakos-Papastavridis, E.; Caldwell, D.G.; Tsagarakis, N.G. Exploiting the natural dynamics of compliant joint robots for cyclic motions. In Proceedings of the 22nd International Conference on Methods and Models in Automation and Robotics (MMAR), Miedzyzdroje, Poland, 28–31 August 2017; pp. 676–681.
105. Khoramshahi, M.; Nasiri, R.; Shushtari, M.; Ijspeert, A.J.; Ahmadabadi, M.N. Adaptive Natural Oscillator to exploit natural dynamics for energy efficiency. *Robot. Auton. Syst.* **2017**, *97*, 51–60. [CrossRef]
106. Noorani, M.S.; Ghanbari, A.; Jafarizadeh, M.A. Using the adaptive frequency nonlinear oscillator for earning an energy efficient motion pattern in a leg-like stretchable pendulum by exploiting the resonant mode. *Amirkabir Int. J. Model. Identif. Simul. Control* 2013, *45*, 47–54.
107. Ruder, S. An overview of gradient descent optimization algorithms. *arXiv* **2016**, arXiv:1609.04747.
108. Morrison, D.D.; Riley, J.D.; Zancanaro, J.F. Multiple Shooting Method for Two-point Boundary Value Problems. *Commun. ACM* **1962**, *5*, 613–614. [CrossRef]
109. Buchli, J.; Righetti, L.; Ijspeert, A.J. A dynamical systems approach to learning: A frequency-adaptive hopper robot. In Proceedings of the European Conference on Artificial Life, Canterbury, UK, 5–9 September 2005; Springer: Berlin, Germany, 2005; pp. 210–220.
110. Buchli, J.; Iida, F.; Ijspeert, A.J. Finding resonance: Adaptive frequency oscillators for dynamic legged locomotion. In Proceedings of the IEEE/RSJ International Conference on Intelligent Robots and Systems (IROS), Beijing, China, 9–15 October 2006; pp. 3903–3909.
111. Ijspeert, A.J. Central pattern generators for locomotion control in animals and robots: A review. *Neural Netw.* **2008**, *21*, 642–653. [CrossRef]
112. Righetti, L.; Buchli, J.; Ijspeert, A.J. Dynamic hebbian learning in adaptive frequency oscillators. *Phys. D Nonlinear Phenom.* **2006**, *216*, 269–281. [CrossRef]
113. Righetti, L.; Buchli, J.; Ijspeert, A.J. Adaptive frequency oscillators and applications. *Open Cybern. Syst. J.* **2009**, *3*, 64–69. [CrossRef]
114. Palomba, I.; Richiedei, D.; Trevisani, A. Reduced-order observers for nonlinear state estimation in flexible multibody systems. *Shock Vib.* **2018**, *2018*. [CrossRef]

© 2019 by the authors. Licensee MDPI, Basel, Switzerland. This article is an open access article distributed under the terms and conditions of the Creative Commons Attribution (CC BY) license (http://creativecommons.org/licenses/by/4.0/).

Article

Study of Personal Mobility Vehicle (PMV) with Active Inward Tilting Mechanism on Obstacle Avoidance and Energy Efficiency

Tetsunori Haraguchi [1,*], Ichiro Kageyama [1] and Tetsuya Kaneko [2]

1. College of Industrial Technology, Nihon University, 1-2-1 Izumi-cho, Narashino, Chiba 275-8575, Japan; kageyama.ichiro@nihon-u.ac.jp
2. Department of Mechanical Engineering for Transportation, Osaka Sangyo University, 3-1-1 Nakagaito, Daito 574-8530, Japan; kaneko@tm.osaka-sandai.ac.jp
* Correspondence: haraguchi@nagoya-u.jp; Tel.: +81-47-474-2337

Received: 5 October 2019; Accepted: 23 October 2019; Published: 6 November 2019

Featured Application: CarMaker of IPG Automotive in Germany was used as a vehicle dynamics simulation tool.

Abstract: Traffic congestion and lack of parking spaces in urban areas etc. may lessen the original benefits of cars, and now new ultra-small mobility concepts called personal mobility vehicles (PMVs) are receiving attention. Among them, PMVs with an inward tilting mechanism in order to avoid overturning on turning, as in motorcycles, look realistic in new, innovative traffic systems. In this study, PMVs with an active inward tilting mechanism with three wheels, double front wheels and single rear wheel, were studied regarding front inner wheel lifting phenomena, capability of obstacle avoidance and energy balance of active tilting mechanism. Based on a comprehensive study of inner wheel lifting and obstacle avoidance, PMVs with a front wheel steering system as the most realistic specification were compared on capability of obstacle avoidance with passenger cars and motorcycles with those that have current market experience, and showed better capability. Although the energy consumption of an active inward tilting mechanism might be in conflict with the energy efficiency of small PMV concepts, the energy needed to tilt PMVs was very little compared with the general energy consumption of driving. It was clarified that the new PMV concepts with inward tilting mechanism have sufficient social acceptability from both mandatory points of safety and efficiency.

Keywords: personal mobility vehicle; active tilting; inner wheel lifting; obstacle avoidance; energy efficiency

1. Introduction

Although automobiles have improved the quality of life of human beings, too big an increase in the number of vehicles in use has the potential to spoil the original benefits of automobiles, due to problems such as traffic congestion and lack of parking space in urban areas. Thus, new ultra-small mobility concepts called personal mobility vehicles (PMVs) are now attracting attention [1,2].

PMVs with narrow total width are likely to be of the inward tilting type, the same as motorcycles, in order to avoid overturning on turning. Among them, tricycles with two front wheels and one rear wheel seem to be a good idea because of the simplicity of the vehicle configuration and security against overturning during braking, as shown in Figure 1. A general understanding of the advantages and disadvantages of the different types of PMV are shown in Table 1.

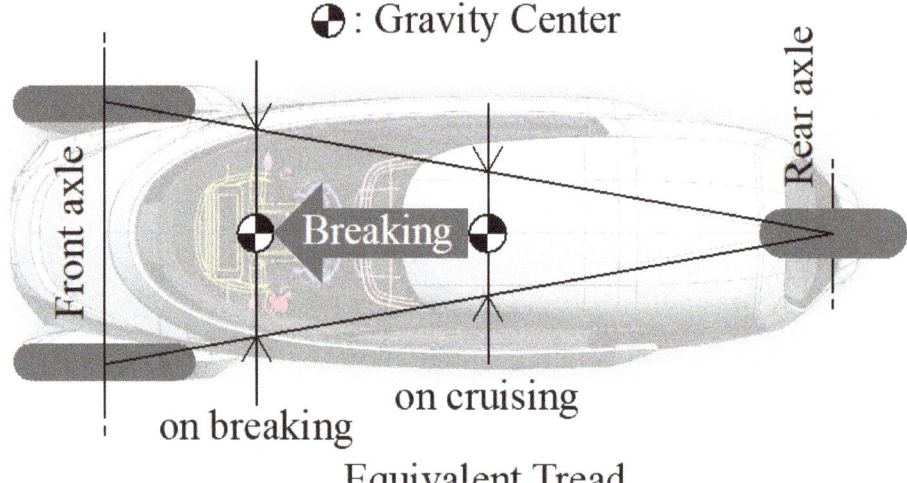

Figure 1. Equivalent tread is wider on breaking.

Table 1. Advantages (✔)/Disadvantages (▼) of different types of personal mobility vehicles (PMVs).

	Simple Construction	Slow Speed	Stability on Braking	Obstacle Avoidance
Single front wheel+Double rear wheels	✔		▼Unstable	▼U.S.
Double front wheels + Single rear wheel			✔	
Front steering + Rear traction				
Passive tilting mechanism	✔	▼Falling		▼Delay
Active tilting mechanism	▼Active	✔		✔
Front traction + Rear steering	▼Steer-by-wire			▼Delay
Double front wheels + Double rear wheels	▼4 wheels		✔	
Front steering + Rear traction				
Passive tilting mechanism	✔	▼Falling		▼Delay
Active tilting mechanism	▼Active	✔		✔

The following concerns should be clarified in order to implement new PMV concepts with inward tilting mechanisms in future mobility society.

The following is the main concern regarding PMVs with a passive tilting mechanism:

- Self-standing ability from stop to very low speed

The following are the main concerns regarding PMVs with an active tilting mechanism:

- Inner wheel lifting on sudden steering input
- Capability on obstacle avoidance
- Energy consumption on active tilting system

1.1. Self-Standing Ability from Stop to Very Low Speed

In order to achieve both a passive tilting mechanism and self-standing ability from stop to very low speed, it should be a simple, mechanical mechanism. Although the details are omitted this time, we devised a self-standing mechanism and confirmed the effectiveness of that mechanism using a motorcycle with two front wheels, as shown in Figure 2.

Figure 2. Self-standing mechanism was devised.

1.2. Study of PMVs with Active Tilting Mechanism

In this report, using the multibody dynamics models described in the next chapter, we study three concerns regarding PMVs with active tilting mechanisms and discuss the social acceptability of inward tilting type PMVs.

2. Multibody Dynamics Model

2.1. Vehicle Model

The specifications of a PMV are shown in Figure 3 and Table 2. Although overall length and width are about half of that of a passenger car, overall height is similar to a passenger car. Therefore, inward tilting is necessary.

CarMaker [3] of IPG Automotive in Germany was used as a vehicle dynamics simulation tool. In order to construct a tilting model with the car simulation tool, the following measures were taken.

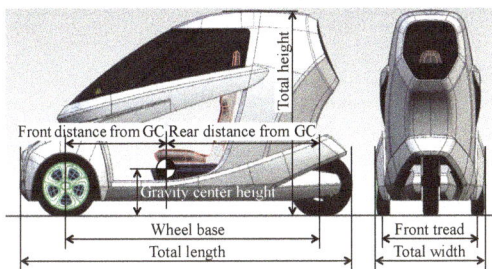

Figure 3. Dimensions of model vehicle.

Table 2. Specifications of model vehicle.

Item	Unit	Value	Item	Unit	Value
Total length	m	2.645	Total mass	kg	369.8
Total width	m	0.880	Front mass distribution	kg	222.1
Total height	m	1.445	Rear mass distribution	kg	147.7
Wheel base	m	2.020	Roll moment of inertia	kgm^2	58.8
Front distance from gravity center	m	0.807	(Roll moment of inertia of sprung mass)	kgm^2	43.0
Rear distance from gravity center	m	1.213	Pitch moment of inertia	kgm^2	197.3
Front tread	m	0.850	(Pitch moment of inertia of sprung mass)	kgm^2	118.0
Gravity center height	m	0.358	Yaw moment of inertia	kgm^2	187.3
Steering gear ratio	-	16.0	(Yaw moment of inertia of sprung mass)	kgm^2	102.3

2.1.1. Suspension

As shown in Figure 4, forced torsional torque is applied to the stabilizer bar. In this method, the vehicle generates internal torque. Although the mechanism might be different from PMVs having direct suspension stroke control mechanisms, the motion characteristics and external forces acting on this vehicle are same.

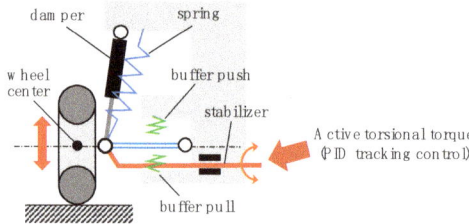

Figure 4. Torsional torque is applied to stabilizer bar.

2.1.2. Target Roll Angle

Target tilting (roll) angle is given as shown in Equation (1), to balance against virtual lateral acceleration, which is simply calculated from tire steered angles, wheel base and vehicle speed. Torsion torque of the active stabilizer to get the target roll angle is given by general proportional-integral-derivative (PID) tracking control. Initial values of control gain are shown in Equation (2).

$$TRA = A \tan^{-1}\left(\frac{\sin(\delta)v^2}{lg}\right) \quad (1)$$

$$P = 4000, I = 100, D = 0 \quad (2)$$

TRA: target roll angle
A: user amplification factor
δ: tire steered angle
v: vehicle speed
l: wheel base
P: proportional gain
I: integral gain
D: derivative gain

2.1.3. Tire

PMVs with tilting mechanisms are strongly influenced by tire camber characteristics. Therefore, we used a motorcycle tire model in CarMaker as shown in Figure 5 to consider the side force on tire camber angle.

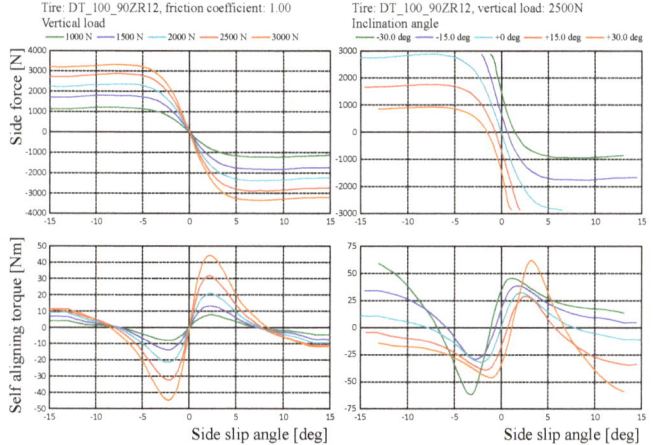

Figure 5. Motorcycle tire model is used in CarMaker.

2.2. Driver Model

IPGDriver from IPG Automotive attached to CarMaker is used as driver model in this study. IPGDriver understands the vehicle characteristics and controls the vehicle speed and steering wheel angle to trace the given course based on a forward quadratic prediction model.

3. Inner Wheel Lifting Phenomena

An active tilting mechanism forcibly applies strokes to suspension. In the case where the roll response of the sprung body is delayed, a difference of vertical loads on the front two wheels occurs. In worse cases, inner wheel lifting phenomena [4,5] occur. Although lifting itself is not a problem, divergent phenomena cause overturning. Therefore, it is better to avoid repeated liftings.

3.1. Understanding Mechanism of Lifting Phenomena

The mechanism of front inner wheel lifting caused by sudden steering inputs seems to be as follows.

1. Inward roll response to steering input cannot occur in time due to roll moment of inertia.
2. Transitional front wheel slip angle due to vehicle response delay caused by yaw moment of inertia causes outward roll moment. This roll moment further delays inward roll response.
3. Too much rebound stroke of front outer wheel caused by inward roll response delay lifts up front body. Pitch moment of inertia delays recovery from front body lifting.

3.1.1. Influence of PID Factors on Roll Angle Tracking Control

In order to study the influence of three axes moments of inertia on front inner wheel lifting phenomena during sudden steering inputs, it is inconvenient if the vehicle has unstable roll phenomena (for example, roll vibration) without intentional steering input. As shown in Figure 6, when the roll moment of inertia of sprung mass (Ixx^*) is increased, the vehicle becomes unstable in roll direction. Roll vibration starts with doubled Ixx^* even on straight running ahead. Overturning occurs with quadrupled Ixx^* without ability to continue straight running, as shown in Figure 7.

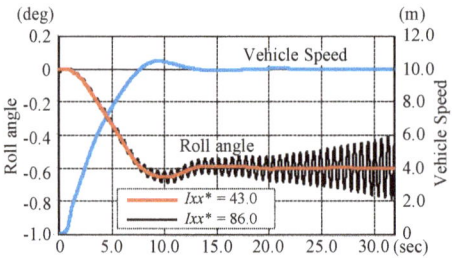

Figure 6. Unstable roll vibration with high Ixx^*.

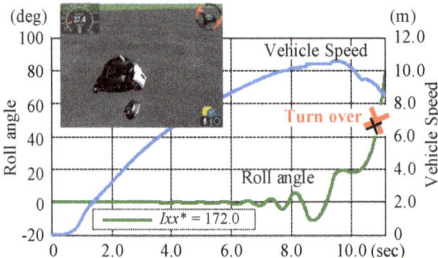

Figure 7. Turning over with too high roll moment of inertia of sprung mass (Ixx^*).

As shown in Figure 8, in case I gain was reduced by half (100→50), it was confirmed that roll vibration is suppressed even under previous conditions (doubled Ixx^* and quadrupled Ixx^*). There is no roll vibration even the roll moment of inertia is quadrupled. We proceeded with I gain = 50 in order to eliminate the disturbing effect of roll tracking control.

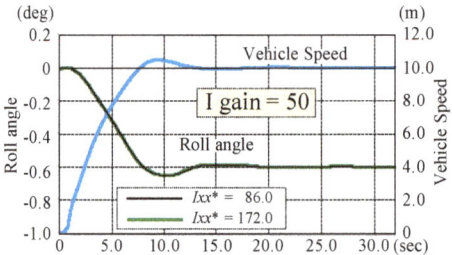

Figure 8. No roll vibration phenomena are observed (I = 100→50).

3.1.2. Steering Input Condition

The lateral displacement of motorcycles in the severe obstacle avoidance scene was about 2 m. The steering input angle was set as the initial condition to get this lateral displacement. Lateral displacement LD is proportional to steering angle MA and the square of vehicle speed v, and inversely proportional to the square of steering input frequency f, as in Equation (3).

Sinusoidal input of about 0.5 Hz ± 60 deg gives about 2 m lateral displacement LD at a vehicle speed of 36 km/h (10 m/s). An input frequency of 0.5 Hz is equivalent to quick lane change (L/C) operation of standard drivers on public roads.

$$LD \propto \frac{MA\ v^2}{f^2} \tag{3}$$

LD: lateral displacement
MA: steering wheel angle
v: vehicle velocity
f: steering input frequency

3.1.3. Relation between Steering Angle and Input Frequency

Lateral displacement *LD* is expressed using Equation (3). Steering input angle *MA* proportioned to the square of steering input frequency *f* gives maintained *LD*, as shown in Figure 9. However, an extreme increase of steering input angle *MA* means fairly severe conditions for inner wheel lifting phenomena. Therefore, increasing steering input frequency *f* results in a sudden turnover.

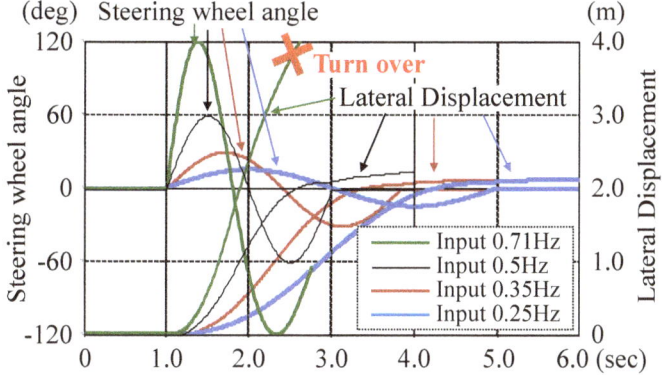

Figure 9. Steering angle proportioned to square of frequency results in sudden turnover (I = 50).

Only steering input frequency *f* was changed, while maintaining steering input angle *MA* = ±60 deg, without maintaining *LD*, as shown in Figure 10. The minimum residual load of the front left wheel (inner wheel) is almost zero at *f* = 0.71 Hz, and the inner wheel lifts clearly at *f* = 1.0 Hz. It can be understood that front inner wheel lifting is caused not only by the large steering angle input but also directly by quick steering input.

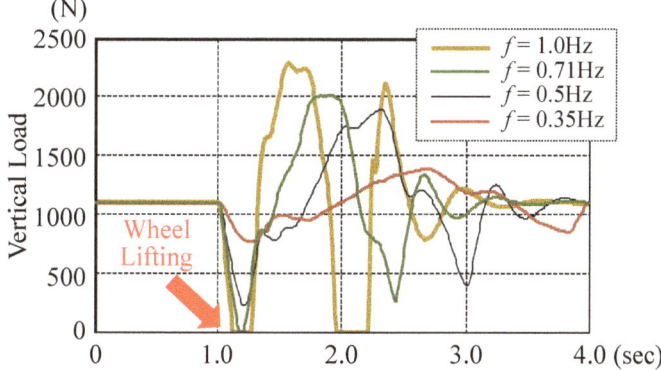

Figure 10. Quick steering input directly causes wheel lifting (I = 50).

Quick and large steering input is actually very difficult for human drivers. The effects of three axis moments of inertia on front inner wheel lifting phenomena by sudden steering input were examined with 0.5 Hz (quick L/C) as the upper limit of standard drivers' ability.

3.1.4. Vehicle Speed

Steering input angle MA inversely proportioned to the square of vehicle speed v gives maintained LD, as shown in Figure 11. However, in order to avoid obstacles completely, steering input frequency f needs to be increased in proportion to the increase of vehicle speed v. Thus, MA is not decreased proportionally to the square of vehicle speed v on actual obstacle avoidance, as shown in Figure 12. The effects of three axis moments of inertia on front inner wheel lifting phenomena by sudden steering input were examined with 36 km/h (10 m/s).

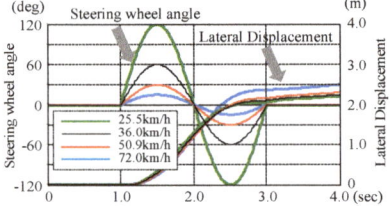

Figure 11. Required steering input angle (I = 50).

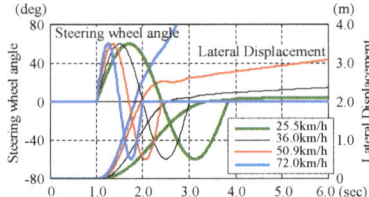

Figure 12. Influence of vehicle speed (I = 50).

3.2. Influences of Three Axis Moments of Inertia

The effects of moments of inertia of sprung mass around three axes Ixx^*, Iyy^* and Izz^* were examined.

3.2.1. Influence of Roll Moment of Inertia

The influence of roll moment of inertia was examined by parameterizing roll moment of inertia of sprung mass Ixx^* out of whole vehicle roll moment of inertia $Ixx = 58.8$ kg/m^2. Figure 13 (I = 50, $f = 0.5$ Hz) shows vehicle behavior at vehicle speed $v = 36$ km/h (10 m/s), steering input frequency $f = 0.5$ Hz, sinusoidal steering angle $MA = \pm 60$ deg, and roll tracking gain I = 50.

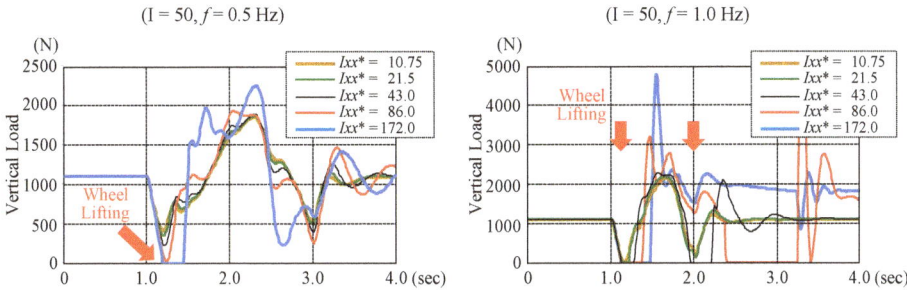

Figure 13. Influence of roll moment of inertia on vertical load of front left wheel (I = 50).

Larger Ixx^* clearly increase front inner wheel lifting. Although no phenomena occur when $Ixx^* = 43.0$ kgm^2 (standard), almost zero vertical load of inner wheel occurs when this is doubled to $Ixx^* = 86.0$ kgm^2, and inner wheel lifting phenomenon clearly occurs when this amount is doubled again to $Ixx^* = 172.0$ kgm^2.

The effect of roll moment of inertia of sprung mass Ixx^* was confirmed also at $f = 1.0$ Hz (sudden L/C). As shown in Figure 13 (I = 50, f = 1.0 Hz), front inner wheel lifting occurs even when $Ixx^* = 43.0$ kgm^2 (standard).

3.2.2. Influence of Pitch and Yaw Moments of Inertia

Influences of pitch moment of inertia of sprung mass Iyy^* and yaw moment of inertia of sprung mass Izz^* were examined as shown in Figure 14. However, no significant influences were observed on front inner wheel lifting phenomena. Therefore, we decided to concentrate on roll direction mechanism 1, as proposed in Section 3.1.

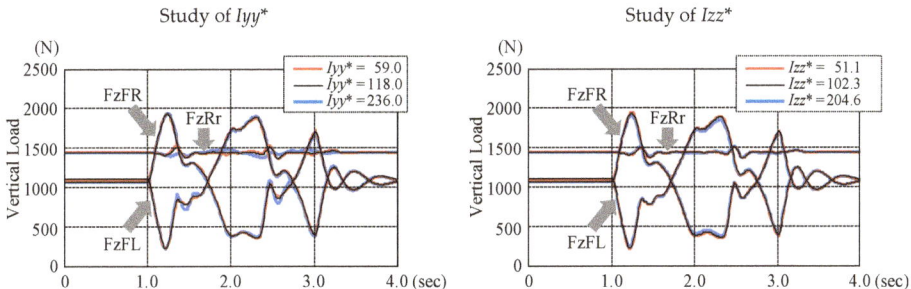

Figure 14. No significant influences were observed with pitch and yaw moments of inertia (I = 50).

3.2.3. Roll Resonance Frequency $f\varphi$ and Influence of Roll Dynamics

PMVs have half total width and same total height (equivalent to nearly half roll moment of inertia), and half front tread (equivalent to one quarter front roll stiffness) compared with passenger cars. PMVs have no roll stiffness on the rear wheel (equivalent to one eighth total roll stiffness). Thus, as shown in Equation (4), roll resonance frequency $f\varphi$ is judged to be further lower (equivalent to 1/2).

Therefore, we grasp roll resonance frequency $f\varphi$, experimentally at first. While driving straight at 36 km/h (10 m/s), a roll pulse was input by a sharp steering pulse (±30 deg, 8 Hz, one cycle of sinusoidal input), as shown in Figure 15. After the pulse input, remained roll vibration was observed, and its frequency was about 1.67 Hz. This roll resonance frequency $f\varphi$ is on suspension roll stiffness, not related to the tilting mechanism.

As shown in Figure 16, vehicle mass, spring/stabilizer-bar stiffness of suspension, damping force characteristics, and roll tracking control gains P, I, D were all doubled in order to get vehicle characteristics with equivalent roll dynamics. Subsequently, front inner wheel lifting phenomena showed almost the same characteristics as the reference vehicle. From this, it was confirmed that front inner wheel lifting phenomena are almost dominated by the roll dynamics of PMVs.

$$f\varphi \propto \left(\frac{K\varphi}{Ixx}\right)^{\frac{1}{2}} \quad (4)$$

$f\varphi$: roll resonance frequency
$K\varphi$: roll stiffness
Ixx: roll moment of inertia

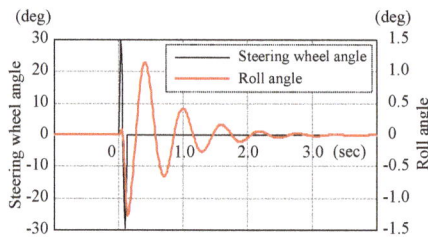

Figure 15. Roll resonance frequency is about 1.67 Hz (I = 100).

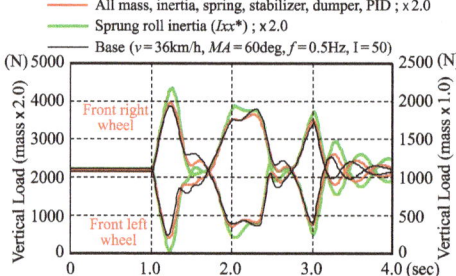

Figure 16. Equivalent roll dynamics gives same wheel lifting phenomena.

3.3. Summary of This Section and Setting Specifications for Further Study

- Lateral displacement LD is proportional to steering angle MA and the square of vehicle speed v, and inversely proportional to the square of steering input frequency f.
- Increase of steering input frequency f itself is disadvantageous for front inner wheel lifting. Increase of input steering wheel angle MA to maintain lateral displacement LD on increase of steering input frequency f causes further severe conditions of front inner wheel lifting.
- Increase of vehicle speed v requires quicker steering input frequency f, which is disadvantageous for front inner wheel lifting.
- Inner wheel lifting phenomenon is only that vehicle roll is unavailable to follow target roll angle TRA due to roll moment of inertia of sprung mass Ixx^* and is almost dominated by the roll dynamics of PMVs. Smaller Ixx^* and larger $K\varphi$, i.e., higher roll resonance frequency $f\varphi$ improves front inner wheel lifting phenomena.
- Unstable roll phenomena without intentional steering input related to front inner wheel lifting phenomena is suppressed by decrease of I gain of PID tracking control to active roll angle. In order to study realistic roll moment of inertia Ixx^* and roll stiffness $K\varphi$, PID control parameters in further study are set as Equation (5).

$$P = 4000, I = 50, D = 0 \tag{5}$$

4. Capability on Obstacle Avoidance

4.1. Comparison Front Wheel Steering (FWS) and Rear Wheel Steering (RWS)

Recently, Toyota Motor Corporation introduced an active tilting type PMV concept vehicle with RWS, considering the package advantageous. However, the RWS system is not generally available on the market. A comparison of FWS and RWS from the point of view of obstacle avoidance is necessary [6].

4.1.1. Understanding of Vehicle Posture on Avoidance Behavior

Although there is no difference between FWS and RWS in Equation (6), which describes vehicle condition on steady circle, RWS vehicles show a clearly inward posture (rear end pushed out), as shown in Figure 17.

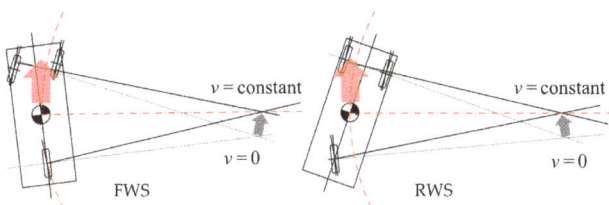

Figure 17. Vehicle postures on steady-state cornering.

An image of vehicle posture on left direction avoidance is shown on Figure 18. The rear vehicle end moves to right at first in RWS, and this causes a delay in the lateral displacement *LD* of RWS vehicles.

$$\rho = \left(1 - \frac{m}{2l^2} \frac{l_f K_f - l_r K_r}{K_f K_r} v^2 \right) \frac{l}{\delta} \tag{6}$$

ρ: turning radius
m: vehicle mass
l: wheel base
l_f: front wheel base
K_f: front cornering power
l_r: rear wheel base
v: velocity
δ: tire steer angle

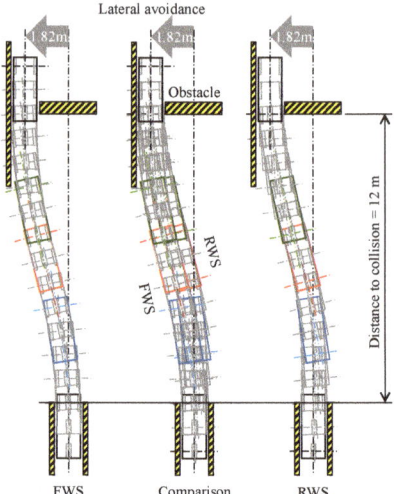

Figure 18. Tracks image of front wheel steering (FWS) and rear wheel steering (RWS) to avoid obstacle.

4.1.2. Open Loop Simulation

First, open loop simulations were conducted using steering angle input as shown on the left in Figure 19. Vehicle speed was fixed at 36 km/h and lateral displacement was 1.82 m, achieved with sinusoidal steering input. The delay in the lateral displacement LD in RWS assumed in the former section is also shown in comparison using a dynamic simulation tool, shown on the right in Figure 19. There is an approximately 2 m delay in longitudinal distance under this condition.

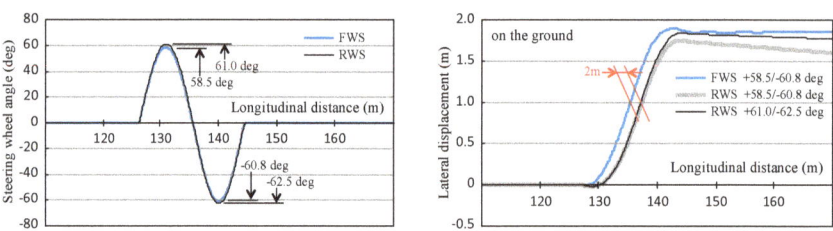

Figure 19. Steering wheel angle and lateral displacement on open loop simulation.

4.1.3. Closed Loop Simulation

A pylon-controlled single lane change course was set for closed loop simulation, as shown in Figure 20. Maximum steering angle, maximum steering speed, and maximum steering acceleration are all sufficiently large for avoidance operation.

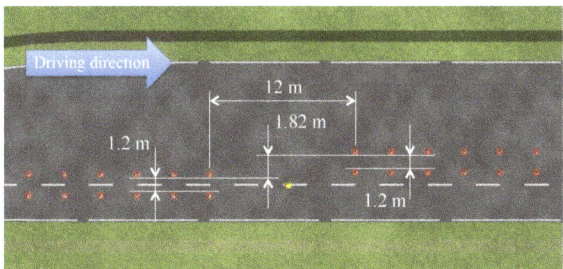

Figure 20. Single lane change course as obstacle avoidance.

In a closed loop, the driver recognizes the given lane change course in advance. Therefore, it makes it easier to pass the obstacle by steering a little to the opposite side before starting avoidance, as shown on the left in Figure 21. However, it is impossible to avoid oncoming obstacles in this way in the real world.

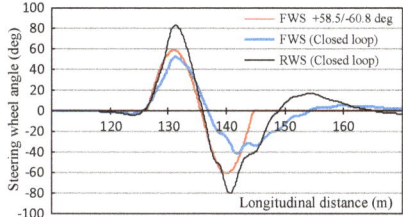 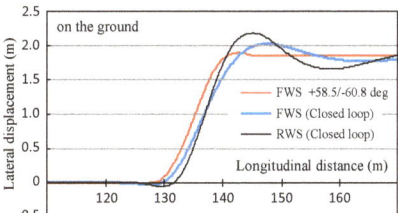

Figure 21. Steering wheel angle and lateral displacement.

As shown in Figure 21, in the case of FWS, the rise of the steering angle after starting avoidance is quite gentle, and the steering angle is considerably smaller and longer in the latter half of avoidance than in open loop. In the case of RWS, a large steering angle is required to recover from the delay of open loop avoidance. It is about a 1.6 times larger angle in the first half of avoidance and about twice as much in the second half. Overshoot at end of avoidance is also noticeable, indicating the difficulty of the avoidance operation in RWS.

4.1.4. Judgment of FWS and RWS on Implementation

The rear end of the RWS vehicle is pushed out at the first time of avoidance, and this causes a delay of lateral displacement. Superiority of FWS to RWS in obstacle avoidance performance was thus shown.

4.2. Comparison with Passenger Cars and Motorcycles

The obstacle avoidance capabilities of motorcycles which are accepted in the market are generally lower than those of passenger cars. When considering the social acceptability of completely new PMVs with tilting mechanisms, it is reasonable to compare them with those with market experience [7].

4.2.1. Vehicle Models of Passenger Cars and Motorcycles

CarMaker of IPG Automotive GmbH in Germany was used also as a passenger car simulation tool and an accompanying general passenger car model was used as a vehicle model. For the motorcycle, we used BikeSim from Mechanical Simulation, USA. Table 3 shows the basic specifications of these vehicles.

Table 3. Specifications of passenger car and motorcycle.

Passenger Car			Motorcycle		
Item	Unit	Value	Item	Unit	Value
Total length	m	4.150	Total length	m	2.140
Total width	m	1.700	Total width	m	0.637
Total height	m	1.600	Total height	m	1.318
Wheel base	m	3.185	Wheel base	m	1.576
Front tread	m	1.508	Front mass distribution	kg	122.8
Rear tread	m	1.494	Rear mass distribution	kg	75.5
Total mass	kg	1463	Total mass	kg	198.3
Gravity center height	m	0.580	Gravity center height	m	0.350

4.2.2. Driving Course

The vehicle models drive into a single lane change course from the approach section. Course severity was set as maximum passing speed, which would be normal driving speed. Three levels, 1.5 m, 2.0 m and 2.5 m, of lateral displacement in a 10 m transition section were set by the parameter study in advance, as shown in Figure 22. The situation is that vehicles run at the right end of the lane and the obstacle is at the right end of lane. This means there is a severer situation for narrower vehicles, PMVs and motorcycles.

Generally, the lane width is set within a certain margin in relation to vehicle width. Therefore, course widths are not the same between PMVs and motorcycles, and passenger cars. However, lateral displacement levels are the same, and the capabilities of obstacle avoidance were judged on maximum passing speeds.

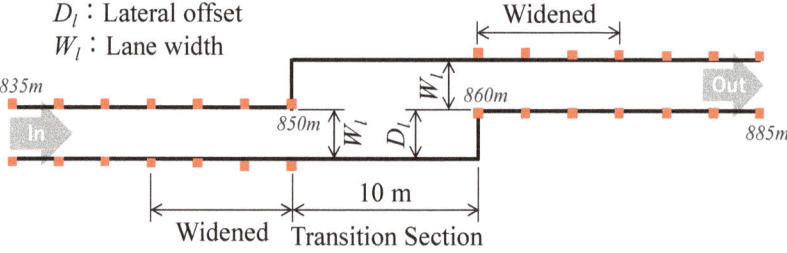

PMV, Motorcycle : W_l=1.5m D_l=1.5m, 2.0m, 2.5m
Passenger Car   : W_l=2.32m D_l=1.5m, 2.0m, 2.5m

Figure 22. Three levels of severity on driving course were set by parameter study.

4.2.3. Passing Speed Comparison and Understanding Social Acceptability

In the case of CarMaker, longitudinal speed is controlled consistently and automatically, but lateral control is switched from straight running model to automatic steering control model just before the transition section. The straight running model runs in the middle of the lane, but the automatic driving model takes the most advantageous course (feint motion) as far as possible, even if a narrow course width is set. For this reason, after entering a single lane change course, the model is switched from straight running model to automatic driving model. Switching timing was set as early so as not to miss the inner pylon and as late so as not to take the most advantageous course. Driver parameter and maneuver are shown in Table 4.

Table 4. Driver maneuvers were set to avoid feint motions.

Driver			Maneuver	
Item	Unit	Value	Item	Model
cornering cutting coefficient (ccc)		0.1	Longitudinal dynamics	IPGDriver
Max. longitudinal acceleration	m/s²	3.0	Lateral dynamics < 841.5 m	Follow course
Max. longitudinal deceleration	m/s²	-4.0	(D_l = 1.5 m) > 841.5 m	IPGDriver
Max. lateral acceleration	m/s²	20	Lateral dynamics < 843.0 m	Follow course
PylonShiftFdCoef		0.15	(D_l = 2.0 m) > 843.0 m	IPGDriver
			Lateral dynamics < 844.0m	Follow course
			(D_l = 2.5 m) > 844.0 m	IPGDriver

In PMVs and passenger cars, vehicle speed is automatically reduced when entering speed is quicker than passable speed. Therefore, in general, entrance speed (avoidance speed) to the transition section and exit speed from the transition section are different. In this study, entrance speed to the transition section was used as passing speed. In the case of PMV, only one occurrence of front inner wheel lifting was accepted and repeated liftings were judged to be failed.

In the case of BikeSim, autonomous maximum speed driving in the given course is not possible. A target trajectory was given to the rider model to drive a similar course to the PMV. The preview time was set according to the model switching distance in the PMV and the riding manner was adjusted by the behavior gain of tilting. The speed was gradually increased and it was judged from actual trajectory whether it was successful or not. The target trajectory was adjusted to get maximum passing speed in necessity.

A comparison of the maximum avoidable speeds is shown in Figure 23. As is generally understood, it was confirmed that there is a large difference between passenger cars and motorcycles. The capability of obstacle avoidance for PMVs is not only clearly superior to motorcycles, but is also equal to or higher than passenger cars. Therefore, we may understand that PMVs have sufficient social acceptability in terms of capability of obstacle avoidance.

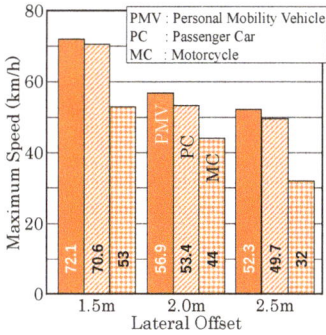

Figure 23. Comparison of capabilities on obstacle avoidance.

4.2.4. Mechanism of Superiority of PMV

As shown in Figure 24, for the dynamics of passenger cars, centrifugal force acting on gravity center height (G.C.H.) and supporting force acting on ground height cause roll moment, and lateral transfer of vertical load causes opposite the direction roll moment. These two moments are balanced with each other and the balance angle is expressed by Equation (7).

$$\tan \varphi_e = \tan \varphi_t = \frac{d_t}{\text{G.C.H.}} \quad (7)$$

$$\tan \varphi_e = \tan \varphi_a = \frac{d_a}{\text{G.C.H.}} \quad (8)$$

φ_e: equivariant roll angle
d_t: moment arm length on lateral transfer of load
G.C.H.: gravity center height
φ_a: actual roll angle
d_a: moment arm length on actual roll

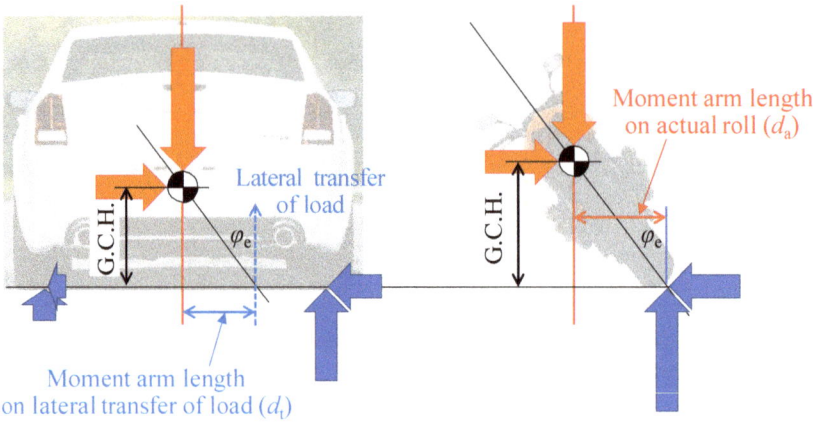

Figure 24. Roll moment balance on car and motorcycle.

In motorcycles, there is no lateral transfer of the vertical load. The roll moment due to the centrifugal force acting on the G.C.H. is balanced by the roll moment due to the combining of the vertical forces. The moment arm length moving in a lateral direction is caused by the actual vehicle roll angle. This is the balance angle of single-track vehicles such as bicycles and motorcycles and is expressed by Equation (8).

In PMVs with active tilting mechanisms, the strokes on both wheels are given to get the roll angle. This is equivalent to positively generating the lateral transfer of the vertical load. After the roll angle is obtained as equivariantly the same as motorcycles, the vertical loads of both wheels settle equally as a result. However, while the response of the roll angle is delayed, lateral transfer of the vertical load occurs between both wheels as in passenger cars.

As shown in Figure 25, the roll moment due to lateral transfer of the vertical load is the same as that of passenger cars, and the roll moment due to the actual roll angle of the vehicle body is the same as that of motorcycles. This is expressed by Equation (9). In the dynamics of passenger cars, lateral force and lateral acceleration on the front body occur without any delay due to the front tires slip angle by steering angle input. However, since the roll moment of inertia balances dynamically with the roll moment due to lateral acceleration, there is a slight delay in the lateral transfer of the vertical loads. In motorcycles, lateral acceleration can only be obtained when inward roll occurs in the vehicle body. This is a factor that causes a large delay in the lateral acceleration of motorcycles (slow avoidance motion), while delay in the lateral acceleration of a passenger car is not seen in avoiding obstacles.

$$\tan \varphi_e = \tan \varphi_a + \tan \varphi_t = \frac{d_a + d_t}{\text{G.C.H.}} \tag{9}$$

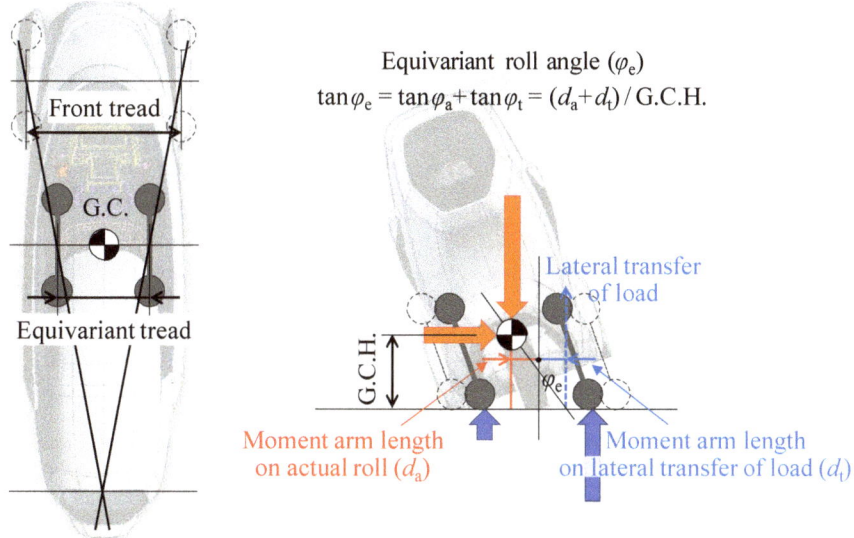

Figure 25. Roll moment balance on PMV with active tilting mechanism.

In the case of PMVs with an actively applied roll angle, as shown on the left of Figure 26, lateral transfer of vertical loads occurs with no delay according to the steering angle input. In comparison with motorcycles, it is equivalent to giving an additional virtual roll angle during delay of the actual roll angle by the tracking control and the roll moment of inertia. These two roll angles are shown on the right of Figure 26. The actual roll angle reaches its peak near the center of the transition section, but an additional virtual roll angle already balances with the lateral acceleration in the opposite direction at the same time.

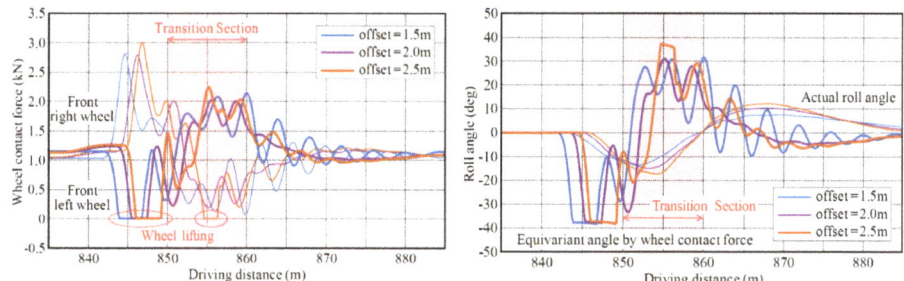

Figure 26. Lateral load transference and equivariant roll angle on load transference.

As shown in Figure 27, the sum of the actual roll angle of the vehicle body and the virtual roll angle of PMVs corresponds equivalently to the inward roll angle of motorcycles. Since this equivalent roll angle balances with lateral acceleration, PMVs are able to provide lateral acceleration without response delay, the same as with passenger cars.

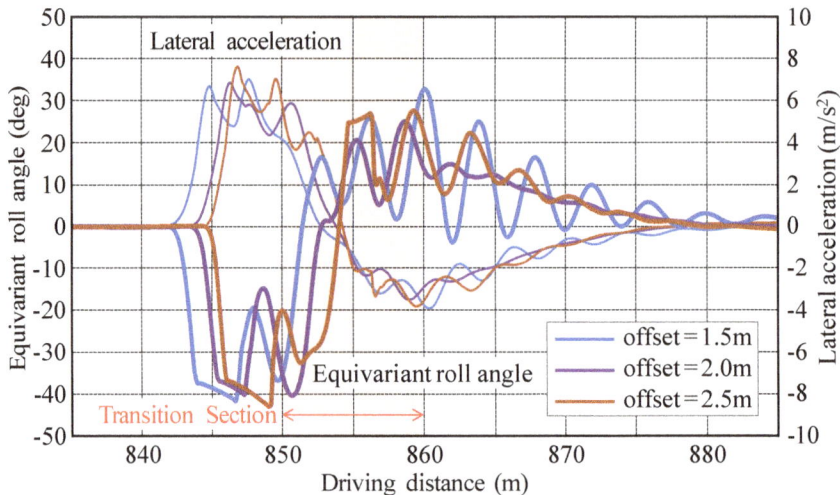

Figure 27. Balance of lateral acceleration and equivariant roll angle.

4.3. Summary of This Section

In a comparison of the capabilities of FWS and RWS on obstacle avoidance, it was observed that RWS has the behavior of the rear end being pushed out at the first timing and causes a delay of lateral displacement. Using dynamic simulation, it was shown to be especially difficult for the driver to steer to compensate for the delay and avoid the obstacle in RWS. Therefore, it can be judged that FWS is superior to RWS in obstacle avoidance performance.

PMVs with active tilting mechanisms have obstacle avoidance ability equal to or higher than that of passenger cars, because they have a much smaller roll moment of inertia than passenger cars and responsiveness equivalent to passenger cars. Thus, sufficient social acceptance of the capability of obstacle avoidance of PMVs with active tilting mechanisms could be confirmed also from their dynamic mechanism.

5. Energy Consumption on Active Tilting System

Although PMVs with a passive tilting mechanism have the negative point of a lack of self-standing ability in stopping and in very low speed the same as motorcycles, PMVs with an active tilting mechanism do not have such a concern. On the other hand, there is a concern about the contradiction with efficiency improvement inherent in PMVs, since energy consumption is inevitably unavoidable in active tilting mechanisms [8].

5.1. Mechanism of Energy Balance

5.1.1. Energy Consumption in Active Tilting Mechanism

Although it is originally necessary to describe the twist energy of the stabilizer bar in relation to the sprung body independently of the bounce strokes of both front wheels, the consumed twist energy can be expressed by the product of the difference of the vertical loads from the mean value and the difference of the bounce strokes from the mean value, as shown in Equations (10) and (11) and Figure 28.

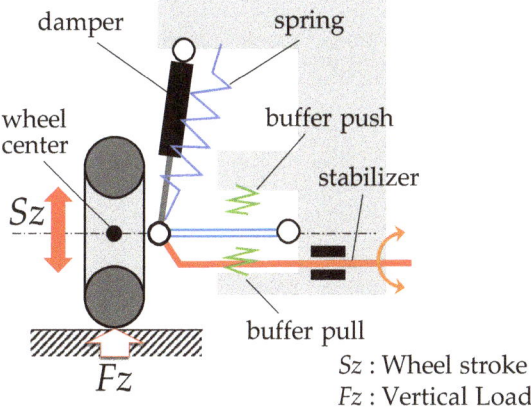

Figure 28. Energy consumption mechanism.

Energy efficiencies in actuating and recovering should be considered. In this study, the actuating efficiency was set to 0.5 and the recovering efficiency was set to 0.1 as an electrical general value.

$$\Delta E = F_{ZL} \times \Delta S_{ZL} + F_{ZR} \times \Delta S_{ZR} \tag{10}$$

IF $\Delta E \geq 0$, $\Delta E^* = \Delta E \times 2$ (actuator energy efficiency rate = 0.5)

IF $\Delta E < 0$, $\Delta E^* = \Delta E \times 0.1$ (actuator energy efficiency rate = 0.1)

$$E = \Sigma \Delta E^* \tag{11}$$

E: energy

5.1.2. Energy Consumption by Cornering Drag in Case of No Tilting

Generally, centripetal forces for vehicles are given as lateral force (SF) by tire slip angle (SA) when vehicles turn. Useful cornering force (Y) for turning is represented as $SF \times \cos(SA)$. Orthogonal $SF \times \sin(SA)$ is the cornering drag component, and energy is consumed wastefully.

Although the vertical load of each tire is always changing, for simplicity, the static load is substituted and the tire cornering power (K) is defined at this load, as shown in Figure 29. Y is obtained from a dynamic simulation and SA is obtained by dividing Y by K.

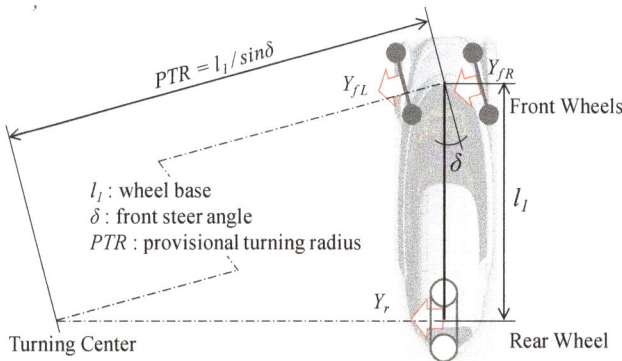

Figure 29. Energy consumption caused by cornering drag.

Total consumed energy rate caused by cornering drag is represented by the product of the cornering drag component and the longitudinal distance of the vehicle, as shown in Equation (12). Then, the accumulated consumed energy is the sum of the consumed energy rate, as shown in Equation (13).

$$\Delta E = Y_{fL} \times \sin\left(\frac{Y_{fL}}{K_{fL}}\right) \times v \times \Delta t + Y_{fR} \times \sin\left(\frac{Y_{fR}}{K_{fR}}\right) \times v \times \Delta t + Y_r \times \sin\left(\frac{Y_r}{K_r}\right) \times v \times \Delta t \quad (12)$$

$$E = \Sigma \Delta E \quad (13)$$

E: energy
Y: cornering force
K: cornering power
v: vehicle velocity
t: time

5.2. Energy Consumption in Typical Driving Modes

Although there are various explanations for the principle of side force generation due to the camber angle, it is usual that the cornering drag component of the side force by the camber angle is not considered, because it has no slip angle.

In this section, under the assumption that there is no cornering drag component of lateral force by camber angle, energy consumption for the active roll angle is compared with energy saving using the camber angle. Then, we consider the social acceptability of PMVs with active inward tilting mechanisms.

5.2.1. Energy Consumption on Steady Circle

In PMVs with inward tilting like motorcycles, during turning on the steady circle, the roll moment due to the centrifugal force is canceled by the roll angle and no additional roll moment is necessary. In other words, energy consumption due to cornering drag can continue to be avoided without energy consumption for tilting.

The energy consumption due to tilting and cornering drag of PMVs while entering the circle course, with a 50 m turning radius, from the straight course are shown. The vehicle starts on the straight course, accelerates while entering the circle course and reaches a steady speed of 60 km/h, as shown in Figure 30. Tilting energy is consumed only for a short time; therefore, the accumulated energy consumption for tilting may be considered essentially zero, as shown on the left of Figure 31. On the other hand, cornering drag energy is saved continuously on the circle course, as shown in right of Figure 31.

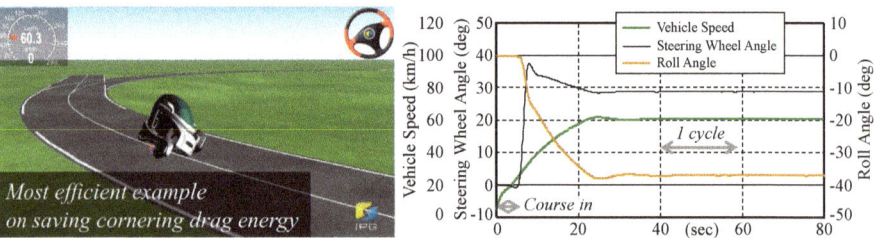

Figure 30. Driving on steady circle.

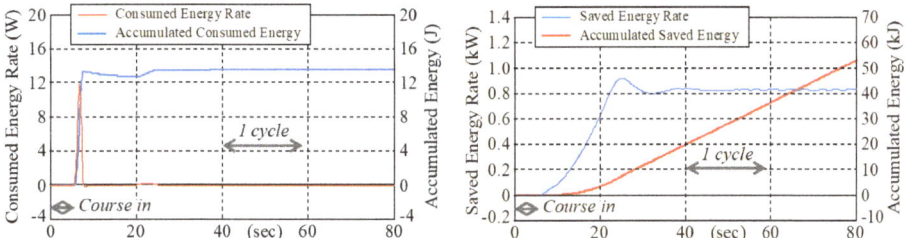

Figure 31. Consumed and saved energy on steady circle.

5.2.2. Energy Consumption in Slalom

In contrast to the steady circle, the pylon slalom with repeated tilting is a typical example of high energy consumption for tilting. We set the slalom course with 10 pylons at 18 m interval, as shown in Figure 32. The vehicle reaches 60 km/h before the slalom section and passes through the slalom section at a constant speed.

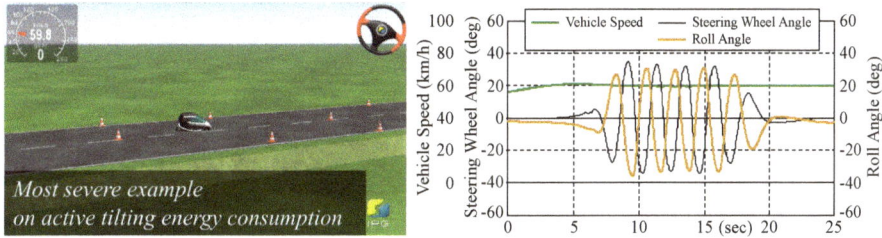

Figure 32. Driving on slalom course.

IPGDriver prepares the "corner cutting coefficient (ccc)" as one of the model parameters of driver characteristics in virtual driving. ccc = 0 means driver runs middle of given course and ccc = 1.0 means driver runs fully in-cut course. In other words, ccc = 1.0 is set to follow the easiest course. Under this condition, with ccc = 0.4, the driver could not pass the given course because of overturning. Therefore, this course setting may be considered to be almost at limit condition.

Although three levels of ccc were compared, which are ccc = 0.5, 0.7 and 0.9, as shown in Figure 33, no major difference was found in the steering wheel angles as input and lateral vehicle accelerations as output. Therefore, we decided to proceed with this study of energy balance at ccc = 0.7, which is considered to be a general driver characteristic.

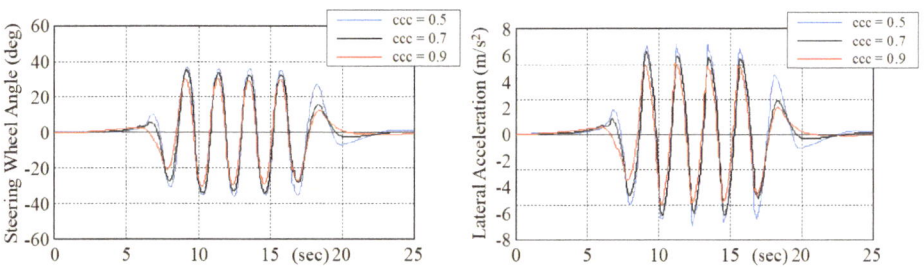

Figure 33. Influence of driver parameter is not significant.

In the slalom, a large amount of energy is consumed to give a quick change in roll angle, as shown on the left of Figure 34. During this time, although energy saving caused by avoidance of cornering drag also occurs significantly, as shown on the right of Figure 34, this cannot offset energy consumption for tilting.

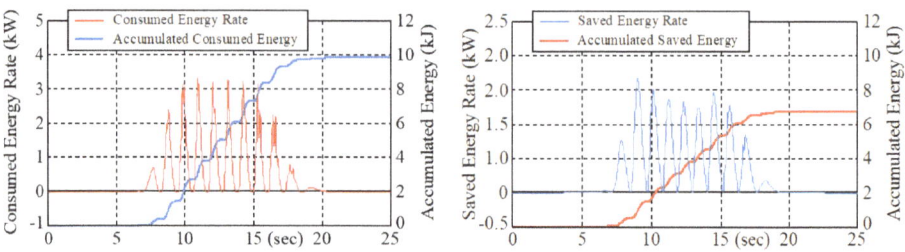

Figure 34. Consumed and saved energy on slalom course (ccc = 0.7).

5.3. Energy Consumption in Real World Condition

Driving modes for homologation of fuel consumption are basically on a straight course in all countries, and do not assume the steering wheel operation or lateral acceleration of the vehicle. However, German "auto motor und sport (AMS)" magazine prepares its own evaluation course, which is a round route on a public road in South Germany.

5.3.1. Typical European Evaluation Course (AMS)

The AMS evaluation course with a total length of 92.5 km and elevation difference of 290 m is shown in Figure 35. It takes about 5000 s to drive this PMV. Maximum speed is determined by speed regulations and the limit of the vehicle's performance. There are many stop intersections on this course. The steering wheel angle is particularly large when a vehicle stops, starts and turns at these intersections.

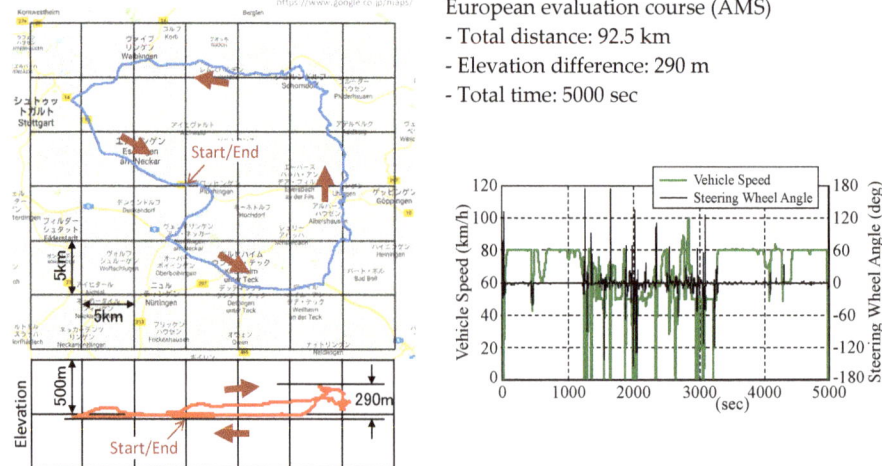

Figure 35. AMS evaluation course as typical European driving route.

5.3.2. Energy Consumption in AMS Magazine Evaluation Course

As shown on the left of Figure 36, the energy consumption rate has sharp peaks at intersections and at tight corners, and the accumulated consumed energy is less than 6 kJ in 5000 s. This amount is fairly small.

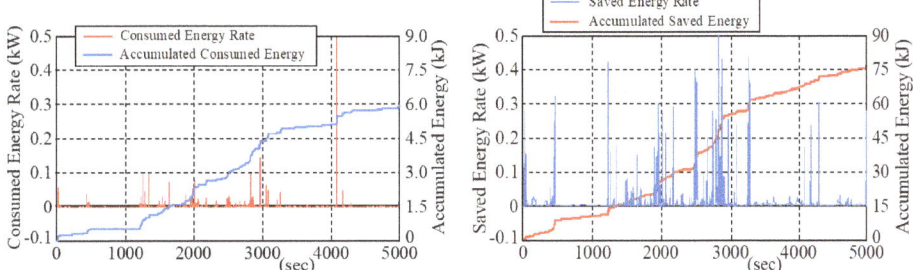

Figure 36. Consumed and saved energy on AMS evaluation course.

As shown on the right of Figure 36, energy saving caused by avoidance of cornering drag occurs much more frequently than energy consumption for tilting. This means that the steering wheel holding time on cornering sections is much longer than the steering input and return.

There is a difference in both energy rates and a larger difference in accumulated energy amounts. The accumulated energy saving against cornering drag reaches 76 kJ in 5000 s.

As shown in Figure 37, there is about 13 times difference between the two accumulated energies, and it shows no need to worry about energy consumption for active tilting mechanisms. High precision in the simulation model is not required at all for this definite result.

As shown in Table 2, the total mass of the vehicle including one passenger of this PMV is about 370 kg, and the total vertical load is about 3600 N. Assuming the tire rolling resistance coefficient (RRC) is approximately 80×10^{-4}, tire rolling resistance is approximately 29 N. With this rolling resistance, 2700 kJ of energy is consumed on the 92.5 km course.

This energy, 2700 kJ, is about 36 times the energy consumption by cornering drag and 450 times the energy consumption for the tilting mechanism, as shown in Figure 37. Energy consumption for driving PMVs is overwhelmingly dominated by tire rolling resistance if acceleration/deceleration and air drag are not considered.

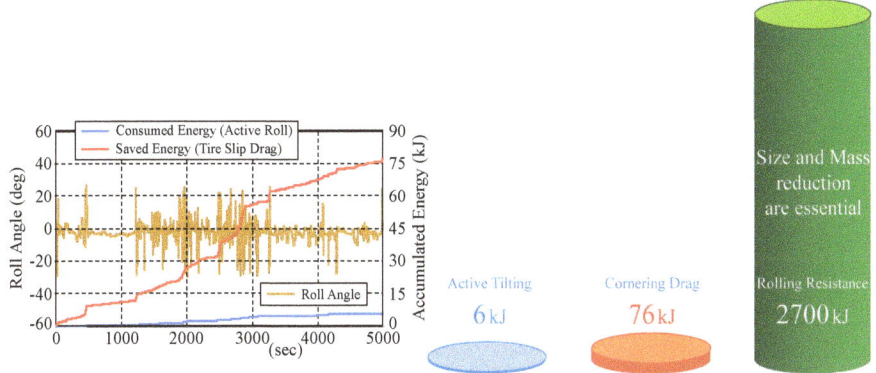

Figure 37. Energy balance on AMS course and comparison with rolling resistance energy.

5.4. Social Acceptance from Viewpoint of Energy Consumption

Energy consumption for tilting is only about one thirteenth of the saving effect of cornering drag; thus, the total efficiency of PMVs with an inward tilting mechanism is better than without a tilting mechanism. Additionally, smaller and lighter PMVs are significantly more efficient than general cars, as shown in Figure 38. These benefits of PMVs with an active inward tilting mechanism show sufficient social acceptability.

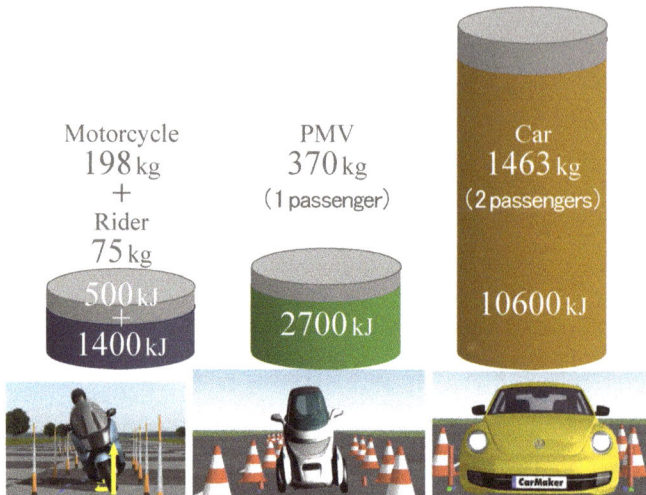

Figure 38. Comparison of energy consumption of general vehicles on rolling resistance.

6. Conclusions

PMVs with an active inward tilting mechanism with three wheels, double front wheels + single rear wheel and front steering + rear traction, are studied regarding front inner wheel lifting phenomena, capability of obstacle avoidance and energy balance of active tilting mechanism.

From the following results, it was shown that PMVs with an active inward tilting mechanism have sufficient social acceptability.

- Front inner wheel lifting phenomenon is only that vehicle roll is unavailable to follow target roll angle TRA due to roll moment of inertia of sprung mass Ixx^*. Smaller Ixx^* and larger $K\varphi$ improve front inner wheel lifting phenomena.
- It is possible to suppress unstable roll phenomena without intentional steering input related to front inner wheel lifting phenomena by decreasing the I gain of roll tracking control (PID).
- PMVs with an active tilting mechanism have two advantages of vehicle dynamics. One is lateral transfer of vertical loads on front both wheels, in the same way as general cars, and the other is inward tilting on turning, the same as motorcycles.
- PMVs with an active tilting mechanism have an obstacle avoidance ability equal to or higher than that of passenger cars, because they have a much smaller roll moment of inertia than passenger cars and responsiveness equivalent to passenger cars.
- Energy consumption for an active tilting mechanism is much smaller than energy saving due to avoiding cornering drag by using the camber angle. This shows there is no need to worry about energy consumption with active tilting mechanisms.
- From a general energy efficiency point of view, smaller and lighter PMVs are significantly more efficient than general cars, because rolling resistance caused by vehicle mass and the tire rolling resistance coefficient is the major resistance in these vehicles.

Author Contributions: Conceptualization, T.H.; formal analysis, T.H.; methodology, T.H. and T.K.; supervision, I.K.; validation, T.H.; writing—original draft, T.H.; writing—review and editing, T.H.

Funding: This research received no external funding.

Conflicts of Interest: The authors declare no conflict of interest.

References

1. Kaneko, T.; Kageyama, I.; Haraguchi, T.; Kuriyagawa, Y. A Study on the Harmonization of a Personal Mobility Vehicle with a Lean Mechanism in Road Traffic (First Report). In Proceedings of the JSAE Annual Congress (Spring), Yokohama, Japan, 25 May 2016; pp. 1350–1354.
2. Haraguchi, T. Are PMVs with inward tilting mechanism like motorcycles accepted in future mobility society? In Proceedings of the Forum text of JSAE Annual Congress (Spring), Yokohama, Japan, 24 May 2019.
3. CarMaker: Virtual Testing of Automobiles and Light-Duty Vehicles, IPG Automotive GmbH. Available online: https://ipg-automotive.com/products-services/simulation-software/carmaker/ (accessed on 20 October 2019).
4. Haraguchi, T.; Kageyama, I.; Kaneko, T. Inner Wheel Lifting Characteristics of Tilting Type Personal Mobility Vehicle by Sudden Steering Input. *Trans. Soc. Automot. Eng. Japan Inc.* **2019**, *50*, 96–101.
5. Kaneko, T.; Kageyama, I.; Haraguchi, T. A Study on Characteristics of the Vehicle Response by Abrupt Operation and an Improvement Method for Personal Mobility Vehicle with Leaning Mechanism. *Trans. Soc. Automot. Eng. Japan Inc.* **2019**, *50*, 796–801.
6. Haraguchi, T.; Kaneko, T.; Kageyama, I.; Kobayashi, M.; Murayama, T. Obstacle Avoidance Maneuver of Personal Mobility Vehicles with Lean Mechanism-Comparison between Front and Rear Wheel Steering-. In Proceedings of the JSAE Annual Congress (Spring), Yokohama, Japan, 24 May 2017; pp. 494–499.
7. Haraguchi, T.; Kaneko, T.; Kageyama, I. Study on Steering Response of Personal Mobility Vehicle (PMV) by Comparison of PMV with Passenger Cars and Motorcycles on the Obstacle Avoidance Performance. In Proceedings of the JSAE Annual Congress (Autumn), Nagoya, Japan, 17 October 2018.
8. Haraguchi, T.; Kaneko, T.; Kageyama, I. Market Acceptability Study on Energy Balance of Personal Mobility Vehicle (PMV) with Active Tilting Mechanism. In Proceedings of the JSAE Annual Congress (Spring), Yokohama, Japan, 24 May 2019.

© 2019 by the authors. Licensee MDPI, Basel, Switzerland. This article is an open access article distributed under the terms and conditions of the Creative Commons Attribution (CC BY) license (http://creativecommons.org/licenses/by/4.0/).

Article

Study on Low-Speed Stability of a Motorcycle

Sharad Singhania [1,*], Ichiro Kageyama [1] and Venkata M Karanam [2]

1. College of Industrial Technology, Nihon University, 1-2-1 Izumi-cho, Narashino-shi, Chiba 275-8575, Japan; kageyama.ichiro@nihon-u.ac.jp
2. Research and Development, TVS Motor Company Ltd., Harita, Hosur 635109, Tamil Nadu, India; Venkata.Raju@tvsmotor.com
* Correspondence: cisa16002@g.nihon-u.ac.jp; Tel.: +81-47-474-2337

Received: 29 April 2019; Accepted: 28 May 2019; Published: 3 June 2019

Abstract: The increased number of vehicles and poor road conditions in many countries result in slow moving traffic. At low-speeds, riding a motorcycle requires continuous input from a rider to achieve stability, which causes fatigue to the rider. Therefore, in this research, the low-speed stability of a motorcycle is studied using a theoretical and experimental approach to identify the parameters that can reduce the rider's effort. Initially, a linear mathematical model of the motorcycle and rider system is presented; wherein, the equation of motion for the stability of the system in roll direction is derived. The open-loop and closed-loop poles from the equation are calculated to determine the regions for the low-speed stability. Subsequently, experiments are conducted on the motorcycle instrumented with the required sensors, on a straight path at speeds below 10 km/h. The input and output parameters from the experimental data are analyzed using a statistical method. Steering angle and steering torque are the input parameters; roll and yaw angles and their corresponding velocities are the output parameters selected for the analysis. Correlation and lead time between the input and output parameters are compared to identify the parameters useful for the rider to attain the low-speed stability. The results obtained from the experimental analysis validate the mathematical model. In addition, these findings also validate that the input parameters required to control the motorcycle to achieve low-speed stability can be estimated using the identified output parameters.

Keywords: low-speed stability; balancing; motorcycle dynamics; rider control

1. Introduction

Motorcycles are a preferred mode of daily transportation in many countries because of congested traffic conditions, and also they are the most economically viable option. The traffic congestion and poor road conditions constrain the speed of motorcycles. Balancing a motorcycle at low-speeds is challenging, especially for new riders because it is unstable below the certain critical speed [1,2]. The rider becomes very cautious at such speeds as it requires continuous input to balance the motorcycle [3]; moreover, the required steering input or the gain value for the steering input increases as speed reduces [4]. This research focuses on the low-speed stability of the motorcycle due to the above-mentioned reasons.

The layout of a motorcycle influences its stability [5–7]. However, it cannot be tuned entirely for stability because it also determines other performance requirements such as manoeuvrability, ride comfort, ergonomics, acceleration feel, braking, etc. Hence, it is necessary to explore other methods to improve stability. In research [8–10], the relations between input and output parameters of a motorcycle are examined to construct a rider robot. Wherein, the speeds are relatively higher, and inputs are derived from the control engineering requirements. Similarly, in research studies [11,12], the low-speed stability of a bicycle is proposed using a theoretical approach. In such cases, frequencies of input to the bicycle are different from that provided by the riders.

The steering input required to improve the low-speed stability can be reduced by adding an extra degree of freedom to a motorcycle using a mechanism [13]. The low-speed stability of a bicycle or motorcycle can also be achieved by adding a device that provides gyroscopic moments [14–16]. In research [17], the low-speed stability of the motorcycle is achieved by providing steering to both front and rear wheels, unlike a typical motorcycle. However, these changes may not retain the conventional form and dynamics of the motorcycle.

A small humanoid robot can balance and steer a scaled-down bicycle by providing input to the handlebar, using the lateral dynamics of the bicycle [18]. A mathematical model of the bicycle and motor integrated by a controlled algorithm can reduce the step of estimating the steering angle for balancing it [19]. In research [20], the bicycle is balanced using both, flywheel and balancer; however, the study of system behaviour coupled with the steering input by the rider is undone. In research [21], the bicycle is balanced by controlling its steering, using the dynamic model derived from the equilibrium of gravity and centrifugal force. In these research studies, the balancing of the bicycle or the motorcycle is attained without assessing the inputs from an actual rider. Whereas, in the present study, the stability of the motorcycle at low-speeds is studied using experiments by the actual riders.

There are many research studies on improving and assessing the stability (weave and wobble) and handling characteristics of the motorcycle at high speeds [22–24]. Whereas, there are limited studies on low-speeds stability of motorcycle by evaluating the rider inputs [25,26], which is the scope of the present research.

In this research, the low-speed stability of a scooter-type motorcycle is studied using theoretical and experimental methods. The details of the motorcycle and the research methodology are given in Section 2. Equation of motion for the theoretical model of the motorcycle is presented in Section 3; wherein, the regions of stability of the motorcycle at speeds 3 and 5 km/h are shown. The experiments and method of analysis are discussed in Section 4. In this section, the theoretical model of the motorcycle is validated, and the parameters important for the low-speed stability are identified and evaluated. The conclusions of the research are given in Section 5.

2. Experimental Motorcycle and Methodology

2.1. Motorcycle Details

The layout, mass and inertia of a scooter-type motorcycle chosen for this research are provided in Table 1. The same table also shows the symbols corresponding to the parameters used in this paper.

Table 1. Layout, mass and inertia of the motorcycle including a rider weighing 65 kg.

Parameters	Symbols	Values	Unit
Total mass	M	165.70	kg
Wheelbase	l	1.236	m
Roll inertia at centre of gravity	I_g	18.79	kgm^2
Height of centre of gravity from ground	h	0.545	m
Horizontal distance of centre of gravity from rear axle	l_r	0.450	m
Caster angle	ϵ	25.8	degree
Fork offset	d_1	0.004	m
Front steering system mass	M_f	17.72	kg
Height of front steering system centre of gravity from ground	h_f	0.540	m
Shortest distance between steering system centre of gravity and steering axis	d	0.005	m
Front wheel rolling radius	r_f	0.214	m
Front wheel spin inertia	I_{fw}	0.122	kgm^2
Rear wheel rolling radius	r_r	0.205	m
Rear wheel spin inertia	I_{rw}	0.112	kgm^2
Acceleration of gravity	g	9.81	m/sec^2

The motorcycle-rider system used for studying low-speed stability is shown in Figure 1. The figure shows that the rider estimates the required input using the state feedback of the motorcycle. The disturbances; road irregularities and rider disruptions shown in the model are components of the output and the input parameters respectively. In general, any disturbance to the motorcycle affects its output parameters which in turn results in a change in the rider input. The objective of the rider is to keep the motorcycle stable by maintaining the reference value of the output parameters (i.e., roll angle (ϕ) to be zero in this case).

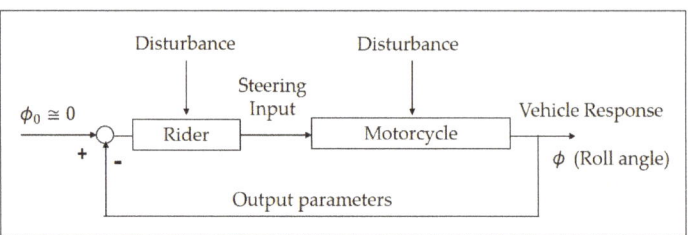

Figure 1. Block diagram for the low-speed stability of a motorcycle and rider system.

2.2. Methodology

In this research, the stability of the motorcycle at low-speed is studied using theoretical and experimental methods. At first, the equations for open-loop and closed-loop systems of the motorcycle were derived for the stability in roll direction using a linear motorcycle model. The regions of stability were determined by solving the equations. Subsequently, experiments were conducted on an actual motorcycle. The motorcycle was ridden on a straight path at speeds ranging from 3 to 10 km/h by professional riders, having riding experience of more than 15 years. The input and output parameters of the motorcycle from the experiments were analyzed using a statistical method. The correlation and lead time between the input and the output parameters were compared to identify the parameters contributing to the low-speed stability. Results obtained from the experiments validated the theoretical model and also confirm that the identified input parameter required to balance the motorcycle at low speeds can be estimated accurately using the identified output parameters.

3. Linear Motorcycle Model

A linear mathematical model for the motorcycle is described in this section. An equation of motion for the stability of the motorcycle is determined in the roll direction. Open-loop and closed-loop systems for the motorcycle are defined using the same equation. In this model, it is assumed that the tire thickness is zero, and all the angles are small.

The layout of a typical motorcycle is shown in Figure 2. The descriptions of the symbols used in the figure are given in Table 1 from Section 2. The points A and O are contact points of the front and rear wheels with ground respectively, and B is the point where the steering axis intersects the ground.

Figure 3 depicts the schematic of the motorcycle, which shows its different states while balancing. The symbols used in the figure are listed in Table 2. The figure shows that the motorcycle-rider system (mass M) first steered by an angle (δ) as shown by legend 2, and then rotated about the x-axis by a roll angle (ϕ) as shown by legend 3. These steering angle and roll angle are generated due to disturbances to the motorcycle. Figure 3a,b show rear-view and top-view of the motorcycle respectively. Various forces acting on the motorcycle are due to weight (Mg), normal reaction (N) and lateral force (F_y) as shown in the Figure 3a. The Figure 3b shows the effect of steering angle (δ) and roll angle (ϕ) on the motorcycle state. Marginal changes in δ and ϕ, which result in yaw angle (ψ), make the motorcycle to follow a circle of radius R. The significance of the steering mechanism on the stability of the motorcycle is described in this section.

Table 2. Input and output parameters used in linear model.

Parameters	Symbols
Roll angle	ϕ
Steering angle	δ
Kinematic steering angle	δ_R
Yaw angle	ψ
Lateral displacement at O	y_o
Speed	v
Instantaneous turning circle radius	R

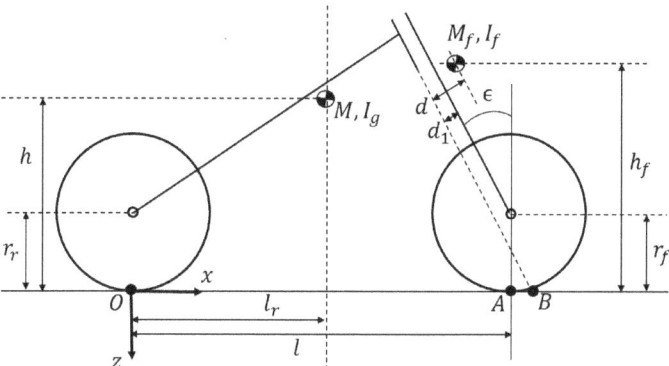

Figure 2. A typical layout of a motorcycle.

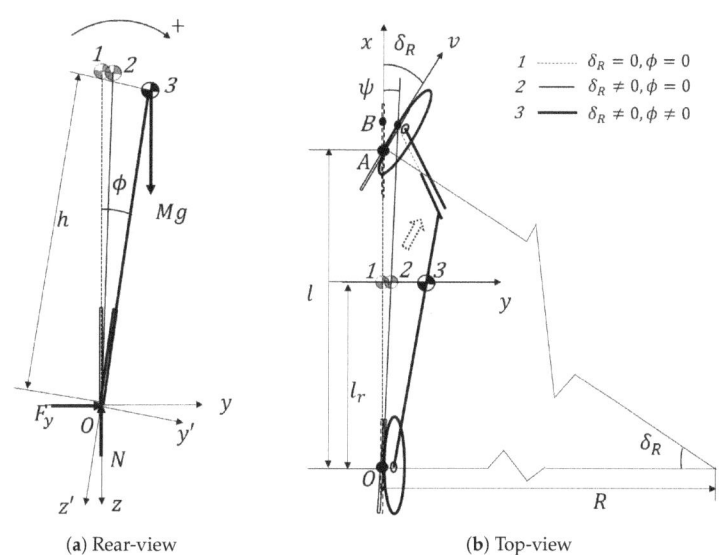

(a) Rear-view   (b) Top-view

Figure 3. Schematic of linear model of motorcycle showing its different states.

The equation of motion for the stability of the motorcycle in roll direction about the x-axis (i.e., the axis intersecting vehicle plane and ground plane at point O) is defined as follows:

$$I_o\ddot{\phi} + M\ddot{y}_o h = Mgh\phi + M_{gyro} + M_{steering} + M_{front\ normal\ reaction}, \qquad (1)$$

where:

The inertia of the motorcycle with respect to point O is as follows:

$$I_o = I_g + Mh^2. \tag{2}$$

The lateral acceleration of the motorcycle $\ddot{y}$ about point O is a function of the centrifugal force and lateral velocity. These parameters are derived from motorcycle speed v and kinematic steering angle δ_R as shown below:

$$\ddot{y}_o = \frac{v^2}{l}\delta_R + v\frac{l_r}{l}\dot{\delta}_R. \tag{3}$$

Gyroscopic moments by both front and rear wheels on the motorcycle, and by the front wheel on the front steering system can be described as:

$$M_{gyro} = -\frac{v^2}{l}\left(\frac{I_{fw}}{r_f} + \frac{I_{rw}}{r_r}\right)\delta_R - \frac{v}{r_f}I_{fw}\dot{\delta}_R. \tag{4}$$

Moment due to the front steering system vibration is as follows:

$$M_{steering} = -M_f dh_f \ddot{\delta}. \tag{5}$$

Moment due to the front wheel normal reaction force is as follows:

$$M_{front\ normal\ reaction} = Mg\frac{l_r}{l}d_1\delta. \tag{6}$$

The following equation is achieved by substituting the values from Equations (2)–(6) in Equation (1).

$$(I_g + Mh^2)\ddot{\phi} + M\left(\frac{v^2}{l}\delta_R + v\frac{l_r}{l}\dot{\delta}_R\right)h = \\ Mgh\phi - \frac{v^2}{l}\left(\frac{I_{fw}}{r_f} + \frac{I_{rw}}{r_r}\right)\delta_R - \frac{v}{r_f}I_{fw}\dot{\delta}_R - M_f dh_f\ddot{\delta} + Mg\frac{l_r}{l}d_1\delta. \tag{7}$$

Further, the kinematic steering angle can be defined by the following equation:

$$\delta_R = \tan^{-1}\left(\frac{\cos(\epsilon)\sin(\delta)}{\cos(\phi)\cos(\delta) - \sin(\phi)\sin(\epsilon)\sin(\delta)}\right). \tag{8}$$

Equation (8) can be approximated in linear form as follows:

$$\delta_R \approx \delta\cos(\epsilon). \tag{9}$$

Substituting the value of kinematic steering angle from Equation (9) in Equation (7) gives the following expression:

$$(I_g + Mh^2)\ddot{\phi} - Mgh\phi = \\ -M\left(\frac{v^2}{l}\delta + v\frac{l_r}{l}\dot{\delta}\right)\cos(\epsilon)h - \frac{v^2}{l}\left(\frac{I_{fw}}{r_f} + \frac{I_{rw}}{r_r}\right)\cos(\epsilon)\delta - \frac{v}{r_f}I_{fw}\cos(\epsilon)\dot{\delta} - M_f dh_f\ddot{\delta} + Mg\frac{l_r}{l}d_1\delta. \tag{10}$$

The open-loop transfer function for the low-speed stability of the motorcycle system can be described from Equation (10) by the following expression:

Open-loop transfer function =

$$\frac{\phi(s)}{\delta(s)} = \frac{-M\left(\frac{v^2}{l} + v\frac{l_r}{l}s\right)\cos(\epsilon)h - \frac{v^2}{l}\left(\frac{I_{fw}}{r_f} + \frac{I_{rw}}{r_r}\right)\cos(\epsilon) - \frac{v}{r_f}I_{fw}\cos(\epsilon)s - M_f dh_f s^2 + Mg\frac{l_r}{l}d_1}{(I_g + Mh^2)s^2 - Mgh}. \tag{11}$$

The transfer function defines the stability of the motorcycle at all the speeds.

Root-locus plot for Equation (11) is shown in Figure 4. The figure shows that one of the poles of the system is always positive at speeds 3, 5 and 10 km/h. Hence, the low-speed stability for the open-loop system of the motorcycle cannot be achieved.

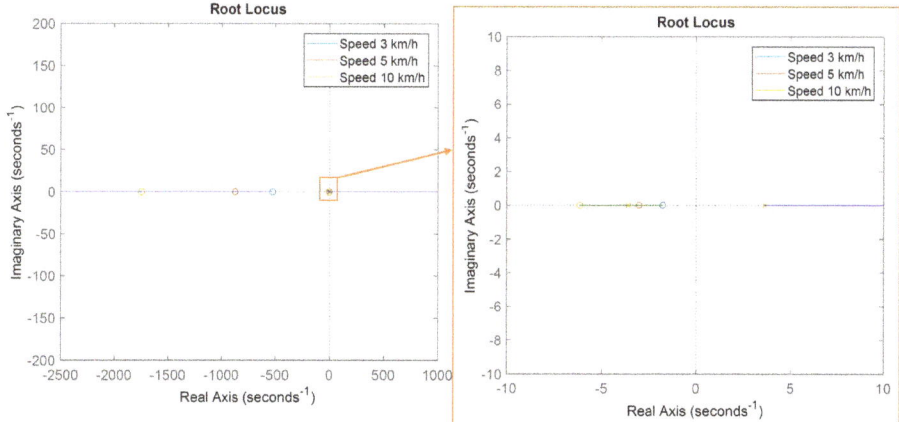

Figure 4. Root-locus plot for open-loop motorcycle system.

Further, the closed-loop feedback of the motorcycle can be defined using the following relationship between the steering angle (δ) and roll angle (ϕ):

$$\delta(t) = a.\phi(t - \tau), \qquad (12)$$

where a is roll angle gain and τ is a lead time for roll angle with respect to the steering angle. Equation (12) can be further simplified for $\tau < 1$ as follows:

$$\delta = a.\phi - a\tau\dot{\phi}. \qquad (13)$$

Substituting the value from Equation (13) in the Equation (10) following equation is obtained:

$$C_1\dddot{\phi} + C_2\ddot{\phi} + C_3\dot{\phi} + C_4\phi = 0, \qquad (14)$$

where,

$$C_1 = -M_f h_f da\tau \qquad (15)$$

$$C_2 = Mh^2 - \frac{Mhl_r v \cos(\epsilon) a\tau}{l} + I_g + M_f h_f da - \frac{I_{fw} v \cos(\epsilon) a\tau}{r_f} \qquad (16)$$

$$C_3 = Mhv\cos(\epsilon)a\left(\frac{l_r}{l} - \frac{v\tau}{l}\right) + \frac{I_{fw} v \cos(\epsilon) a}{r_f} - \frac{v^2 \cos(\epsilon) a\tau}{l}\left(\frac{I_{fw}}{r_f} + \frac{I_{rw}}{r_r}\right) + \frac{Mgl_r d_1 a\tau}{l} \qquad (17)$$

$$C_4 = \frac{v^2 \cos(\epsilon) a}{l}\left(\frac{I_f w}{r_f} + \frac{I_r w}{r_r}\right) - Mgh - Ma\left(\frac{gdl_r + hv^2 \cos(\epsilon)}{l}\right). \qquad (18)$$

The constants of the closed-loop characteristic Equation (14): C_1, C_2, C_3 and C_4 are shown in Equations (15)–(18). The eigenvalues of the characteristic equation are calculated for different values of the roll angle gain a and the lead time τ. The motorcycle is stable at a point where all the real parts of the eigenvalues are negative for a particular value of a and τ. The shaded zones in Figure 5a,b show regions for the stable motorcycle corresponding to the values of a and τ at speeds 3 and 5 km/h respectively. The rider must be operating inside the shaded zone shown in Figure 5 to achieve the low-speed stability.

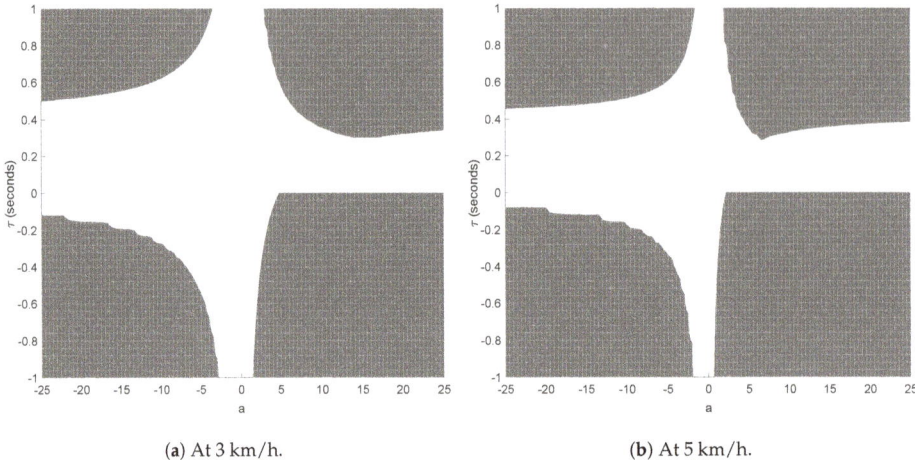

(a) At 3 km/h.   (b) At 5 km/h.

Figure 5. Regions of stability for different values of roll angle gain a and lead time τ for closed-loop system of motorcycle.

In the next section, the theoretical results are validated from experimental results.

4. Experiment and Analysis

4.1. Experiments Details

The motorcycle was instrumented with two analogue sensors namely a potentiometer (range of linearity ±50°) and a piezoelectric sensor (range ±200 Nm and sensitivity −175 pC/Nm) to measure the steering angle and the steering torque respectively. An inertia measurement unit (IMU) (angular rate range ±400°/s with an angle measurement accuracy of 0.2°) was used for measuring roll, pitch and yaw angles, corresponding velocities and accelerations. A GPS antenna was used for measuring longitudinal displacement, lateral displacement and speed of the motorcycle. The data was logged at a sample rate of 100 Hz [27].

The sensors were mounted on the motorcycle for the experiments. The rotating shaft of the potentiometer was mounted on top of the handlebar, and its non-rotating body was fixed to the frame. The handlebar and front steering system were disintegrated, and two flanges were fixed to them such that the faces of the flanges are perpendicular to the steering axis. The piezoelectric sensor was mounted between these flanges under high preload. The IMU was mounted on top of the pillion seat, and the GPS antenna was kept on a metal plate fixed to the rear frame of the motorcycle at the highest point on the motorcycle.

Experiments were conducted on a proving ground using the motorcycle instrumented with aforementioned sensors at speeds 3, 5 and 10 km/h. The motorcycle was ridden on the same dry asphalt road to inhibit the road variations in the results. The riders had instructed to follow a straight line marked on the proving ground. The experiments were conducted mainly to focus on balancing the motorcycle. Three professional riders, having riding experience of more than 15 years, were selected for the experiments to ensure the quality of results. Each rider had repeated the experiments two times to ensure the repeatability of the results.

4.2. Analysis of Experiments

The maximum correlation coefficient (MCC) and the lead time between the input and the output parameters are calculated from the experimental data. Steering angle and steering torque are the input parameters; roll and yaw angles and their corresponding velocities are the output parameters selected for the analysis. The MCC defined herein as a cost function to determine the maximum possible correlation between the input and the output parameters by shifting the output parameter with respect to the input parameter. The time step at which the cost function becomes maximum was defined as the lead time. The MCC is a measure of the dependency of the input parameter on the output parameter, and the lead time indicated the usability of the output parameter by the rider for predicting the steering control.

For example, Figure 6 shows time series data for roll angle and steering angle of a professional rider, normalized in [−1,1] range. The correlation between the roll angle and the steering angle was calculated while shifting the roll angle curve in the time domain at a step size of 0.01 s in the direction as shown by the arrow in the figure. Maximum correlation was observed after shifting the roll angle curve by 0.53 s. This correlation was named as the MCC and the time by which the roll angle was shifted to arrive at it is the lead time (i.e., 0.53 s). The MCC and the lead time are determined from the experimental data for all the riders.

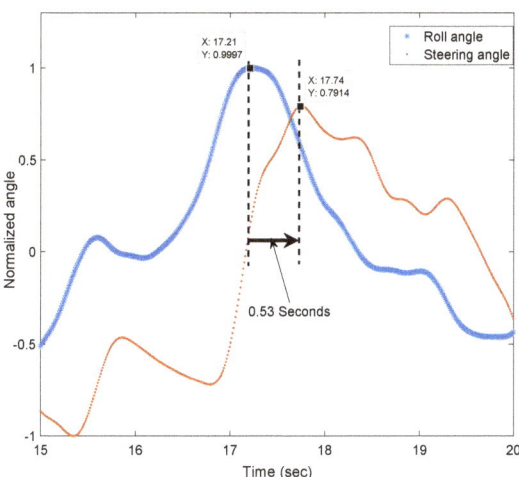

Figure 6. Steering angle and roll angle curve of a rider normalized in [−1,1] range @ 3.1 km/h.

In this research, the experimental data were analyzed using a statistical method. All the experimental data were divided into the sample size of 300 (i.e., 3 s) to examine the instantaneous input and output parameters while balancing the motorcycle at low speeds. The average (avg) and standard deviation (SD) from the experimental data were calculated and compared to ensure the repeatability and reliability of the experiments.

The analysis was done in four stages. Firstly, the repeatability and reliability of the experiments were determined. Secondly, the theoretical results were validated with the experimental results using the relationship between the steering angle and the roll parameters in Equations (12) and (13). Thirdly, the MCC and the lead time were calculated for output parameters with the input parameters. These were calculated to identify the output parameters important for the low-speed stability. Steering angle and steering torque were the input parameters; roll and yaw angles and their corresponding velocities were the output parameters selected for the analysis. Finally, multiple regression analysis (MRA) was performed between the identified output parameters as independent variables and the input parameters as dependent variables, to estimate and identify the most important input parameter between the steering angle and the steering torque. Further, the estimated input parameter from MRA was validated by comparing the same with the measured data.

4.2.1. Repeatability and Reliability of Experiments

Table 3 shows the avg and SD for both the motorcycle speed (*v*) and roll angle gain (RA gain *a*, for the positive value regions) for each experiment. The avg and SD were calculated for three professional riders who have ridden the motorcycle at target speeds of 3, 5 and 10 km/h. Each rider repeated the experiments two times (named as 'set' in the table). The table shows that the avg and SD of the motorcycle speeds and the roll angle gains match closely when the same rider repeated the experiment. This ensures the repeatability of the experiments.

The avg and SD of the RA gain increased sharply when the target speed reduced from 5 to 3 km/h than the same from 10 to 5 km/h; although, the SD of the speeds were similar. This was due to the fact that the effort and correction in steering input required to attain the low-speed stability increase sharply as speed reduces. The avg and SD in the RA gains were similar for all the rider at the particular target speed, which ensured the reliability of the experiments.

Table 3. Averages and standard deviations of motorcycle speed (v) and roll angle gain (a).

Target Speed			3 km/h			5 km/h				10 km/h			
		Speed		RA Gain		Speed		RA Gain		Speed		RA Gain	
Rider	Set	avg	SD	avg	SD	avg	SD	avg	SD	avg	SD	avg	SD
1	1	2.46	0.32	14.49	4.26	5.54	0.44	6.21	1.34	10.16	0.34	2.59	0.67
1	2	2.56	0.48	14.18	4.20	5.49	0.63	6.55	1.72	10.32	0.35	2.42	0.66
2	1	3.12	0.34	16.31	4.11	5.69	0.71	7.00	1.73	10.62	0.52	2.22	0.50
2	2	2.95	0.40	17.42	5.31	5.81	0.36	7.52	1.44	10.48	0.40	2.63	0.54
3	1	3.12	0.46	9.20	3.92	5.27	0.79	6.46	1.58	11.07	0.59	1.88	0.66
3	2	3.02	0.38	10.78	4.02	5.60	0.57	5.22	1.10	10.36	0.35	2.22	0.51

4.2.2. Validation of Theoretical Model

The roll angle gains *a* and the lead time τ were calculated as per Equation (12) from the experimental data. The results obtained from the experiments were compared with the theoretical results at speeds 3 and 5 km/h as shown in Figure 7a,b respectively. The figures show that the values of *a* and τ calculated from the experiments are in the stable regions determined from the theoretical results thereby validating the theoretical model. It also confirms that the model can be used to predict the stable zone for a motorcycle at low speeds.

Further, the relationship between the steering angle and roll parameters defined by Equation (13) in the Section 3 was examined. Linear MRA between the steering angle as a dependent variable; and, the roll angle and the roll rate as independent variables was performed from the experimental data using the following equation:

$$\delta = a_1 \phi + a_2 \dot{\phi}, \tag{19}$$

where a_1 and a_2 are the regression coefficients of the roll angle and the roll rate respectively. The regression coefficient a_2 was compared with the corresponding coefficient $(-a\tau)$ of roll rate derived from the experimental data as per the Equation (13). The correlation coefficients between the coefficients a_2 and $(-a\tau)$ were strong and more than 0.7. Figure 8 shows that both the coefficients match closely which shows the strong relationship between the theoretical model and experimental results.

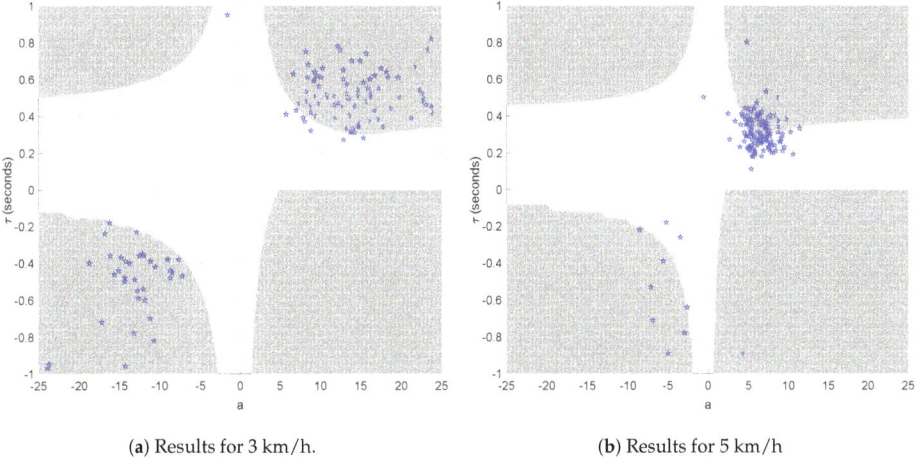

(a) Results for 3 km/h. (b) Results for 5 km/h

Figure 7. Validation of theoretical results from experiments.

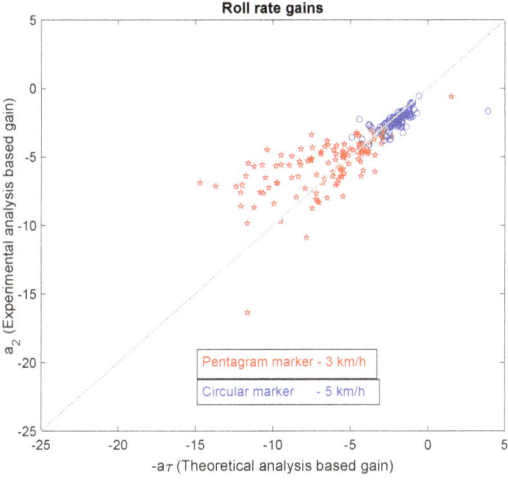

Figure 8. Validation of roll rate gain values based on theory and experiments.

4.2.3. Correlation Analysis

In this section, the experimental data are analyzed to identify the important input and output parameters for the low-speed stability of the motorcycle.

The avg and SD for the lead time and the MCC, for the output parameters with steering angle and steering torque are shown in Figures 9 and 10 respectively using error bars. The MCC is calculated for the leading side of the output parameters such that it can be used for estimating the steering input to achieve the low-speed stability. The results of the analysis are examined in two steps: Firstly, the MCCs between the input and the output parameters are compared to identify the output parameters which can estimate the input parameter. Secondly, the lead time between useful output parameters and input parameters are compared to assess their usability by the riders.

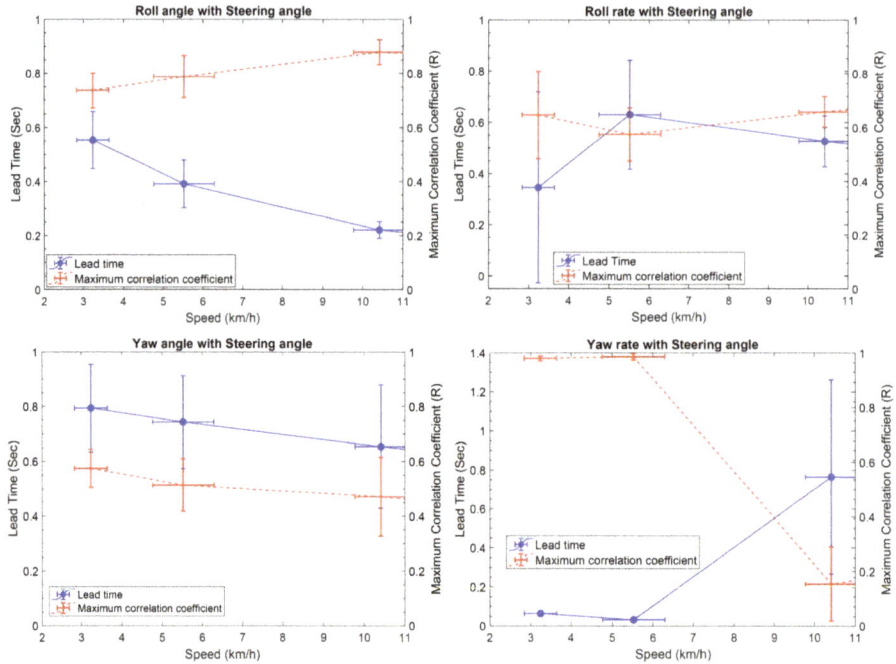

Figure 9. MCC (R) and lead time for output parameters with the steering angle.

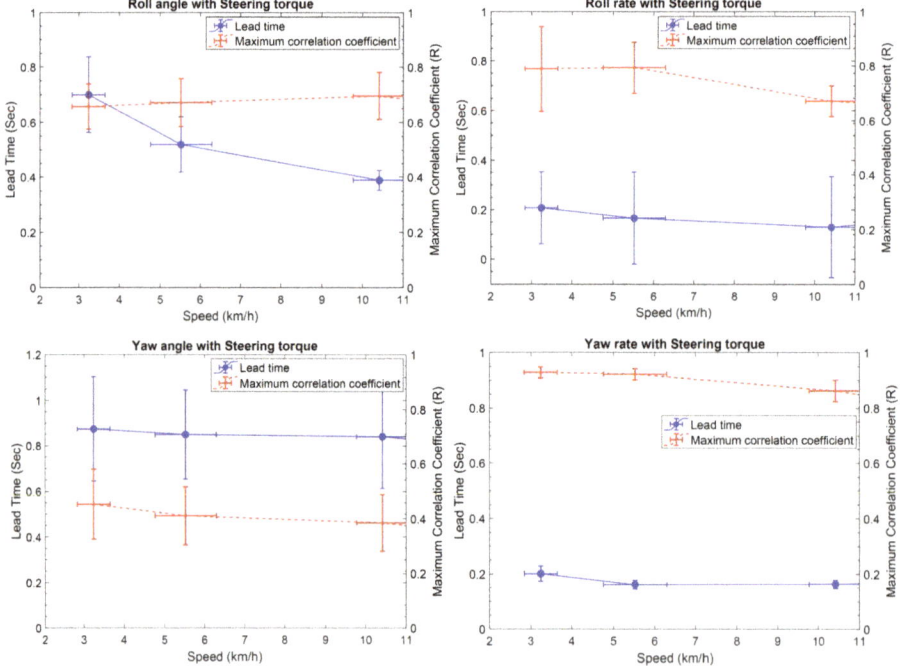

Figure 10. Maximum correlation coefficient (MCC) (R) and lead time for output parameters with steering torque.

Figure 9 shows that the roll angle and the roll rate are the output parameters that can estimate the steering angle because the MCCs for these parameters with steering angle are strong; and, the lead time is also sufficient to be used by the riders. Whereas, the correlation between the yaw angle and steering angle is weak. The correlation between steering angle and yaw rate was strong (above 0.9) at 3 and 5 km/h. Although, it was not the parameter useful for the riders because its lead time was too low and beyond the human limit to use it for estimating the steering angle.

Similarly, Figure 10 shows that the steering torque can be estimated from roll angle, roll rate and yaw rate because the MCCs between the steering torque and these parameters are strong. Whereas, the MCC between the yaw angle and steering torque was weak. The lead time for the yaw rate with the steering torque is below 0.2 s, which was lower than typical human response time. Thus, the yaw rate cannot be used by the rider for estimating steering torque, and it is a result of the change in other parameters. Therefore, the roll angle and the roll rate are important output parameters used by the riders to estimate the steering input to balance the motorcycle.

Further, the appropriate input parameter between the steering torque and the steering angle is selected based on two requirements. Firstly, the multiple correlations between the input parameter and output parameters i.e., roll angle and roll rate should be strong. Secondly, it is preferred to select the input that can be applied as late as possible to accommodate the delay in the system.

The MRA was performed individually for steering angle and steering torque as a dependent parameter using Equation (19). Figure 11a shows that the regression correlation between the identified output parameters (i.e., the roll angle and the roll rate) and the steering angle was stronger than the same with the steering torque. This shows that the steering angle can be estimated more accurately than the steering torque. The lead time and the MCC for the steering torque with the steering angle are shown in Figure 11b. The MCC between the parameters is strong. The negative lead time (approximately -0.15 s) shows that the steering torque had a delay with respect to the steering angle, which made it a suitable parameter for accommodating the delay in the system. Although, the steering angle was selected as an appropriate input parameter as the priority is given to the accurate estimation of the input parameter from the output parameters.

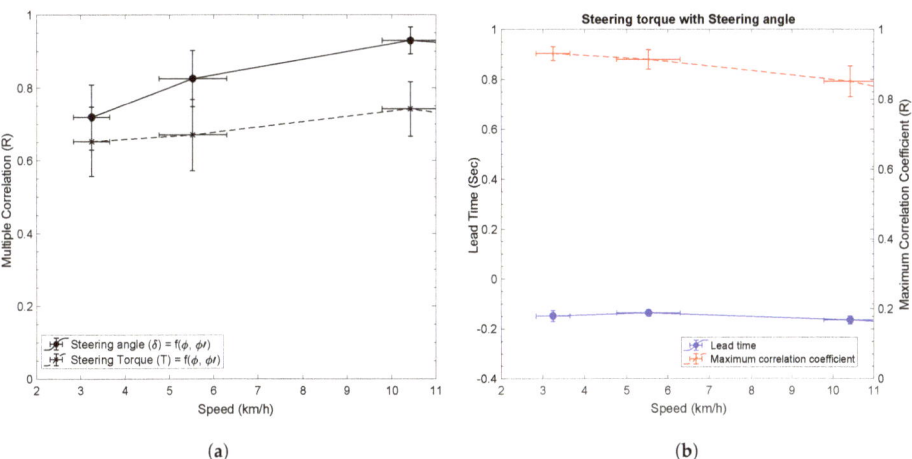

Figure 11. Analyses to select an appropriate input parameter between the steering angle and the steering torque. (**a**) Regression correlation for the steering angle and the steering torque; (**b**) Lead time and MCC (R) for the steering torque with the steering angle.

4.2.4. Output Parameter Evaluation

The steering angle was estimated from the roll angle and roll rate using the MRA Equation (19). The mean values of the regression coefficients a_1 and a_2 are shown in Figure 12, which can be defined by Equations (20) and (21), respectively as a function of motorcycle speed v.

$$a_1 = 0.011v^3 - 0.27v^2 + 2.5v - 5 \tag{20}$$

$$a_2 = -0.044v^2 + 0.92v - 6.4 \tag{21}$$

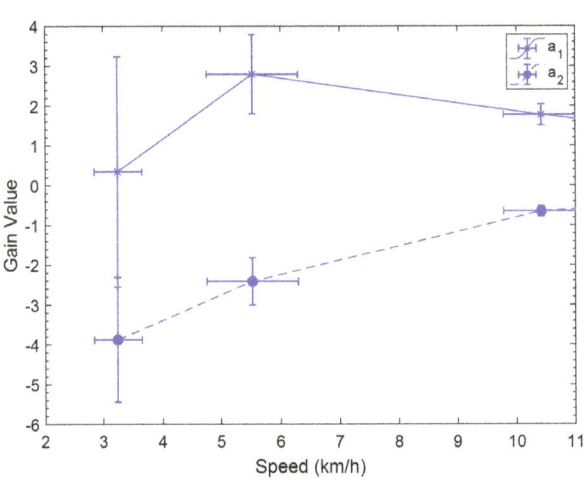

Figure 12. Regression coefficients for roll angle (a_1) and roll rate (a_2).

The steering angle estimated from the regression model is compared with its measured data from the experiments as shown in Figure 13. Figure 13a,b show that the estimated steering angle matches closely with the measured steering angle at speeds 3 and 5 km/h respectively. These results validate that the regression model can be used for estimating the steering angle, using the roll angle and roll rate accurately.

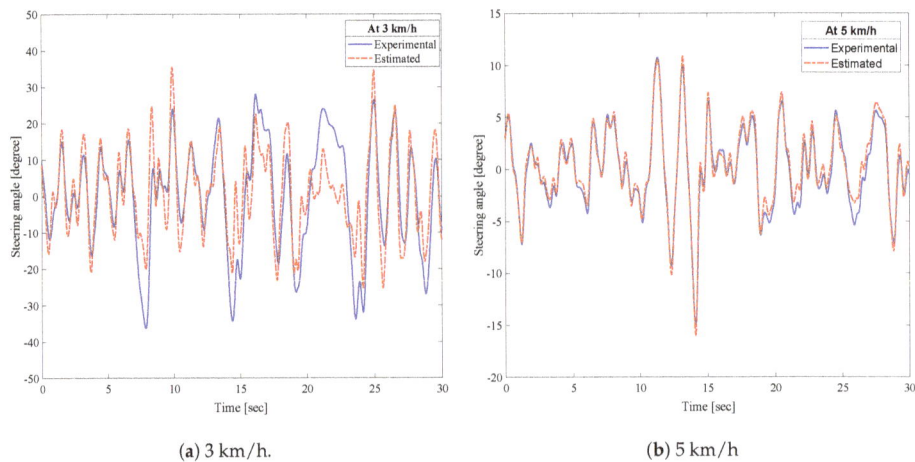

(a) 3 km/h. (b) 5 km/h

Figure 13. Validation of estimated steering angle from the experimental measurements.

5. Conclusions

In this research, the low-speed stability of a scooter-type motorcycle is studied using theoretical and experimental methods. The regions for the stability of the motorcycle are determined from the theoretical method. Subsequently, experiments are conducted on a straight path with professional riders at speeds 3, 5 and 10 km/h. A detailed analysis of experimental data is performed using statistical methods. Maximum correlation coefficient and lead time between input and output parameters are calculated and compared to identify the important output parameters to achieve the stability of the motorcycle at low speeds. Further, the appropriate input parameter is identified from a linear multiple regression analysis between the identified output parameters and the input parameters. Following conclusions are drawn from the study:

1. The experimental results validate that the theoretical method of analysis can be used to find the regions of low-speed stability of a motorcycle.
2. The roll angle and roll rate are important output parameters to be assessed to achieve low-speed stability.
3. The steering angle is the appropriate input parameter over steering torque for low-speed stability because the regression correlation for the steering angle is significantly stronger than the steering torque, with the roll angle and the roll rate.

In future, the results obtained from this research are to be used for developing a controller that can enable a motorcycle in attaining the low-speed stability, which is to be validated by simulations and experiments. Also, the theoretical model is to be improved by including other forces and parameters such as overturning moments, tire thickness and pitch motion of the motorcycle.

Author Contributions: Conceptualization, S.S., I.K. and V.M.K.; methodology, S.S. and I.K.; validation, S.S. and I.K.; formal analysis, S.S. and I.K.; writing—original draft preparation, S.S.; writing—review and editing, S.S., I.K. and V.M.K.; supervision, I.K. and V.M.K.

Funding: This research received no external funding.

Conflicts of Interest: The authors declare no conflict of interest.

Abbreviations

The following abbreviations are used in this manuscript:

MCC	Maximum Correlation Coefficient
MRA	Multiple Regression Analysis
avg	Average
SD	Standard Deviation

References

1. Astrom, K.J.; Klein, R.E.; Lennartsson, A. Bicycle dynamics and control: Adapted bicycles for education and research. *IEEE Control Syst.* **2005**, *25*, 26–47. [CrossRef]
2. Sharp, R.S. The stability and control of motorcycles. *J. Mech. Eng. Sci.* **2005**, *13*, 316–329. [CrossRef]
3. Zhang, Y.; Li, J.; Yi, J.; Song, D. Balance control and analysis of stationary riderless motorcycles. In Proceedings of the 2011 IEEE International Conference on Robotics and Automation, Shanghai, China, 9–13 May 2011; pp. 3018–3023. [CrossRef]
4. Yokomori, M.; Higuchi, K.; Ooya, T. Rider's Operation of a Motorcycle Running Straight at Low Speed. *JSME Int. J.* **1992**, *35*, 553–559. [CrossRef]
5. Kooijman, J.D.G.; Meijaard, J.P.; Papadopoulos, J.M.; Ruina, A.; Schwab A.L. A Bicycle Can Be Self-Stable Without Gyroscopic or Caster Effects. *Sci. Mag.* **2005**, *332*, 339–342. [CrossRef] [PubMed]
6. Jones D.E.H. The stability of the bicycle. *Phys. Today* **1970**, *23*, 34–40. [CrossRef]
7. Cossalter, V. *Motorcycle Dynamics*, 2nd ed.; Lulu.com: Morrisville, NC, USA, 2007.
8. Miyagishi, S.; Kageyama, I.; Takama, K.; Baba, M.; Uchiyama, H. Study on construction of a rider robot for two-wheeled vehicle. *JSAE Rev.* **2003**, *24*, 321–326. [CrossRef]

9. Takenouchi, S.; Sekine, T.; Okano, M. Study on condition of stability of motorcycle at low speed. In Proceedings of the Small Engine Technology Conference & Exposition, Kyoto, Japan, 29–31 October 2002; No. 2002-32-1797; SAE Technical Paper: Warrendale, PA, USA, 2002.
10. Schlipsing, M.; Salmen, J.; Lattke, B.; Schröter, K.G.; Winner, H. Roll angle estimation for motorcycles: Comparing video and inertial sensor approaches. In Proceedings of the IEEE Intelligent Vehicles Symposium, Alcala de Henares, Spain, 3–7 June 2012; pp. 500–505. [CrossRef]
11. Schwab, A.L.; Kooijman, J.D.G.; Meijaard, J.P. Some recent developments in bicycle dynamics and control. In Proceedings of the Fourth European Conference on Structural Control (4ECSC), St. Petersburg, Russia, 8–12 September 2008.
12. Zhang, Y.; Yi, J. Dynamic modeling and balance control of human/bicycle systems. In Proceedings of the 2010 IEEE/ASME International Conference on Advanced Intelligent Mechatronics, Montreal, QC, Canada, 6–9 July 2010; pp. 1385–1390. [CrossRef]
13. Kimura, T.; Ando, Y.; Tsujii, E. Development of new concept two-wheel steering system for motorcycles. *SAE Int. J. Passeng. Cars Electron. Electr. Syst.* **2014**, *7*, 36–40. [CrossRef]
14. Beznos, A.V.; Formal'sky, A.M.; Gurfinkel, E.V.; Jicharev, D.N.; Lensky, A.V.; Savitsky, K.V.; Tchesalin, L.S. Control of autonomous motion of two-wheel bicycle with gyroscopic stabilisation. In Proceedings of the 1998 IEEE International Conference on Robotics and Automation, Leuven, Belgium, 16–20 May 1988; Volume 3, pp. 2670–2675. [CrossRef]
15. Lot, R.; Fleming, J. Gyroscopic stabilisers for powered two-wheeled vehicles. *Veh. Syst. Dyn.* **2018**. [CrossRef]
16. Karnopp, D. Tilt control for gyro-stabilized two-wheeled vehicles. *Veh. Syst. Dyn.* **2002**, *37*, 145–156. [CrossRef]
17. Yang, C.; Murakami, T. Full-speed range self-balancing electric motorcycles without the handlebar. *IEEE Trans. Ind. Electron.* **2016**, *63*, 1911–1922. [CrossRef]
18. Huang, C.; Tung, Y.; Yeh, T. Balancing control of a robot bicycle with uncertain center of gravity. In Proceedings of the IEEE International Conference on Robotics and Automation (ICRA), Singapore, 29 May–3 June 2017; pp. 5858–5863. [CrossRef]
19. Vatanashevanopakorn, S.; Parnichkun, M. Steering control based balancing of a bicycle robot. In Proceedings of the 2011 IEEE International Conference on Robotics and Biomimetics, Karon Beach, Phuket, Thailand, 7–11 December 2011; pp. 2169–2174. [CrossRef]
20. Keo, L.; Yoshino, K.; Kawaguchi, M.; Yamakita, M. Experimental results for stabilizing of a bicycle with a flywheel balancer. In Proceedings of the 2011 IEEE International Conference on Robotics and Automation, Shanghai, China, 9–13 May 2011; pp. 6150–6155. [CrossRef]
21. Tanaka, Y.; Murakami, T. Self sustaining bicycle robot with steering controller. In Proceedings of the AMC '04 8th IEEE International Workshop on Advanced Motion Control, Kawasaki, Japan, 25–28 March 2004; pp. 193–197. [CrossRef]
22. Sharp, R.S.; Watanabe, Y. Chatter vibrations of high-performance motorcycles. *Veh. Syst. Dyn.* **2013**, *51*, 393–404. [CrossRef]
23. Massaro, M.; Lot, R.; Cossalter, V.; Brendelson, J.; Sadauckas, J. Numerical and experimental investigation of passive rider effects on motorcycle weave. *Veh. Syst. Dyn.* **2012**, *50* (Suppl. 1), 215–227. [CrossRef]
24. Sakai, H. *Theoretical and Fundamental Consideration to Accord Between Self-Steer Speed and Rolling in Maneuverability of Motorcycles*; No. 2018-32-0049; SAE Technical Paper: Warrendale, PA, USA, 2018. [CrossRef]
25. Karanam, V.M. Studies in the Dynamics of Two and Three Wheeled Vehicles. Ph.D. Thesis, Indian Institute of Science, Bangalore, India, 2012.
26. Popov, A.A.; Rowell, S.; Meijaard, J.P. A review on motorcycle and rider modelling for steering control. *Veh. Syst. Dyn.* **2010**, *48*, 775–792. [CrossRef]
27. Data Logger. Available online: http://www.racelogic.co.uk/_downloads/vbox/Datasheets/Data_Loggers/RLVB3i_Data.pdf (accessed on 27 April 2019).

© 2019 by the authors. Licensee MDPI, Basel, Switzerland. This article is an open access article distributed under the terms and conditions of the Creative Commons Attribution (CC BY) license (http://creativecommons.org/licenses/by/4.0/).

Article

A Multi-Physics Modeling-Based Vibration Prediction Method for Switched Reluctance Motors

Xiao Ling, Jianfeng Tao *, Bingchu Li, Chengjin Qin and Chengliang Liu

The State Key Laboratory of Mechanical System and Vibration, School of Mechanical Engineering, Shanghai Jiao Tong University, Shanghai 200240, China; lingxiao@sjtu.edu.cn (L.X.); bchli@sjtu.edu.cn (B.L.); qinchengjin@sjtu.edu.cn (C.Q.); chlliu@sjtu.edu.cn (C.L.)
* Correspondence: jftao@sjtu.edu.cn; Tel.: +86-021-3420-6053

Received: 19 September 2019; Accepted: 23 October 2019; Published: 25 October 2019

Abstract: Currently, vibration has been one crucial factor hindering the application of switched reluctance motor (SRM). Hence, it is of crucial importance to predict and suppress this undesirable vibration. This paper proposes a multi-physics analysis-based vibration prediction approach for SRM. It consists of three modules: digital controller and drive circuit module, electromagnetic field module, and mechanical module. In the mechanical module, it not only includes the influence of the stator, but also fully considers the influence of rotor, end cover, bearing, and other components of the motor on the system modal. Moreover, the vibration data under different control strategies are obtained in real time, and data dynamic interaction between the three segments can also be achieved. By combining the electromagnetic forces and the system structure modal, the vibration of SRM can be predicted. Finally, the effectiveness of the proposed method was verified on 12/8 poles, 1.5KW SRM drive system test bench. The results demonstrate that the modal simulation method based on static pre-calculation achieves high accuracy, and the vibration spectrums predicted by the proposed method shows good agreement with the experimental results.

Keywords: switched reluctance motor; vibration prediction; multi-physics modelling

1. Introduction

Due to its simple structure, low cost and high reliability, a switched reluctance motor (SRM) has been highlighted for many industrial applications [1,2]. For instance, SRM is considered as a promising candidate for the distributed-driven electric vehicle [3,4]. Based on the analytical Fourier fitting method, Sun et al. [5] presented a convenience method for modeling an in-wheel switched reluctance motor. To alleviate the harmful effects of vibration induced by the unbalanced electromagnetic force, Qin and co-workers [6] developed a vibration mitigation method based on a dynamic vibration absorbing structure. Despite its many advantages, SRM suffers from inherent vibration and acoustic noise, which limits its application in the industrial potential market. Therefore, researchers have investigated the vibration and acoustic noise problem for SRM from various aspects: stator and rotor pole shaping [7], motor topology optimization [8,9], switching angle adjustment [10], insert of zero voltage loop during switching [11], and current waveform profiling [12]. In order to predict the vibration of SRM, accurate modal analysis and radial force calculation are of vital significance.

The natural frequencies and vibration modes of SRM can be calculated by modal analysis, which are used to analyze the vibration characteristics. Cai et al. [13] showed that the vibration is largest when the harmonic frequency of electromagnetic force coincides with the modal frequency. Callegaro et al. [14] recommended a phase radial force shaping method based on harmonic content analysis, which was experimentally validated with experiments at different speed and load conditions. Kimpara and co-workers [15] analyzed the mechanical coupling path of stator and rotor and its influence on vibration modes. Taking into account the electromagnetic characteristics, mechanical vibration, and acoustic noise, Liang et al. [16] proposed a numerical modeling method for classical SRM and mutually coupled SRM. In [17], FEA methods were used to determine the modal shapes and natural frequencies for SRM, and the influence of end-bell and self-weight were also considered. However, the above methods only analyzed the modal of the motor stator, and did not consider the influence of windings, end cover, bearings, and other components on the system modal. Radial electromagnetic force is the main excitation source for the vibration and noise of SRM [18]. Calculation of electromagnetic force can be categorized into analytical methods and finite element analysis (FEA) methods. Anwar et al. [19] and Husain et al. [20] derived the electromagnetic field distribution of SRM by dimension parameters and calculated the radial force according to the virtual displacement principle. However, the analytical methods require precise system parameters, and the corresponding derivation process of the radial force is quite complicated. In [21,22], the finite element-based multi-physics analysis method was adopted to calculate the electromagnetic force, and the electromagnetic force was used as the excitation of the vibration model. Fiedler et al. [23] and Srivas et al. [24] proposed a multi-physical field numerical simulation method, in which only the vibration characteristics of the stator were considered.

With the obtained electromagnetic force and modal information, the vibration of the motor can be determined quantitatively. By utilizing the modal information and electromagnetic force, Fiedler et al. [23] calculated the stator vibration response by a modal superposition method. Tang and co-workers [25] established the transfer function model of vibration based on impulse excitation experiment and employed the transfer function model to calculate and analyze the vibration. Anwar et al. [19] and Lin et al. [26] computed the SRM vibration by convolution method. When the number of switching actions in a cycle is relatively large, the computational time will also increase. With the aid of a piecewise first-order linear function, Guo et al. [27] proposed an efficient vibration calculation method. However, the saturation effect is not taken into consideration.

In-depth analysis shows that the above vibration prediction methods only analyzed the vibration characteristics of the stator and ignored the influence of other components on the system vibration. To predict its vibration more accurately, it is necessary to consider the influence of more components including stator, rotor, housing winding, end cover, and bearing on the system modal. Consequently, this paper proposes a novel vibration prediction method for SRM based on multi-physics modeling. It includes three core segments: control algorithm integration, electromagnetic field dynamic analysis, electromagnetic force calculation, and modal analysis. The highlight of the proposed method is that it allows modification of system inputs, such as geometry dimensions, material parameters, or control strategies, at each step of the simulation process. The vibration data under different control strategies is obtained in real time, and data dynamic interaction between the three segments can be also achieved. Therefore, from the perspective of reducing vibration, it can be used to optimize the structural design and the control strategy of the switched reluctance motor. In addition, in the mechanical module, it not only includes the influence of the stator, but also fully considers the influence of rotor, end cover, bearing, and other components of the motor on the system modal. The remainder of this paper is organized as follows. In Section 2, the operational principle and the origin of vibration of the SRM are presented. Section 3 develops the multi-physics modeling-based vibration prediction method. In Section 4, the implementation of the prediction model is presented, and experiments are conducted to validate the proposed method. Finally, the conclusions are drawn in Section 5.

2. Vibration of the Switched Reluctance Motor

SRM mainly consists of a stator unit, a rotor unit and corresponding structural components, as shown in Figure 1a. The stator unit includes iron core, winding, slot wedge, insulating material and motor lead, while the rotor unit contains rotor shaft and rotor core. Both the stator and the rotor use a salient pole structure that was formed by laminating silicon steel sheets. In addition, the windings are installed only at the stator pole. Furthermore, the structural components are mainly composed of housing, end cover, bearing cover, bearing, junction box, and junction column. Figure 1b illustrates the stator and rotor structure for a 12-8 SRM. Four radial relative windings at the stator pole form a phase. If current is applied to a specific phase of the motor (e.g., phase A), the magnetic field force is generated due to the distortion of the electromagnetic field, which makes the rotor rotate to the position where the pole axis of the rotor and the pole axis of the stator coincide. At this time, the magnetic resistance is at a minimum. When the phases B, C, A and B in Figure 1b are sequentially energized, the rotor will continuously rotate in a counterclockwise direction, and, vice versa, in a clockwise direction.

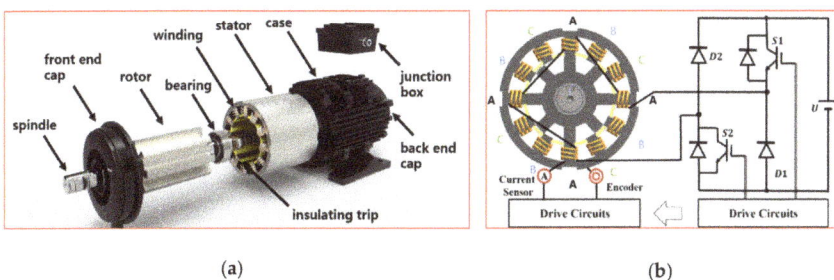

Figure 1. Structure and working principle of switched reluctance motor (SRM): (**a**) structure of SRM; (**b**) working principle of a 12-8 SRM.

Currently, the vibration problem is one of the research hotspots of the SRM. It is mainly caused by the change of the electromagnetic force between air gap magnetic fields of the stator and rotor. When the stator winding of SRM is energized, an induced magnetic field is formed at the stator pole, which further magnetizes the rotor. At this time, electromagnetic forces are generated between the stator and rotor, as shown in Figure 2. The electromagnetic forces acting on the stator and rotor, namely F_s and F_r, are a pair of forces and reaction forces. The electromagnetic force F_s can be decomposed into tangential magnetic force F_{st} and radial magnetic force F_{sr}. When the stator winding is excited, the radial magnetic force F_{sr} increases, stretching the stator along the radial direction. However, if the current is cut, the radial magnetic force F_{sr} is reduced to zero, and the stator rebounds in the radial direction. With the continuous commutation of the excitation winding for the SRM, the radial magnetic force F_{sr} of the different phases repeatedly attracts and releases the stator, causing structural deformation of the motor stator. If the frequency of F_{sr} is close to the natural frequency of the motor, it will excite the resonance of the motor body and cause strong radial vibration. Similarly, the electromagnetic force F_r can be also decomposed into tangential magnetic force F_{rt} and radial magnetic force F_{rr}. Due to the doubly salient construction and fundamental excitation principles, the majority of the flux in SRM is in the radial direction. Although the torque ripple caused by fluctuations of F_{rt} during motor excitation and demagnetization may also induce tangential vibration, it can be neglected compared with radial vibration, which has been also verified by the works of several researchers [1,18,26]. For instance, Cameron et al. [18] designed a series of experiments to investigate the vibration in SRM, and the experimental results showed that radial vibration of the stator frame caused by radial electromagnetic forces is the dominant source of vibration in SRM. The authors in [1,26] also proved that the acceleration response of SRM mainly resulted from the sudden change of radial electromagnetic forces. Consequently, this paper focuses on the establishment of a radial vibration prediction model.

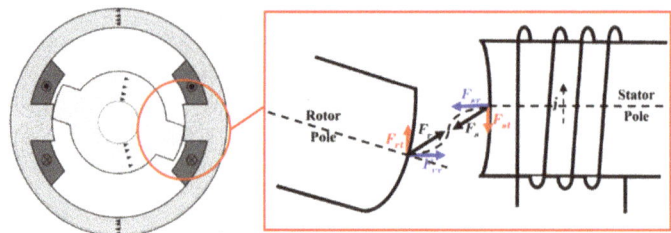

Figure 2. Schematic diagram of an electromagnetic field and electromagnetic force.

3. The Proposed Multi-Physics Modelling-Based Method

The operation of an SRM is a process of interaction of electro-magnetic-mechanical multiple physical fields. The performance parameters of multiple physical fields are constrained by each other. The electromagnetic characteristics of traditional motors are usually linear or quasi linear. For SRM, it has poles both on stator and rotor, and the windings are bounded on the stator, as shown in Figure 1. On the other hand, due to the saturation characteristics of ferromagnetic material, the flux linkage begins to saturate when the current reaches certain values. During the motor running process, the SRM's electromagnetic properties versus position and phase current exhibit nonlinear characteristics and thus are difficult to model accurately compared with traditional motors. Due to its nonlinear electromagnetic characteristics and the special action mode of electromagnetic force, the multi-physics interaction behavior of SRMs is more complicated than that of traditional motors. In addition, the components of SRM, including stator, rotor, shell winding, end cover and bearing, have influence on the system modal. Therefore, for the performance optimization and control method design of SRM, it is of vital importance to model and analyze from the perspective of multi-physics field, and fully take into account the influence of more components on the system modal. This paper presents a numerical simulation method that fully considers the interaction relationship of multi-physical fields. It includes three segments: digital controller and drive circuit module, magnetic field module, and mechanical module. To obtain an excitation current, the gate signals of the controller module are employed to drive the power amplifier. The winding of SRM is energized by the generating current, and an induced magnetic field will be formed. Thereafter, the electromagnetic forces can be acquired. Finally, the vibration of SRM is predicted by combining the electromagnetic forces and the system structure modal.

3.1. The Controller and Drive Circuit Models

As shown in Figure 3, the modeling of SRM controller is divided into four parts: speed controller, current controller, position and speed calculation module, and execution angle calculation module. The speed controller outputs the reference current I_{ref}, while the current controller generates a gate signal G_i that controls the interruption of insulated gate bipolar translator (IGBT). In addition, the position and speed calculation module compute the rotor angle θ_{est} and speed ω_{est}, while the execution angle calculation module obtains the turn on and turn off angles, i.e., θ_{on} and θ_{off}. For SRMs, the modeling of the drive circuit mainly involves the modeling of the switching device IGBT. If the change of the capacitance parameter and the switching process characteristics are ignored, the saturation voltage V_{sat} and the current I_{sat} can be calculated as follows:

$$\begin{cases} V_{sat} = A_{FET} \times (V_{GS} - V_P)^{M_{FET}} \\ I_{sat} = \frac{k}{2} \times (V_{GS} - V_P)^{N_{FET}} \end{cases}, \quad (1)$$

where A_{FET} is the saturation factor, V_{GS} is the gate to source voltage, V_p is the pinch-off voltage, M_{FET} is the saturation exponent, N_{FET} is the transfer characteristic exponent, and k is the transfer constant. Meanwhile, the current I_C of the linear region is expressed as:

$$I_C = B_N \times I_{sat} \times (1 + KLM \times V_{DS}) \times (2 - \frac{V_{DS}}{V_{sat}}) \times \frac{V_{DS}}{V_{sat}}, \quad (2)$$

where B_N is gain of current, the KLM is the channel length modulation factor, and V_{DS} is the drain to source voltage. The coefficient k, the voltage V_p, and the current gain B_N are parameters that vary with temperature and can be corrected based on the actual temperature value.

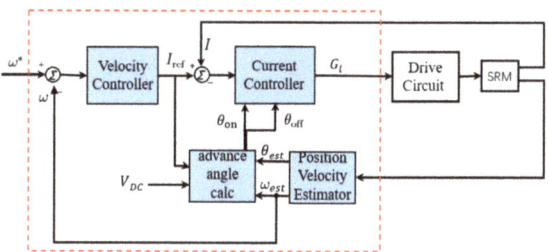

Figure 3. Controller of SRM.

3.2. The Electromagnetic Module

There are three main numerical methods available for the electromagnetic field simulation, including the finite element methods, boundary element methods, and finite difference methods. In this work, we adopt the most widely used finite element method based on the following assumptions [28]:

1. Neglect both magnetic flux leakages from the excitation pole to the stator yoke through the stator pole space and from the excitation pole to the adjacent stator teeth.
2. The eddy current effect is negligible.
3. Neglect the magnetic field and motor end effect outside the boundary formed between the outer diameters of stator and rotor.
4. The equation $B_s = \nabla \times U$ is satisfied when the vector magnetic potential U is introduced in the solution domain.

Considering the above assumptions, the electromagnetic field problem in the plane Ω can be expressed as the following boundary value problem [29]:

$$\begin{cases} \Omega : \frac{\partial}{\partial x}\left(\gamma \frac{\partial U_z}{\partial x}\right) + \frac{\partial}{\partial x}\left(\gamma \frac{\partial U_z}{\partial y}\right) = -J_z \\ \Gamma_1 : U_z = U_0 \\ \Gamma_2 : \gamma \frac{\partial U_z}{\partial n} = -H_t \end{cases}, \quad (3)$$

where γ is the magnetoresistance of the stator and rotor cores, J_z is the current density, U_z is the vector magnetic potential, H_t is the tangential component of the magnetic field, and Γ_1 and Γ_2 are the first and second boundary conditions, respectively. Equation (3) can be equivalent to a problem of the conditional variation as follows:

$$\begin{cases} W(U_z) = \iint_\Omega \left(\int_0^{B_s} \gamma B_s dB_s - J_z U_z\right) dxdy - \int_{\Gamma_2} (-H_t) U_z dl = \min \\ B_s = \sqrt{B_{sx}^2 + B_{sy}^2} = \sqrt{\left(\frac{\partial U_z}{\partial x}\right)^2 + \left(\frac{\partial U_z}{\partial y}\right)^2} \\ \Gamma_1 : U_z = U_0 \end{cases}, \quad (4)$$

After a finite element division of the entire solution domain of the electromagnetic field, the difference function for each small element can be given by:

$$U = N_i U_i + N_j U_j + N_m U_m, \tag{5}$$

where U is the magnetic vector potential at any point in the element, N_i, N_j, and N_m are the shape functions that are related to the element node coordinates; U_i, U_j, and, U_m are the magnetic vector potentials on the element nodes.

The algebraic equations of the nodal function can be obtained through converting the variational problem to another problem of finding the extremum of the energy function W. We can get the nodal magnetic potential by solving the algebraic equations using the Newton–Raphson method. Then, the field quantity can be obtained by a finite element post-processing. According to [30], the flux linkage ψ and inductance L through the phase winding can be expressed as

$$\begin{cases} \psi = l_{st} \int_{x_r}^{x_l} B_{sy} dx = l_{st}(U_r - U_l) \\ L = \frac{\psi}{i} = \frac{l_{st}(U_r - U_l)}{i} \end{cases} \tag{6}$$

where l_{st} is the axial length of the motor stator. U_r and U_l represent the magnetic vector potentials on the left and right sides of the coil, respectively. The radial force F_r and tangential force F_t are calculated by Maxwell tensor method as follows:

$$\begin{cases} F_r = \frac{1}{2\mu_0} \oint (B_{sr}^2 - B_{st}^2) dS = \frac{1}{2\mu_0} \frac{D}{2} l_{st} \sum_{i=1}^{n} [(B_{sri}^2 - B_{sti}^2)(\theta_i - \theta_{i-1})] \\ F_t = \frac{1}{\mu_0} \oint (B_{sr} B_{st}) dS = \frac{1}{\mu_0} \frac{D}{2} l_{st} \sum_{i=1}^{n} [(B_{sri} B_{sti})(\theta_i - \theta_{i-1})] \end{cases}, \tag{7}$$

where D is the arc length of the integral path, N is the number of arc segments, B_{sri} and B_{sti} are the radial and tangential components of flux density B_s, μ_0 is the permeability of vacuum, and $(\theta_i - \theta_{i-1})$ is the radian angle of each arc. Finally, the electromagnetic torque T_e can be calculated by the tangential force F_t [30]:

$$T_e = \frac{D}{2} 2pF_t = \frac{l_{st}}{\mu_0} \frac{D^2}{4} 2p \sum_{i=1}^{n} [(B_{sri} B_{sti})(\theta_i - \theta_{i-1})], \tag{8}$$

where p is the number of motor rotors.

3.3. The Mechanical Module

The mechanical module includes two sub-modules: modal identification and vibration prediction. The modal identification integrates the constrained modal simulation and takes into account the effects of stator, rotor, shell winding, end cover, bearing, gravity, and assembly stress on the stiffness of the installed motor. The modal simulation of SRM is developed based on static pre-calculation, and the dynamic equilibrium equation of its vibration system can be given by

$$\left(\begin{pmatrix} K_{ff} & K_{fr} \\ K_{rf} & K_{rr} \end{pmatrix}_{n \times n} + \begin{pmatrix} S_{ff} & S_{fr} \\ S_{rf} & S_{rr} \end{pmatrix}_{n \times n} - \omega^2 \begin{pmatrix} M_{ff} & M_{fr} \\ M_{rf} & M_{rr} \end{pmatrix}_{n \times n} \right) \begin{pmatrix} \{u_f\} \\ 0 \end{pmatrix}_{n \times 1} = \begin{pmatrix} 0 \\ \{R_r\} \end{pmatrix}_{n \times 1}, \tag{9}$$

where $\begin{pmatrix} K_{ff} & K_{fr} \\ K_{rf} & K_{rr} \end{pmatrix}_{n \times n}$ is the prestress matrix, $\begin{pmatrix} S_{ff} & S_{fr} \\ S_{rf} & S_{rr} \end{pmatrix}_{n \times n}$ is the stress stiffening matrix, $\begin{pmatrix} M_{ff} & M_{fr} \\ M_{rf} & M_{rr} \end{pmatrix}_{n \times n}$ is the mass matrix, and R_r is the binding force vector. The characteristic equation in Equation (9) can be expressed as

$$(K_{ff} + S_{ff} - \omega^2 M_{ff})\{u_f\} = 0 \tag{10}$$

Solving Equations (9) and (10) yields the eigenfrequencies ω_i and eigenvectors $\{u_{fi}\}$ of each order modes, which respectively correspond to the natural frequencies and modal modes of the vibration system. The SRM can be simplified to a second-order system, and its input and output are the radial force F_r and the motor surface displacement x, respectively. The motor surface displacement is the displacement of the measuring point, which is located on the outer surface of the stator core and directly above the salient pole of any phase (e.g., phase A), as shown in Figure 4. The generic modal superposition method [25] is used to predict the system output x, namely

$$x(s) = \sum_{i=1}^{n} Q_i \frac{1}{s^2 \zeta_i \omega_{ni} s + \omega_{ni}^2}, \tag{11}$$

where ω_{ni} is the characteristic frequency, $\zeta_i = c/2m\omega_n$ is the damping ratio, and Q_i is the gain coefficient of the i_{th} order mode.

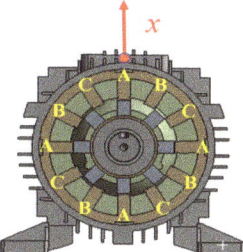

Figure 4. Illustration of the motor surface displacement.

4. Experimental Results and Verification

4.1. Implementation of Simulated and Experimental Platforms

The implementation of the vibration prediction model based on the multi-physics analysis is illustrated in Figure 5. The controller is built with MATLAB/Simulink (R2017a, the MathWorks, Inc., Natick, MA, USA). The information exchange between controller and drive circuit is realized by S-Function. The speed control module is constructed by the existing module of Simulink. The current control module, i.e., the drive circuit switch control module, is programmed using the S-Function. The modeling circuit of the driver circuit module is the same as the actual driver circuit that consists of IGBT. In addition, the inductance and resistance of the driver circuit are added to the model to simulate the characteristics of the real circuit. The drive circuit model of SRM is built by the ANSYS Simplorer (19.0.0, ANSYS, Inc., Canonsburg, PA, USA) and the mechanical module is established by the ANSYS Structure. The digital controller determines the gate signal generated by the driver circuit controller, and the power amplifier is driven by controlling the switch of the Insulated Gate Bipolar Transistor (IGBT). Consequently, an excitation current is generated in the field winding to form an electromagnetic field. Radial and tangential components of the electromagnetic force are then transmitted to the mechanical structure for vibration prediction. It is worth noting that the computing time for the multi-physics simulation depends on the running time we set. On average, it takes about 18 h and 30 min to accomplish the entire simulation process of 0.2 s on a workstation computer (Intel Xeon E5-1650 3.5 GHz, 64.0 GB of DDR3L RAM, Intel, Inc., Santa Clara, CA, USA, Windows 10 OS, Microsoft, Inc., Redmond, WA, USA).

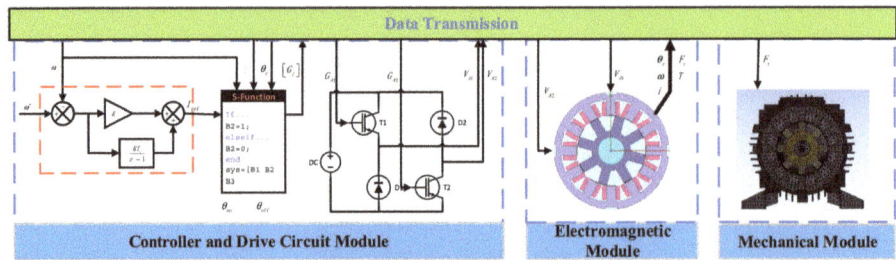

Figure 5. Schematic diagram of the simulation platform.

In general, electromagnetic field simulation of SRMs uses a two-dimensional model. The two-dimensional model ignores the edge effect of the motor, and the accuracy of the simulation results is slightly reduced compared to the three-dimensional model [31]. However, the computational efficiency of the two-dimensional model can be greatly improved compared with that of the three-dimensional model [32]. Therefore, electromagnetic field simulation is conducted with the two-dimensional model in this paper. In the two-dimensional simulation of electromagnetic field, it is assumed that the magnetic field is uniformly distributed along the axis. Meanwhile, the effects of end magnetic field and external magnetic field on the motor are neglected, and the stator outer ring is set as the simulation boundary. The main parameters of the SRM in this simulation is shown in Table 1, and the corresponding material parameters are listed in Table 2. Both stator and rotor of the SRM are laminated by the silicon steel sheet (Grade 50W470). The shaft material is 10# steel, while the winding material is copper. The stator, rotor and shaft of the SRM are modeled in the software and their material are set accordingly. Subsequently, the gap between the stator and rotor is modeled and its material is set to be air. Eventually, electromagnetic field simulation model can be established, as shown in Figure 6.

Table 1. Main dimension parameters of the model.

Name	Value	Name	Value	Name	Value
Stator Pole Number	12	Rotor Pole Number	8	Coil number	100
Stator outer diameter	130 mm	Rotor outer diameter	77.4 mm	Lamination coefficient	0.95
Stator inner diameter	78 mm	Rotor inner diameter	30 mm	Rotor pole arc coefficient	0.355
Stator yoke height	8 mm	Winding diameter	1.2 mm	Winding parallel number	2
Core length	120 mm	Rotor yoke height	7 mm	Stator pole arc coefficient	0.5

Table 2. Material parameters setup in finite element analysis (FEA).

Material	Density (kg/m^3)	Poisson's Ratio	Young's Modulus (Pa)	Shear Modulus (Pa)
Silicon Steel	7700	0.26	2×10^{11}	7.9365×10^{10}
Gray Cast Iron	7200	0.28	1.1×10^{11}	4.2969×10^{10}
Structure steel	7850	0.30	2×10^{11}	7.6923×10^{10}
GCr15	7830	0.30	2.19×10^{11}	8.4200×10^{10}

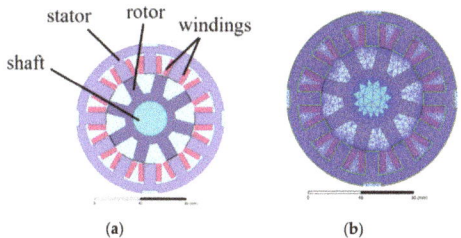

Figure 6. Electromagnetic field model: (**a**) structure (**b**) mesh.

In order to accurately model the structure of the SRM in Figure 1a, the constraints between the components are determined according to their real assembly arrangements. The modeling methods for individual components of the SRM are as follows.

The motor frame is fixed to the workbench through bolts, and the actual installation is completed by the constraints of the bolt holes and the frame surface. Considering the end cover and motor frame fixed to each other, they are handled as an integration. The stator is pressed into the casing through a pressing machine and is integrated with the frame. In addition, the stator is fixed to the motor frame. The winding is fixed to the stator slot by insulating sheets, which has a great influence on the stator mode and cannot be ignored. The equivalent effect of the windings is explored by changing the density of the stator material:

$$\widetilde{\rho}_s = \frac{\rho_s V_s + \rho_c V_W}{V_s}, \qquad (12)$$

where $\widetilde{\rho}_s$ is the equivalent stator density, ρ_s is the stator density (the density of silicon steel sheet), V_S is the stator volume, ρ_c is the winding density (copper wire density), and V_W is the winding volume.

Due to interference fit, the rotor and shaft are fixed to each other. The shaft is fixed to the inner ring of the bearings and thus it remains stationary between the front and rear end covers. The outer rings of both bearings are bonded to the front and rear end covers, and the friction coefficients between the inner and outer rings are set to be 0.0015.

Figure 7a shows the mechanical structure of the SRM and Figure 7b presents its meshing model. In the mechanical module, it not only includes the influence of the stator (containing iron core, winding, slot wedge, insulating material, and motor lead), but also fully considers the influence of rotor(containing rotor shaft and rotor core), end cover, bearing, and other structural components of the motor on the system modal. Meanwhile, the boundary conditions are set as the constraint of the movement and rotation of the base of the machine. Based on the simulation results and experimental data, the material parameters are adjusted to make the mass of the simulation mechanical model consistent with the actual mass of SRM. In addition, the model is tuned many times according to experimental results.

Figure 7. Three-dimensional structure of SRM and its meshing model: (**a**) mechanical structure of the SRM; (**b**) meshing model of the SRM.

To verify the effectiveness of the proposed multi-physics modeling-based method, a 1.5 kW SRM prototype with the same geometric parameters as the simulation model is utilized to build the experimental platform, as shown in Figure 8. The control of SRM is implemented in a Digital Signal Processor (DSP) TMS320F2812 from Texas Instruments (Dallas, TX, USA), which work at a frequency of 150 MHZ. The current and voltage signals are measured in real time through sensors and data acquisition card. The rotor position is obtained by an incremental encoder with resolution of 1024. Meanwhile, the vibration is measured using a piezoelectric accelerometer attached to the housing surface of the stator core on the radial direction of the A-phase tooth as shown in Figure 8. Therefore, the measured shell surface vibration can reflect the actual vibration of the motor to the greatest extent. A 1.5 kW DC motor is used as the mechanical load and can provide a loading ranging from 0 to 20 Nm.

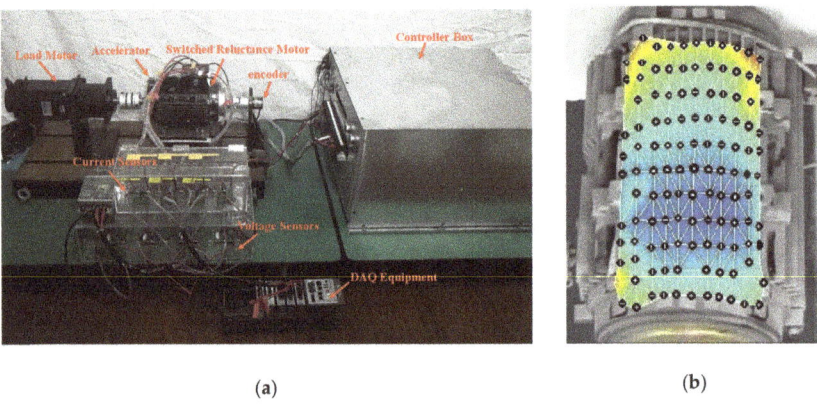

Figure 8. (**a**) The experimental platform of SRM; (**b**) the measurement points.

4.2. Simulation and Experimental Results

4.2.1. Analysis of Flux Linkage Characteristics

Based on the established magnetic field module, the flux linkage characteristics of several rotor angles are analyzed. Considering the accuracy of the position sensor used in the experimental test, the rotor angle is set from 0° (aligned position) to 22.5° (unaligned position) with the step increment of 4.5°. In the Ansys software (Canonsburg, PA, USA), the mesh can be divided intelligently. As can be seen from Figure 6b, the mesh around the air gap is finer. The exaction current is applied to phase A windings, and the value is between 2A and 16A, with the increment of 2A. The current density J_z is obtained by dividing the excitation current and the cross-sectional area of the winding wire. The magnetic vector distribution is solved by the software solver. The flux linkage data can be obtained after post-processing. To verify the simulation results, relevant experiments are carried out. The test procedure is as follows: fix the rotor at the required position, and apply a certain width pulse voltage to the A phase to make the current rise to at least 16A, then record the current and voltage data; tune the rotor to 4.5 degrees, repeat the previous operation until the rotor angle is 22.5 degrees, and then finish the measurement.

The comparison between the simulation results and experimental results is shown in Figure 9. It is seen that the flux linkage curve calculated by the simulated magnetic field is consistent with the experiment results, which proves that the magnetic field simulation model of the SRM built in this paper conforms to the characteristics of the actual motor. Meanwhile, Figure 9 shows that the flux linkage curve of the SRM is nonlinear, which is related to the rotor angle and current magnitude. When the rotor is at the aligned position, the flux linkage is the largest, while the flux linkage is smallest at the unaligned position. Due to the saturation characteristics of ferromagnetic material, the flux linkage begins to saturate when the current reaches to 6A. In addition, the simulated flux linkage curve agrees well with the experimental curve in the rotor position region with smaller flux linkage, but there are some differences at the regions with larger flux linkage. The difference is manifested in the fact that the saturation effect of the flux linkage is greater in the experimental measured curve, that is, the slope of the relationship curve decreases faster.

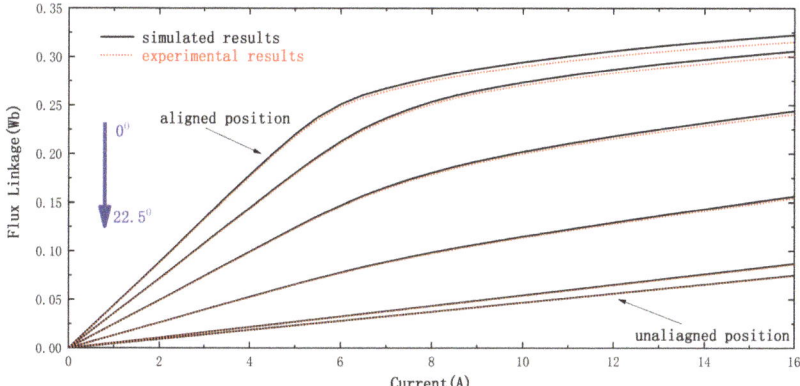

Figure 9. Flux linkage curve.

4.2.2. Comparisons of the Obtained System Modal Results

The system modal of the SRM is obtained by using the simulation method presented in Section 3.3. In addition, the sinusoidal excitation method is adopted to acquire the actual system modal. Specifically, by applying different frequencies of sinusoidal signals to the SRM through exciter, the frequency response characteristics of the motor are obtained, and the vibration characteristics of the motor are determined. In the test, the SRM is fixed on the desktop by elastic cotton, and the sinusoidal excitation force is generated by the exciter (BK 4809, 10N). In addition, the vibration of the shell surface is recorded and analyzed by the laser Doppler Vibrometer (Polytech PSV-500-3D). Then, the exciter generates sinusoidal excitation with varying frequencies and scans the vibration response of the motor. The frequency response of the test points is shown in Figure 10. It is seen that the frequency of the first four orders are 618 Hz, 1215 Hz, 3022 Hz, 3856 Hz, respectively. The experimental characteristic frequencies are compared with those obtained by the proposed finite element method, as shown in Table 3. According to the simulation and experimental results, it is shown that the proposed modal simulation method based on static pre-calculation achieves high accuracy, and the errors between predicted and experimental results are within 3.7%. Compared with the results predicted by only considering the stator, the prediction error of the proposed method is reduced by at least 2.5%.

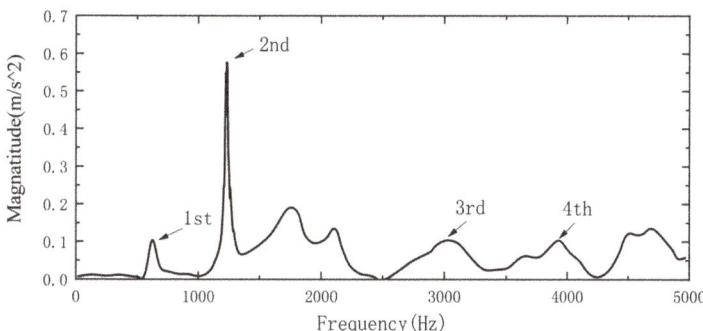

Figure 10. Frequency domain responses at the test point.

Table 3. Natural frequency results of FEA and the experimental method.

Mode Order	Experiment (Hz)	The Proposed FEA Method		FEA Considering Stator Only	
		Predicted Frequency (Hz)	Error (%)	Predicted Frequency (Hz)	Error (%)
1	618	602	2.59	583	5.66
2	1215	1198	1.40	1167	3.91
3	3022	2966	1.85	2885	4.53
4	3856	3999	3.70	3601	6.61

4.2.3. Verification of the Vibration Prediction Model

In order to verify the accuracy of the vibration predictive model, the SRM operates under the condition of constant torque and speed. Vibration tests were carried out both on the simulation platform and experimental setup. The drive circuit adopted asymmetric half-bridge power amplifier. The speed controller adopted a proportional integral (PI) control algorithm, and the current controller used the hysteresis control method. To begin with, the speed was set as 625 rpm and the load torque was 2.5 Nm. For the sake of simplicity, the turn-on angle and turn off angle were chosen as 40° and 160° according to [33]. In addition, the damping coefficient is determined by measuring the attenuation speed of the motor vibration amplitude after the excitation stops, and the gain of the corresponding mode is calculated according to the measured damping coefficient. The obtained parameters for the vibration prediction model are shown in Table 4.

Table 4. Transfer function parameters for Modal 1–4 orders.

Mode Order	Frequency	Damping Ratio	Gain
1	602	0.0109	0.0000453
2	1198	0.0156	0.0000376
3	2966	0.0167	0.0000013
4	3999	0.0172	0.0000292

Since the electromagnetic force acting on the stator is directly related to the phase current, it is essential to accurately predict the current waveform. Figure 11a shows the comparison between the obtained current from the vibration prediction model and experimental results. The vibration spectrum obtained by experiment and simulation is shown in Figure 11b. It can be seen from the current curve that the experimental current value is not zero at the current turn-off point due to the zero drift of the current sensor. Although the simulation data and experimental data do not completely match because of the signal interference of the experimental platform itself, the error of current is relatively small and within an acceptable range. Simultaneously, based on Figure 11b, it can be seen from the vibration spectrum curve that the error of vibration magnitude is small, which validated the effectiveness of the proposed multi-physics modeling-based vibration prediction method.

Then, the speed was set as 1500 rpm and the load torque was 6.0 Nm. The turn-on angle and turn off angle were also chosen as 40° and 160°. Figure 12a presents the comparison between the obtained current from the vibration prediction model and experimental results for this case. The vibration spectrum obtained by experiment and simulation is illustrated in Figure 12b. It can be seen that, with the increase of speed and torque, the errors between the predicted values and the experimental results were still within the acceptable range. Consequently, the presented vibration prediction method based on multi-physical modeling can be utilized for the structural design optimization and control strategy optimization of SRMs.

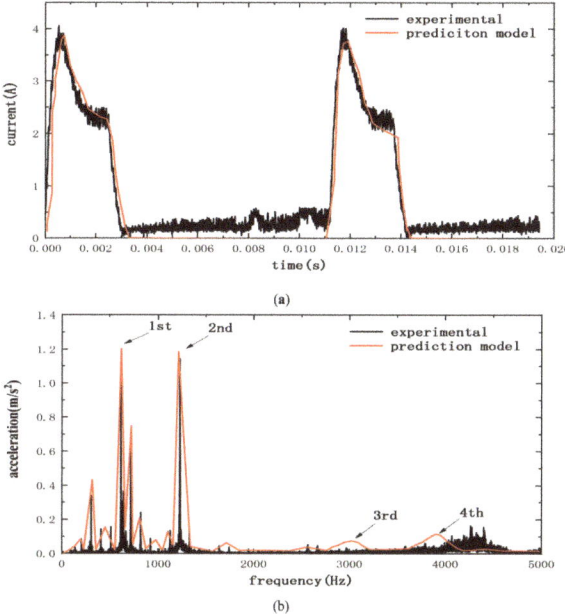

Figure 11. Current and vibration spectrum under simulated and experimental platform with the speed 625 rpm and the load torque 2.5 Nm: (**a**) current and (**b**) vibration spectrum.

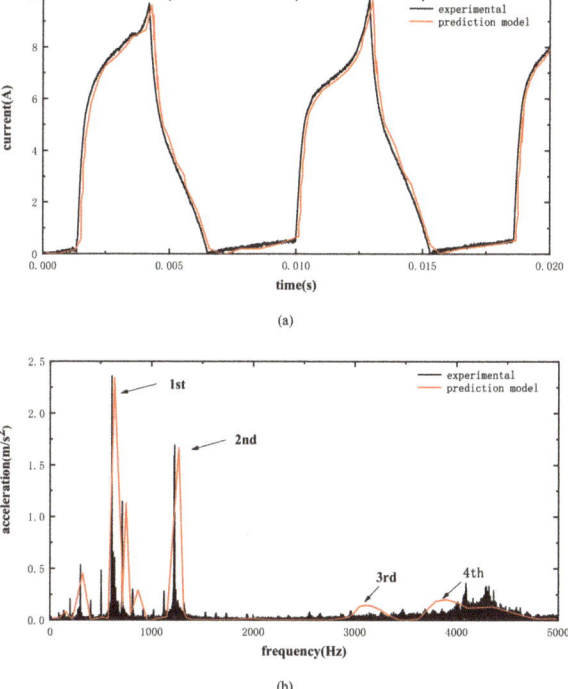

Figure 12. Current and vibration spectrum under simulated and experimental platform with the speed 1500 rpm and the load torque 6.0 Nm: (**a**) current and (**b**) vibration spectrum.

5. Conclusions

In this work, we develop a novel multi-physics modeling-based vibration prediction framework for SRM, in which the influence of components including stator, rotor, shell winding, end cover, and bearing on the system modal and interaction relationship of multi-physical field are fully taken into account. Through the integration of the control algorithm and the electromagnetic field analysis, the dynamic distribution of the electromagnetic field can be obtained to calculate the radial force. In the modal analysis, the influence of the stator, rotor, end cover, bearing and other components of the motor is fully considered. In addition, data dynamic interaction can be realized between the three segments, and the vibration data under different control strategies can be acquired in real time. To verify the proposed multi-physics modeling-based method, comparisons between the numerical and experimental results have been made. It is shown that the proposed vibration prediction method is provided with a high prediction accuracy for SRMs. Compared with the experimental results, the predicted error of vibration magnitude is quite small. Meanwhile, due to consideration of more components, the obtained system modal by the proposed method is rather accurate and the errors between predicted and experimental results are within 3.7%. Therefore, the proposed multi-physics modeling-based vibration prediction method can be applied to the structural design optimization and control strategy optimization of SRMs.

Author Contributions: Conceptualization, X.L. and J.T.; Data curation, X.L., B.L., and C.Q.; Formal analysis, X.L. and J.T.; Funding acquisition, C.L.; Investigation, J.T.; Methodology, X.K., J.T., B.L., and C.Q.; Project administration, C.L.; Resources, X.L. and J.T.; Supervision, C.L.; Validation, X.L., J.T., B.L., and C.Q.; Visualization, C.L.; Writing—original draft, X.L. and J.T.; Writing—review and editing, J.T. and C.L.

Funding: This research was funded by the National Natural Science Foundation of China (NSFC) (Grant No. 51935007 and No. 11202125 and No. 51305258).

Conflicts of Interest: The authors declare no conflict of interest.

References

1. Gan, C.; Wu, J.; Sun, Q.; Kong, W.; Li, H.; Hu, Y. A Review on Machine Topologies and Control Techniques for Low-Noise Switched Reluctance Motors in Electric Vehicle Applications. *IEEE Access* **2018**, *6*, 31430–31443. [CrossRef]
2. Takeno, M.; Chiba, A.; Hoshi, N.; Ogasawara, S.; Takemoto, M.; Rahman, M.A. Test Results and Torque Improvement of the 50-kW Switched Reluctance Motor Designed for Hybrid Electric Vehicles. *IEEE Trans. Ind. Appl.* **2012**, *48*, 1327–1334. [CrossRef]
3. Bostanci, E.; Moallem, M.; Parsapour, A.; Fahimi, B. Opportunities and challenges of switched reluctance motor drives for electric propulsion: A comparative study. *IEEE Trans. Transp. Electrif.* **2017**, *3*, 58–75. [CrossRef]
4. Yang, Z.; Shang, F.; Brown, I.P.; Krishnamurthy, M. Comparative Study of Interior Permanent Magnet, Induction, and Switched Reluctance Motor Drives for EV and HEV Applications. *IEEE Trans. Transp. Electrif.* **2015**, *1*, 245–254. [CrossRef]
5. Sun, W.; Li, Y.; Huang, J.; Zhang, N. Vibration effect and control of In-Wheel Switched Reluctance Motor for electric vehicle. *J. Sound Vib.* **2015**, *338*, 105–120. [CrossRef]
6. Qin, Y.; He, C.; Shao, X.; Du, H.; Xiang, C.; Dong, M. Vibration mitigation for in-wheel switched reluctance motor driven electric vehicle with dynamic vibration absorbing structures. *J. Sound Vib.* **2018**, *419*, 249–267. [CrossRef]
7. Gan, C.; Wu, J.; Shen, M.; Yang, S.; Hu, Y.; Cao, W. Investigation of Skewing Effects on the Vibration Reduction of Three-Phase Switched Reluctance Motors. *IEEE Trans. Magn.* **2015**, *51*, 1–9. [CrossRef]
8. Widmer, J.D.; Mecrow, B.C. Optimized Segmental Rotor Switched Reluctance Machines with a Greater Number of Rotor Segments Than Stator Slots. *IEEE Trans. Ind. Appl.* **2013**, *49*, 1491–1498. [CrossRef]
9. Kurihara, N.; Bayless, J.; Sugimoto, H.; Chiba, A. Noise Reduction of Switched Reluctance Motor with High Number of Poles by Novel Simplified Current Waveform at Low Speed and Low Torque Region. *IEEE Trans. Ind. Appl.* **2016**, *52*, 3013–3021. [CrossRef]

10. Chai, J.Y.; Lin, Y.W.; Liaw, C.M. Comparative study of switching controls in vibration and acoustic noise reductions for switched reluctance motor. *IEE Proc. Electr. Power Appl.* **2006**, *153*, 348–360. [CrossRef]
11. Zhu, Z.Q.; Liu, X.; Pan, Z. Analytical Model for Predicting Maximum Reduction Levels of Vibration and Noise in Switched Reluctance Machine by Active Vibration Cancellation. *IEEE Trans. Energy Convers.* **2011**, *26*, 36–45. [CrossRef]
12. Takiguchi, M.; Sugimoto, H.; Kurihara, N.; Chiba, A. Acoustic Noise and Vibration Reduction of SRM by Elimination of Third Harmonic Component in Sum of Radial Forces. *IEEE Trans. Energy Convers.* **2015**, *30*, 883–891. [CrossRef]
13. Cai, W.; Pillay, P.; Tang, Z.; Omekanda, A.M. Low-vibration design of switched reluctance motors for automotive applications using modal analysis. *IEEE Trans. Ind. Appl.* **2003**, *39*, 971–977. [CrossRef]
14. Callegaro, A.D.; Liang, J.; Jiang, J.W.; Bilgin, B.; Emadi, A. Radial Force Density Analysis of Switched Reluctance Machines: The Source of Acoustic Noise. *IEEE Trans. Transp. Electrif.* **2019**, *5*, 93–106. [CrossRef]
15. Kimpara, M.; Wang, S.; Reis, R.; Pinto, J.; Moallem, M.; Fahimi, B. On the Cross Coupling Effects in Structural Response of Switched Reluctance Motor Drives. *IEEE Trans. Energy Convers.* **2019**, *34*, 620–630. [CrossRef]
16. Liang, X.; Li, G.; Ojeda, J.; Gabsi, M.; Ren, Z. Comparative Study of Classical and Mutually Coupled Switched Reluctance Motors Using Multiphysics Finite-Element Modeling. *IEEE Trans. Ind. Electron.* **2014**, *61*, 5066–5074. [CrossRef]
17. Li, J.; Cho, Y. Investigation into Reduction of Vibration and Acoustic Noise in Switched Reluctance Motors in Radial Force Excitation and Frame Transfer Function Aspects. *IEEE Trans. Magn.* **2009**, *45*, 4664–4667.
18. Cameron, D.E.; Lang, J.H.; Umans, S.D. The origin and reduction of acoustic noise in doubly salient variable-reluctance motors. *IEEE Trans. Ind. Appl.* **1992**, *28*, 1250–1255. [CrossRef]
19. Anwar, M.N.; Husain, O. Radial force calculation and acoustic noise prediction in switched reluctance machines. *IEEE Trans. Ind. Appl.* **2000**, *36*, 1589–1597.
20. Husain, A.R.; Nairus, J. Unbalanced force calculation in switched-reluctance machines. *IEEE Trans. Magn.* **2000**, *36*, 330–338. [CrossRef]
21. Bracikowski, N.; Hecquet, M.; Brochet, P.; Shirinskii, S.V. Multiphysics Modeling of a Permanent Magnet Synchronous Machine by Using Lumped Models. *IEEE Trans. Ind. Electron.* **2012**, *59*, 2426–2437. [CrossRef]
22. Le Besnerais, J.; Fasquelle, A.; Hecquet, M.; Pelle, J.; Lanfranchi, V.; Harmand, S. Multiphysics Modeling: Electro-Vibro-Acoustics and Heat Transfer of PWM-Fed Induction Machines. *IEEE Trans. Ind. Electron.* **2010**, *57*, 1279–1287. [CrossRef]
23. Fiedler, O.; Kasper, K.A.; De Doncker, R.W. Calculation of the Acoustic Noise Spectrum of SRM Using Modal Superposition. *IEEE Trans. Ind. Electron.* **2010**, *57*, 2939–2945. [CrossRef]
24. Srinivas, N.; Arumugam, R. Analysis and characterization of switched reluctance motors: Part II. Flow, thermal, and vibration analyses. *IEEE Trans. Magn.* **2005**, *41*, 1321–1332. [CrossRef]
25. Tang, Z.; Pillay, P.; Omekanda, A.M. Vibration prediction in switched reluctance motors with transfer function identification from shaker and force hammer tests. *IEEE Trans. Ind. Appl.* **2003**, *39*, 978–985. [CrossRef]
26. Lin, C.; Fahimi, B. Prediction of Radial Vibration in Switched Reluctance Machines. *IEEE Trans. Energy Convers.* **2013**, *28*, 1072–1081. [CrossRef]
27. Guo, X.; Zhong, R.; Zhang, M.; Ding, D.; Sun, W. Fast Computation of Radial Vibration in Switched Reluctance Motors. *IEEE Trans. Ind. Electron.* **2018**, *65*, 4588–4598. [CrossRef]
28. Arumugam, R.; Lowther, D.; Krishnan, R.; Lindsay, J. Magnetic field analysis of a switched reluctance motor using a two dimensional finite element model. *IEEE Trans. Magn.* **1985**, *21*, 1883–1885. [CrossRef]
29. Parreira, B.; Rafael, S.; Pires, A.J.; Branco, P.J.C. Obtaining the magnetic characteristics of an 8/6 switched reluctance machine: From FEM analysis to the experimental tests. *IEEE Trans. Ind. Electron.* **2005**, *52*, 1635–1643. [CrossRef]
30. Bilgin, B.; Jiang, J.W.; Emadi, A. *Switched Reluctance Motor Drives: Fundamentals to Applications*, 1st ed.; CRC Press: Boca Raton, FL, USA, 2018.
31. Chen, H.; Nie, R.; Sun, M.; Deng, W.; Liang, K. 3-D Electromagnetic Analysis of Single-Phase Tubular Switched Reluctance Linear Launcher. *IEEE Trans. Plasma Sci.* **2017**, *45*, 1553–1560. [CrossRef]

32. Deshpande, U. Two-dimensional finite-element analysis of a high-force-density linear switched reluctance machine including three-dimensional effects. *IEEE Trans. Ind. Appl.* **2000**, *36*, 1047–1052. [CrossRef]
33. Ye, J.; Malysz, P.; Emadi, A. A Fixed-Switching-Frequency Integral Sliding Mode Current Controller for Switched Reluctance Motor Drives. *IEEE J. Emerg. Sel. Top. Power Electron.* **2015**, *3*, 381–394.

© 2019 by the authors. Licensee MDPI, Basel, Switzerland. This article is an open access article distributed under the terms and conditions of the Creative Commons Attribution (CC BY) license (http://creativecommons.org/licenses/by/4.0/).

Article

Dynamics of a Turbine Blade with an Under-Platform Damper Considering the Bladed Disc's Rotation

Shangwen He [1], Wenzhen Jia [1], Zhaorui Yang [1,*], Bingbing He [2] and Jun Zhao [1]

[1] School of Mechanics and Safety Engineering, Zhengzhou University, Zhengzhou 450001, China; hsw2013@zzu.edu.cn (S.H.); jiawenzhen1993@163.com (W.J.); zhaoj@zzu.edu.cn (J.Z.)
[2] College of Mechanical and Electrical Engineering, Shaanxi University of Science and Technology, Xi'an 710021, China; hebb714@gmail.com
* Correspondence: zryang@zzu.edu.cn; Tel.: +86-1856-760-1945

Received: 16 August 2019; Accepted: 27 September 2019; Published: 7 October 2019

Featured Application: This study supplies a clear thinking to improve the dynamic model of turbine blade with under-platform damper. With this thinking, deeper study can be done to make the damper design more scientific.

Abstract: To simplify the dynamic model, it is generally assumed that the bladed disc is stationary in current studies on the dynamics of the turbine blade with an under-platform damper. With this assumption, convective inertial force in tangential direction and Coriolis inertial force are not be considered in the dynamic model and its equations. To make the dynamic model and relevant analysis more scientific, an approximation method has been developed with the theory of compositive motion. With this method, the response of the blade relative to the rotating bladed disc could be calculated, and the influence of the bladed disc's rotation is considered in this paper. Considering the bladed disc from startup to steady-state, the dynamic characteristics of the system were studied. The influence of damper mass, damper vibration stiffness, and external excitation amplitude on the vibration reduction characteristics of the system were obtained. A method for determining the time when the system reaches steady-state vibration was proposed with the normalized cross-correlation function (NCCF) and the bisection method. The simulation results show that the bladed disc's rotation has an obvious influence on the dynamic characteristics of the system, and some new conclusions were obtained.

Keywords: under-platform damper; bladed disc's rotation; compositive motion; relative displacement; dynamical characteristic

1. Introduction

High-cycle fatigue (HCF) failure of the turbine blades of aero-engines caused by high vibrational stresses is one of the main causes of aero-engine incidents. Due to its insensitivity to temperature and its simple structure, the under-platform damper was widely used to reduce the vibration of the aero-engine turbine blades [1]. To predict the response of the turbine blade with an under-platform damper more and more accurately, recently, there have been quite a lot of developments in the calculations and analyses of under-platform dry friction damper. These studies are mainly about structural dynamic model, dry friction contact model, and methods for solving response of the nonlinear system.

Menq [2] and Sanliturk [3] studied the friction contact and the effect on the vibration reduction of the two-dimensional friction motion. Xia [4] proposed a model for investigating the stick-slip motion caused by dry friction of a two-dimensional oscillator under arbitrary excitations and provided a numerical approach to investigate the system with the Coulomb friction law. Shan and Zhu [5,6] used the numerical tracking method to analyze the complex motion and studied the dynamic response of the plate blade with an under-platform damper. Ma et al. established a dynamic model of rotating shrouded blades considering the effects of the centrifugal stiffening and spin softening of the blade [7]. He B and Ouyang H studied the forced vibration response of a turbine blade with a new kind of under-platform damper, in which the vertical motion of the damper leads to time-varying contact forces and can cause horizontal stick-slip motion [8]. For understanding the actual dynamics of the blade–damper interaction, a novel experimental test rig was developed to extensively investigate the damper's dynamic behavior [9,10]. Umer and Botto [11] explored the contact forces and relative displacement between the damper–blade contact interface with an experimental study for the first time. Liao and Li [12] proposed a two-dimensional friction ball/plate model and established a dynamics model of the rotor with elastic support/dry friction dampers.

Qi and Gao [13,14] established a one-dimensional macro-micro slip friction model to analyze the dynamic characteristics of the damper system, and compared it with the results of the finite element method. The phenomenological macro-slip of dry friction modeling was described in mathematical form by two approaches, and both approaches were illustrated using different acceleration excitations to describe the differences between them [15]. A model was proposed to characterize friction contact of non-spherical contact geometries obeying the Coulomb friction law with constant friction coefficient and constant normal load and the dissipated energies were obtained for different contact geometries [16]. A decrease in vibrational amplitudes was explained by changes in boundary conditions induced by a stick/slip behavior, and the contribution of respective energy dissipation and change of contact state on peak levels was shown [17]. He and Ren [18] studied the reducing vibrational characteristics of the blade by the two-dimensional friction model and finite element model. A purposely developed contact model was tuned on a single-contact test and then included in the numerical model of a curved-flat damper to simulate its cylindrical interface [19]. A microslip model was developed for analyzing the damping effect of under-platform dampers for turbine blades, but the inertia and rotating effects of the damper were ignored for simplicity [20].

Wang and Zhang [21] studied the free vibration and forced vibration of a dry friction oscillator, which was composed of the Iwan model and a mass by harmonic balance method. The multi-harmonic balance method was used to analyze the periodic vibrations of the damper system and to investigate the steady-state solutions of the nonlinear system [22]. A method to predict the nonlinear steady-state response of a complex structure was described, and two differential forms of friction force were given to solve the tangential force of the blades with under-platform dampers accurately [23]. The vertical contact forces and the resultant friction forces acted as moving loads, and the finite element method and the modal superposition method were used to obtain the numerical modes and to solve the dynamic response of the dry friction dampers [24]. Yu and Xu studied the properties of cubic nonlinear systems with dry friction damping and an approximate method was used to get the frequency-response function [25].

In the most studies, it is generally assumed that the bladed disc is stationary, to simplify the dynamic model in the design and analysis of the blade. Bladed disc's rotation is considered in some studies from the aspect of the centrifugal stiffening of the blade. There are few studies about the improvement of the dynamic model considering the bladed disc's rotation. Besides, the study of the dynamic characteristics of the whole process from startup to steady-state has not been included.

In engineering practice, the blades are set up in a circle around the disc, and the under-platform damper can be installed between two adjacent blades. The whole structure can be considered to be in cyclic symmetry. If the normal pressure between contact surfaces is supposed to be distributed equally between two adjacent blades, then the structural model of under-platform damper, as used in this paper, can be described as in Figure 1, where the xyz orthogonal coordinate system (called the moving coordinate system) is defined in accordance with the axial (x, along to the angular velocity direction of the blade), tangential (y), and radial (z) directions. The coordinate system is attached to the bladed disc and rotates with it. A static coordinate system fixed to the ground is defined.

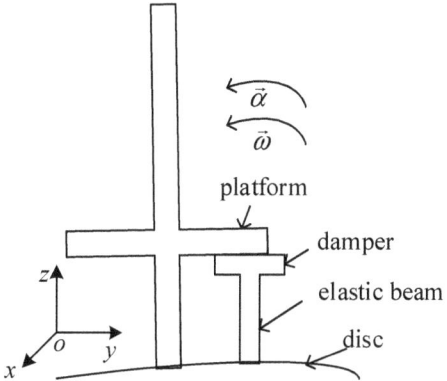

Figure 1. Structure of blade with under-platform damper.

When the rotation of the bladed disc is considered, firstly the vibration stress of the blade mainly depends on its relative displacement (response) to the bladed disc, and the relative displacement should be studied instead of absolute displacement. Secondly, at the rotating state the variation of the convective inertial force and Coriolis inertial force leads to the change of normal pressure and tangential force, which has a significant influence on the damping effects of the damper and the dynamic characteristics of the blade. To study the influence of bladed disc's rotation and improve the accuracy of analysis of the damper, an approximation method for the dynamic response of the blade relative to the bladed disc (moving coordinate system) has been proposed in this paper by combining the theories of compositive motion and dynamics. Compositive motion describes the motion of a moving body relative to different coordinate systems. The response before steady-state could be defined as the transient response. With this method, the convective inertial force and Coriolis inertial force are considered in dynamic equations; the properties of the system solution are derived; and the vibration-damping law of the steady-state response and the transient response of the blade are studied. Some new conclusions about the damping characteristics of the turbine blade with an under-platform damper were obtained, which will be useful in the damper design in engineering practice.

2. Mechanical Model and Dynamics Equations

2.1. Blade-Damper Model

In accordance with Figure 1, the vibration of the first bending mode of the blade in y direction is considered in this paper. The vibrations of the blade in the x and z directions will be neglected, as they are higher order modes and very small relative to the displacement in the y direction. The blade is taken as the moving body; the bladed disc is the moving coordinate system. The convective motion, the motion of the moving coordinate system relative to the static coordinate system, is the bladed disc's rotation about a fixed axis. The motions of the blade and damper relative to the bladed disc can be approximated as the linear motion along the y direction. In Figure 1, $\vec{\omega}$ is the angular velocity of the bladed disc; $\vec{\alpha}$ ($\vec{\alpha} = d\vec{\omega}/dt = \dot{\vec{\omega}}$) is the angular acceleration. The absolute motion of the blade consists of rigid body rotation and vibration relative to the bladed disc. It is assumed that the rigid damper is mounted on an elastic beam and that there is no friction between the damper and the elastic beam. When the bladed disc is rotating, the damper is pressed on the platform by the normal pressure which contains the centrifugal force, Coriolis inertia force and the component of the damper's gravity. The damper and the platform are not separated during the whole process. The sizes of the damper and the blade are ignored for approximate calculation. Based on the above assumption, the structure of under-platform damper (Figure 1) is simplified as a spring-mass model, which is shown in Figure 2.

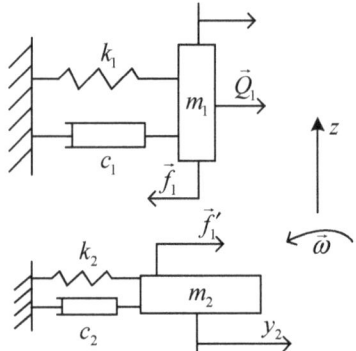

Figure 2. Dynamic model of a blade with an under-platform damper.

Two local coordinate systems parallel to the $oxyz$ coordinate system are defined at the platform and the damper, and are stationary relative to the bladed disc. Displacement of the blade relative to the corresponding local coordinate system is y_1, while relative displacement of the damper is y_2, both of which are in the y direction. The equivalent mass of the blade is m_1. The total equivalent mass of the damper and the elastic beam given by the energy method is m_2. The vibrational stiffnesses of the blade and damper in the y direction are k_1 and k_2. The linear damping coefficients of the blade and damper in the y direction are c_1 and c_2. $\vec{Q}_1$ is aerodynamic excitation force. The dry friction force between the damper and the platform is $\vec{f}_1$, and its reaction force is $\vec{f}_1{}'.(\vec{f}_1 = -\vec{f}_1{}'$, in y direction).

2.2. The Dynamics Equations of Blade-Coupled Vibration

In the yoz plane, the sizes of the damper and the platform are neglected when calculating the convective acceleration. In accordance with Figure 2, the dynamic equation of the system in vector form is established with the theories of compositive motion, which can be written as:

$$\begin{cases} m_1(\vec{a}_{1r} + \vec{a}_{1e} + \vec{a}_{1k}) + c_1\vec{v}_{1r} + \vec{F}_{K1} = -m_1\vec{g} + \vec{Q}_1 + \vec{f}_1 + \vec{N} \\ m_2(\vec{a}_{2r} + \vec{a}_{2e} + \vec{a}_{2k}) + c_2\vec{v}_{2r} + \vec{F}_{K2} = -m_2\vec{g} + \vec{f}_1{}' - \vec{N} \end{cases} \quad (1)$$

where $m_1\vec{g}$ and $m_2\vec{g}$ are gravity vectors, which can be ignored, as they are very small in comparison with the other forces. The relative acceleration vectors of m_1 and m_2 in the corresponding local coordinate systems are $\vec{a}_{1r}$ and $\vec{a}_{2r}$, respectively (in y direction). The corresponding convective acceleration vectors are $\vec{a}_{1e}$ and $\vec{a}_{2e}$. (When $\vec{\omega}$ is constant, the convective acceleration is along the z direction. When $\vec{\omega}$ changes, the tangential component of the convective acceleration is along y direction, and the normal component of the convective acceleration is along the z direction.) The corresponding Coriolis acceleration vectors are $\vec{a}_{1k}$ and $\vec{a}_{2k}$. (The direction of $\vec{\omega}$ is perpendicular to the yoz plane; according to $\vec{a}_k = 2\vec{\omega} \times \vec{v}_r$, the Coriolis acceleration is along the z direction.) The corresponding relative velocity vectors are $\vec{v}_{1r}$ and $\vec{v}_{2r}$ (in y direction). $\vec{F}_{K1}$ and $\vec{F}_{K2}$ are the corresponding stiffness forces vectors in y direction. $\vec{Q}_1$ is the aerodynamic excitation forces' vector in y direction. $\vec{N}$ is the normal pressure vector between the blade and the damper in z direction. Ignoring $m_1\vec{g}$ and $m_2\vec{g}$, the dynamic Equation (1) is projected to the y direction, and the scalar form of it could be written as

$$\begin{cases} m_1\ddot{y}_1 + c_1\dot{y}_1 + k_1 y_1 = Q_1 - f_1 \\ m_2\ddot{y}_2 + c_2\dot{y}_2 + k_2 y_2 = f_1{}' = f_1 \end{cases} \quad (2)$$

$$\begin{cases} m_1(\ddot{y}_1 + a_{1e\tau}) + c_1\dot{y}_1 + k_1 y_1 = Q_1 - f_1 \\ m_2(\ddot{y}_2 + a_{2e\tau}) + c_2\dot{y}_2 + k_2 y_2 = f_1' = f_1 \end{cases} \quad (3)$$

The dynamic equation of the system in y direction is Equation (2) when the rotation speed of the bladed disc is constant, with the rotation speed varying the equation shown as Equation (3). The response characteristics of the blade vibration relative to the bladed disc can be studied at working speed (uniform rotation) and in start-stop condition (non-uniform rotation). In Equation (3), $a_{1e\tau}$ and $a_{2e\tau}$ exist with the bladed disc's rotation considered, and the tangential component of the convective acceleration is described as

$$\begin{cases} a_{1e\tau} = -l_1\alpha = -l_1\dot{\omega} \\ a_{2e\tau} = -l_2\alpha = -l_2\dot{\omega} \end{cases} \quad (4)$$

where l_1 and l_2 are, respectively, the rotation radius of the platform and the damper rotating around the center of the bladed disc, and could be supposed to be constant values during the motion. The scalar quantity of angular acceleration and angular velocity are α and ω respectively; the counterclockwise direction is taken as their positive direction.

2.3. Normal Pressure and Dry Friction Force

Studies have shown that normal pressure is a very critical parameter in the design of dry friction dampers [26,27]. According to Figures 1 and 3, the normal pressure of the under-platform damper consists of the normal component of the convective inertia force (centrifugal force), the Coriolis inertial force, and the gravity of damper. The Coriolis inertial force is generated by the movement of the damper in the y direction relative to the bladed disc and varies with the relative velocity to the disc and the rotational speed of the disc. In addition, the component of the gravity of the damper in the normal pressure's direction changes with the rotation of the bladed disc, and it can be ignored, as it is small enough to be, relative to the centrifugal force and the Coriolis inertial force.

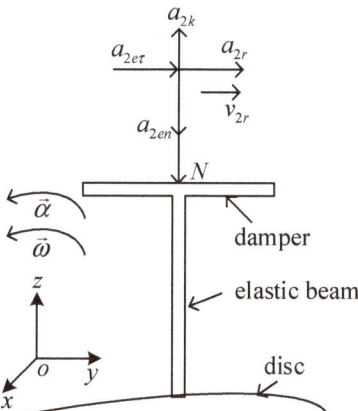

Figure 3. Acceleration synthesis and the normal load diagram of the damper.

In Figure 3, the motion of the damper relative to the bladed disc can be approximated as a linear motion along the y direction. The relative velocity, relative acceleration and Coriolis acceleration of each point on the damper are equal, and the direction of Coriolis acceleration changes with the direction of relative velocity. When calculating the total normal pressure, the mass-center acceleration is used to approximately replace the acceleration of each point on the damper.

2.3.1. Normal Pressure

The mass of the damper is m_3, less than m_2 which is the total equivalent mass of the damper and the elastic beam. Calculating the normal pressure N, the projection of the damper gravity in the z direction is very small in comparison with the convective inertia force and Coriolis inertia force, thus it can be ignored. As can be seen in Figure 3:

$$m_3(a_{2en} - a_{2k}) = N \qquad (5)$$

where

$$\begin{cases} a_{2en} = l_2\omega^2 \\ a_{2k} = 2\omega v_{2r} \end{cases} \qquad (6)$$

In Equation (5), a_{2k} and a_{2en} are the Coriolis acceleration and the normal component of the convective acceleration of the damper, respectively. Obviously, normal pressure is caused by the normal component of the convective inertial force (centrifugal force) and the Coriolis inertial force of the damper. The Coriolis inertial force being equal to $-ma_{2k}$ exists with consideration of the bladed disc's rotation. According to Equation (6), the centrifugal force $m_3 l_2 \omega^2$ contributes positively to the normal pressure, while the direction of relative velocity v_{2r} determines the positive or negative contributions of the Coriolis inertial force to the normal pressure.

2.3.2. Dry Friction Force

The dry friction force between two moving bodies is calculated by a bilinear hysteresis model with which stick-slip-separation transition is considered and captured by the bisection method.

As shown in Figure 4, the contact of two moving bodies is considered in this study. The friction force is simulated by a spring which has no initial length and can yield. The contact stiffness is k_d, N is the normal pressure, and μ is the coefficient of kinetic friction. Point 1 is attached to the platform and remains attached at all times, while point 2 is attached to the damper and remains attached at all times. Point b is the sliding contact point, which is initially attached to point 2 with a limiting friction force μN. Initially the sliding contact point b coincides with point 1 and point 2. The displacements of the platform and the damper relative to the bladed disc are y_1 and y_2, respectively; the displacement of the sliding contact b relative to the bladed disc is y_b. With two bodies moving, point b keeps static with the damper when $|y_1 - y_b|$ is less than $\mu N/k_d$; otherwise, point b keeps static with point 1 and the distance of the two points equals to $\mu N/k_d$.

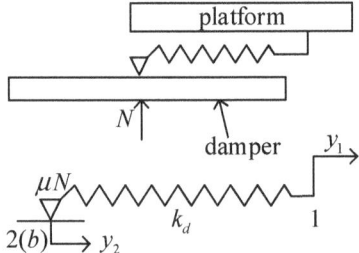

Figure 4. Friction contact model between the damper and the platform.

Corresponding to the two local coordinate systems of y_1 and y_2, assuming that $y_1 = y_2 = y_b = 0$ in the initial state of the system, then the friction force can be determined by Equation (7).

$$f_1 = k_d(y_1 - y_b) \qquad (7)$$

Obviously, the friction force is positive when $y_1 \geq y_b$; otherwise, it is negative.

3. Numerical Simulation

Referring to the engineering, the angular acceleration of the bladed disc increases from 0, and then decreases to 0 after it reaches the top value. The fourth-order Runge–Kutta algorithm was used to compute the relative vibration responses and study the influence of bladed disc's rotation on the dynamic characteristic of the system. The vibration reduction effect is illustrated. The smooth function is used to describe the angular acceleration as $\alpha(t) = \omega_0[\cos(2\pi t/3 + \pi) + 1]/3$. The working angular velocity of the bladed disc is ω_0, and the angular acceleration α decreases to 0 after the angular velocity reaches ω_0; then, the bladed disc rotates at a uniform angular velocity ω_0. The steady excitation frequency is f_e, which equals $\omega_0/2\pi$. F_0 is the external excitation amplitude. This paper is a mechanistic study. The parameters of the system can be taken from Table 1. The other parameters are given in the following simulations.

Table 1. The parameters of the system.

Parameters	Values	Parameters	Values
m_1	0.5 kg	c_1	1 N s/m
k_1	4×10^5 N/m	c_2	1 N s/m
k_d	1×10^6 N/m	l_1	0.01 m
ω_0	600 rad/s	l_2	0.01 m
μ	0.2	Q_1	$F_0 \sin(\omega(t) \times t)$

3.1. The Analysis of the Vibration Response's Characteristics

1. The parameters are taken from Table 2. Figures 5 and 6 show the simulation results.

Table 2. The parameters of the system.

Parameters	Values
m_2	0.049 kg
m_3	0.04 kg
k_2	6×10^5 N/m
F_0	400 N

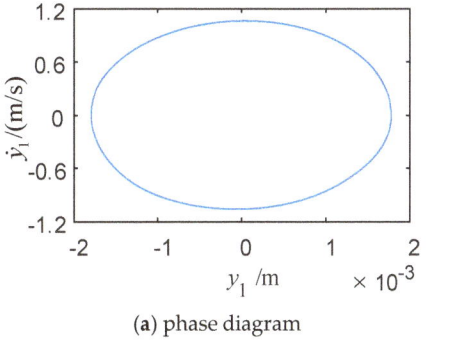
(a) phase diagram

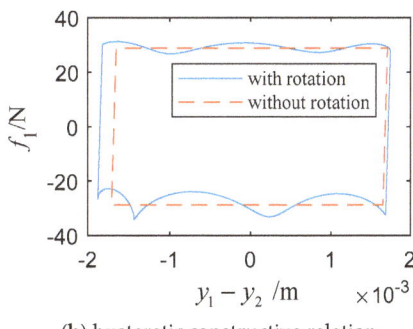
(b) hysteretic constructive relation

Figure 5. Phase diagram (**a**) and the hysteretic constructive relationship of f_1 and $y_1 - y_2$ (**b**) when the system reaches steady-state.

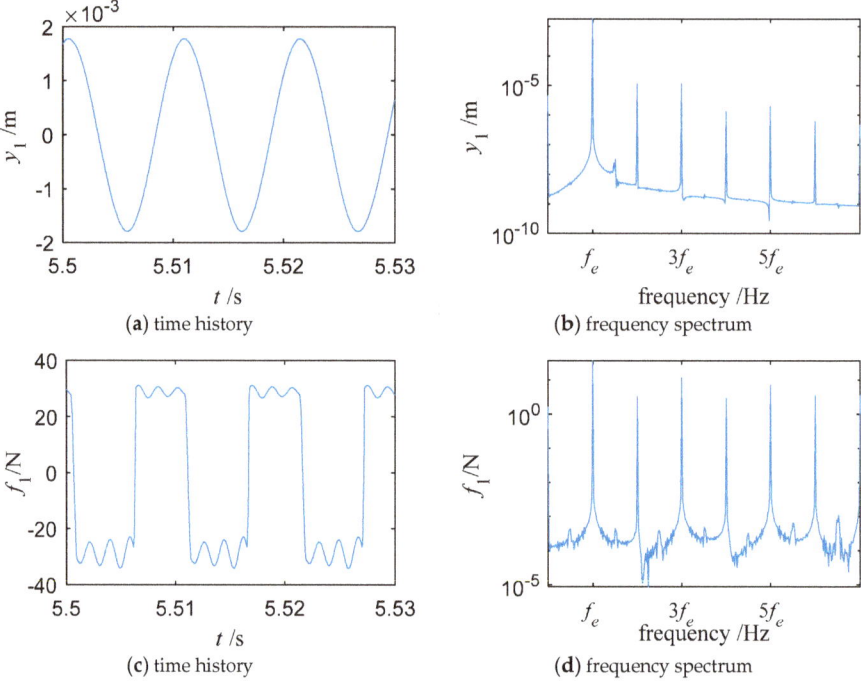

Figure 6. Steady-state dynamic characteristics of the response (**a**,**b**) and dry friction force (**c**,**d**).

2. The parameters are taken from Table 3. Figures 7 and 8 show the simulation results.

Table 3. The parameters of the system.

Parameters	Values
m_2	0.079 kg
m_3	0.07 kg
k_2	8×10^5 N/m
F_0	600 N

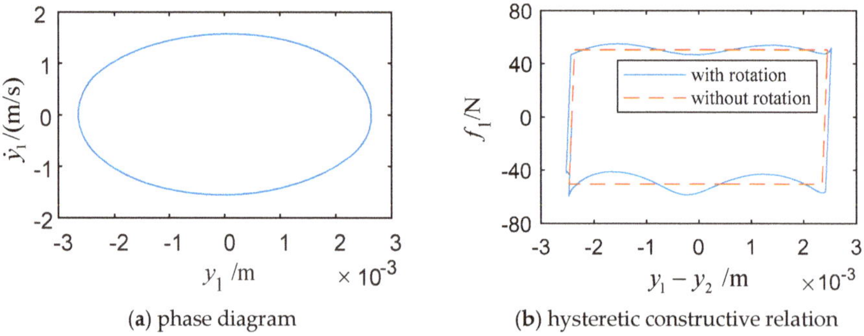

Figure 7. Phase diagram (**a**) and the hysteretic constructive relationship of f_1 and $y_1 - y_2$ (**b**) when the system reaches steady-state.

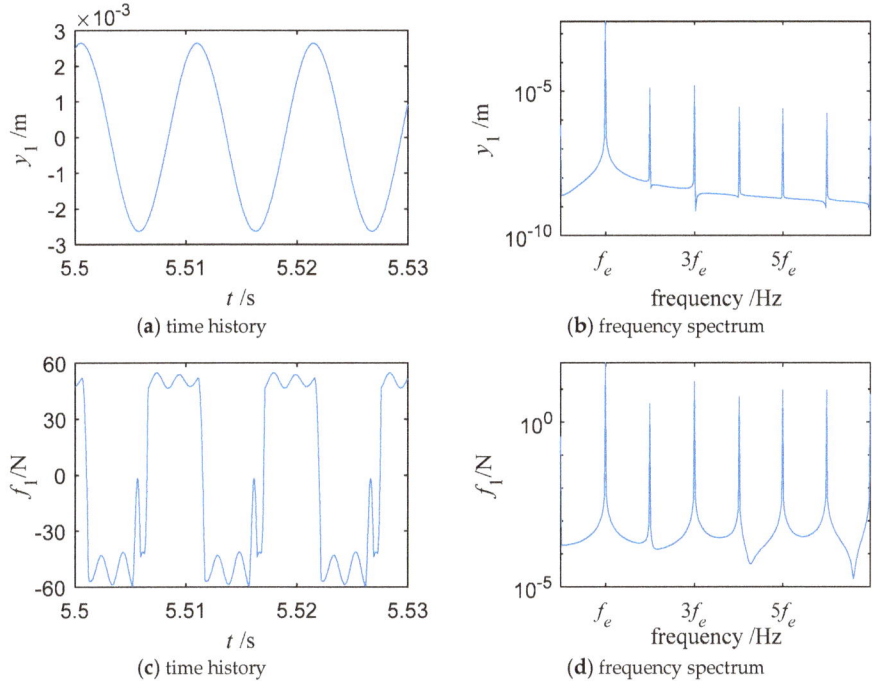

Figure 8. Steady-state dynamic characteristics of the response (**a**,**b**) and dry friction force (**c**,**d**).

The two simulations above are typical among the simulating results in this study. When the blade is at steady-state, the motion is periodic from Figures 5a and 7a. Figure 5b and b show the comparison of hysteretic constructive relation of friction force and relative displacement with and without the Coriolis inertial force, and the two constructive relations in the same figure are obviously different. Considering the bladed disc's rotation, the Coriolis inertial force exists and changes the normal pressure; therefore, the hysteresis loop is not symmetrical. As the normal pressure is a very critical parameter in the dry friction damper's design, the dynamic characteristics of the system will be different with that without considering the Coriolis inertial force. In Figures 6 and 8, f_e is the steady excitation frequency, and only odd multiple frequencies of y_1 and f_1 can be observed. When the mass of the damper, the vibration stiffness of the damper, and the amplitude of external excitation change, there are no fractional frequencies, nor any bifurcation or chaos with the friction contact surface not being separated. The motion of the system is periodic, and the minimum period of the steady-state response T is equal to that of the external excitation, $T = 2\pi/\omega_0$.

3.2. The Decision of Steady-State of the Blade

To supply more reference to dry friction platform damper design in engineering, the blade's vibrational reduction of the steady-state response and the transient response will be studied in the next section; therefore, it is necessary to get the moment t_0 when the system reaches steady-state. In this section, a method for deciding the stable state of the system is proposed: combining the normalized cross-correlation function (NCCF) and the bisection method. The principle of this method is as followings: choosing steady-state response of the last period as a reference sequence, and a response before the last period as the target sequence. Window size h is the length of target sequence which is taken out each time. The correlation coefficient maximum C of reference sequence and target sequence is calculated by the NCCF. The closer that the value of C is to 1, the more that target sequence is in agreement with reference sequence. $flag$ is a given parameter, and when C is greater than its value the response can be considered steady-state. The step length s is the moving length of a target sequence each time, and is changeable via the bisection method, which is used to improve the calculation's accuracy and efficiency. Comparing target sequence from back to front with the reference sequence, the moment t_0 can be obtained when the step length equals to 1. A computational scheme of the mothed is shown in Figure 9.

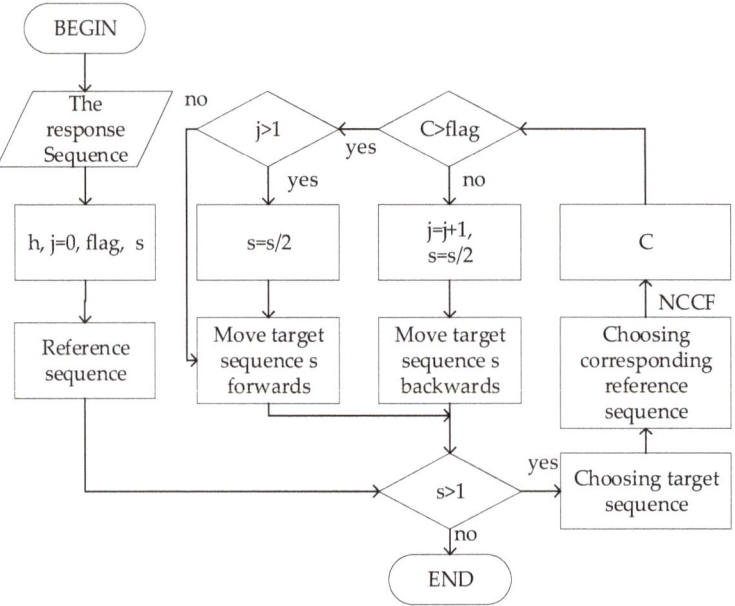

Figure 9. The computational scheme of the method.

When $flag$ is 0.999, 0.998, and 0.997, the other parameters are shown in Table 2. The results are shown in Figure 10. The difference between the two lines has been amplified by three times for clarity. The moment t_0 when the system reaches steady-state was obtained. When $flag$ is 0.999, t_0 satisfies the accuracy requirement.

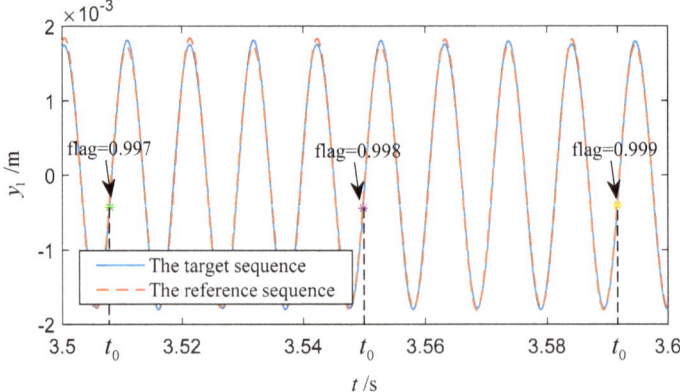

Figure 10. The results with different value of flag.

With t_0, the response of blade before t_0 can be defined as y_{1t}, and after t_0 can be can be defined as y_{1s}; therefore, y_1 is divided into y_{1t} and y_{1s}.

3.3. The Vibration Reduction Characteristics of the System

Based on the analyses done in Sections 3.1 and 3.2, the influence of damper mass, damper vibration stiffness, and external excitation amplitude on the vibration reduction characteristics of the system are studied in this section. Relevant parameters were set as in Table 1. The other parameters are given in the following simulation.

E_{ds} is the vibrational power reduction rate of the steady-state response of the blade. As the steady-state response is periodic, T is the minimum period of the steady-state response, which is equal to the period of the external excitation with the value of $2\pi/\omega_0$. Therefore, E_{ds} can be expressed as Equation (8). E_{dt} is the average power-reduction rate of the transient response of the blade, and can be expressed as Equation (9). The relative displacement of the blade without an under-platform damper is y'_1, and the moment when the system without an under-platform damper reaches steady-state is t'_0. Similarly, the response of the blade before t'_0 can be defined as y'_{1t}, and after t_0 can be defined as y'_{1s}. The response is divided into y'_{1t} and y'_{1s}.

$$E_{ds} = \frac{\int_T y'^2_1 dt/T - \int_T y^2_1 dt/T}{\int_T y'^2_1 dt/T} = \frac{\int_T y'^2_1 dt - \int_T y^2_1 dt}{\int_T y'^2_1 dt} = \frac{\int_T (y'^2_1 - y^2_1) dt}{\int_T y'^2_1 dt} \tag{8}$$

$$E_{dt} = \frac{\int_0^{t'_0} y'^2_1 dt/t'_0 - \int_0^{t_0} y^2_1 dt/t_0}{\int_0^{t'_0} y'^2_1 dt/t'_0} = \frac{\int_0^{t'_0} y'^2_1 dt - \int_0^{t_0} t'_0 y^2_1 dt/t_0}{\int_0^{t'_0} y'^2_1 dt} = \frac{\int_0^{t'_0} (y'^2_1 - t'_0 y^2_1/t_0) dt}{\int_0^{t'_0} y'^2_1 dt} \tag{9}$$

The maximum values of $|y_{1t}|$ and $|y_{1s}|$ are y_{1tm} and y_{1sm} respectively, Similarly, the maximum values of $|y'_{1t}|$ and $|y'_{1s}|$ are y'_{1tm} and y'_{1sm} respectively. A_{ds} is the reduction rate of y_{1sm} and A_{dt} is the reduction rate of y_{1tm}. A_{dt} and A_{ds} are expressed in Equation (10).

$$\begin{cases} A_{ds} = \frac{y'_{1sm} - y_{1sm}}{y'_{1sm}} \\ A_{dt} = \frac{y'_{1tm} - y_{1tm}}{y'_{1tm}} \end{cases} \tag{10}$$

3.3.1. The Effect of Damper Mass on the Vibration Reduction

When the working speed is constant, the damper mass has a great influence on the normal pressure. The numerical simulation parameters are taken from Table 4.

Table 4. The parameters of the system.

Parameters	Values
m_2	$m_3 + 0.009$ kg
m_3	0.04 kg~0.08 kg
k_2	8×10^5 N/m
F_0	400 N

The results are as follows:

In Figure 11, E_{dt} and E_{ds} vary with the increase of the damper mass m_3; some peaks of E_{dt} and E_{ds} are extant. There is a significant reduction of vibrational power with proper damper mass adding to the blade. In Figure 12, A_{dt} increases with the damper mass's increase, while A_{ds} fluctuates while the damper mass increases. y_{1tm} and y_{1sm} reduce significantly with the proper damper mass adding to the blade. From Figures 11 and 12, the laws of E_{ds} and A_{ds} varying with m_3 are basically the same, while the laws of E_{dt} and A_{dt} are obviously different, as the transient response of the blade is complicated. The maximum values of E_{ds} and A_{ds} are smaller than those of E_{dt} and A_{dt} with the same parameters.

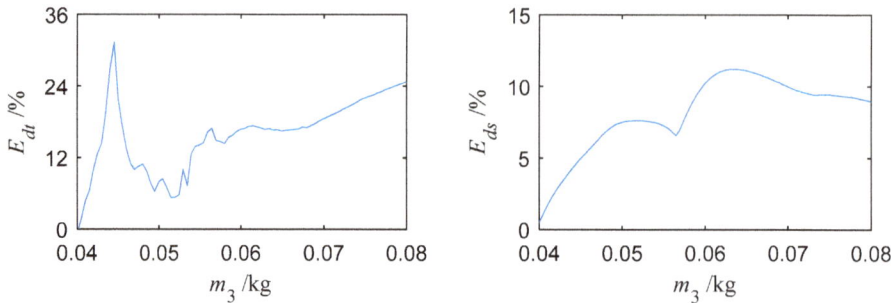

Figure 11. The influence of damper mass on the vibrational power reduction.

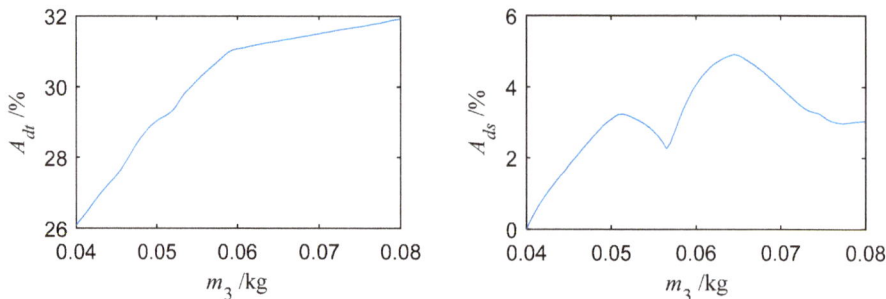

Figure 12. The influence of damper mass on the reduction of the maximum absolute value of vibrational response.

3.3.2. The Effect of a Damper's Vibrational Stiffness on the Vibration Reduction

The effect of a damper's vibration stiffness on the vibration reduction of the blade with an under-platform damper was studied. The parameter values are shown in Table 5.

Table 5. The parameters of the system.

Parameters	Values
m_2	0.059 kg
m_3	0.05 kg
k_2	6×10^5 N/m~1×10^6 N/m
F_0	400 N

The results are as follows:

In Figure 13, E_{dt} and E_{ds} fluctuate with the increasing of the damper stiffness k_2, and there is a significant reduction of vibrational power with proper damper stiffness k_2. In Figure 14, the damper stiffness k_2 has an obvious influence on A_{dt} and A_{ds}; y_{1tm} and y_{1sm} reduce significantly with the proper damper stiffness. From Figures 13 and 14, the laws of E_{ds} and E_{dt} varying with k_2, are basically the same, while the laws of E_{ds} and E_{dt} are obviously different. The maximum values of E_{ds} and A_{ds} are smaller than those of E_{dt} and A_{dt} with the same parameters.

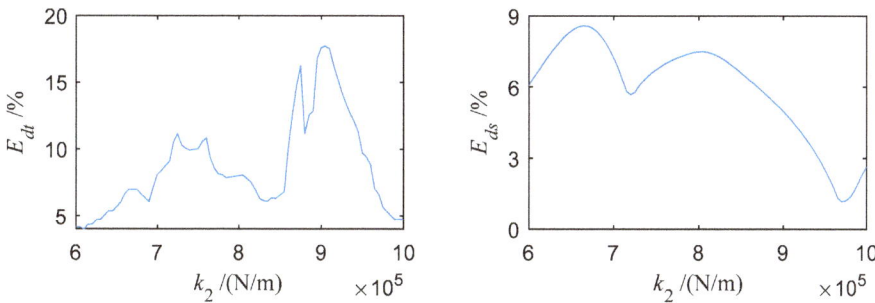

Figure 13. The influence of damper stiffness on the vibrational power reduction.

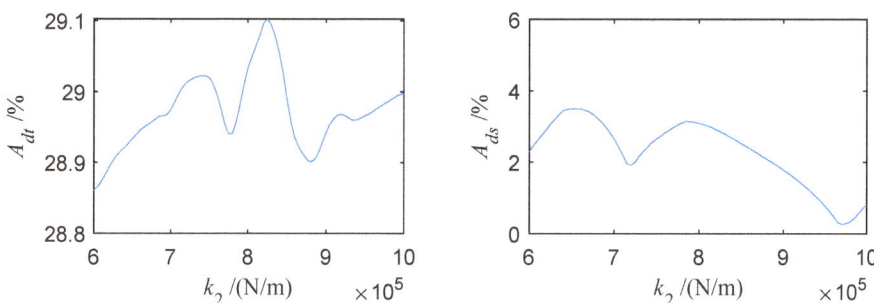

Figure 14. The influence of damper stiffness on the reduction of maximum absolute value of vibrational response.

3.3.3. The Effect of External Excitation Amplitude on the Vibrational Reduction

The parameters are shown in Table 6. The results are shown in Figures 13 and 14.

Table 6. The parameters of the system.

Parameters	Values
m_2	0.059 kg
m_3	0.05 kg
k_2	8×10^5 N/m
F_0	200 N~800 N

In Figure 15, E_{dt} and E_{ds} basically decrease with the increase of external excitation amplitude F_0. In Figure 16, A_{dt} and A_{ds} decrease with increasing F_0. From Figures 15 and 16, the increasing external excitation amplitude causes the vibrational reduction effect of the blade to decrease, obviously, and a larger normal pressure would be needed to make the damper work well. The laws of E_{ds} and E_{dt} varying with k_2, are basically the same, as are E_{ds} and E_{dt}. Besides, E_{dt} could be negative with F_0 increasing, which needs to be considered when engineering damper designs.

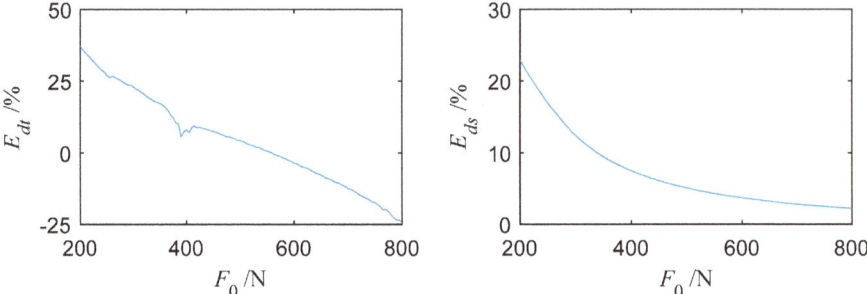

Figure 15. The influence of external excitation amplitude on the vibrational power reduction.

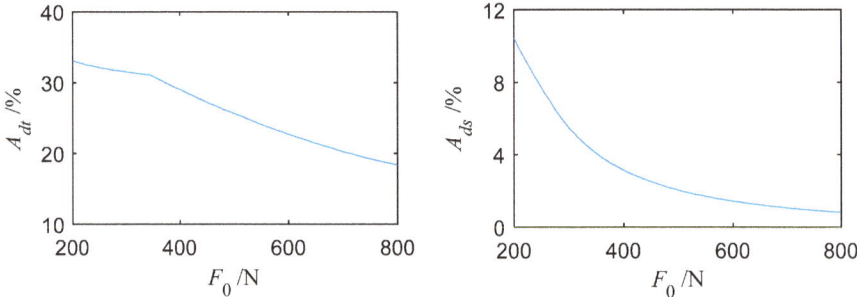

Figure 16. The influence of external excitation amplitude on the reduction of the maximum absolute value of vibrational response.

4. Conclusions

The paper presents a study on the dynamic characteristics of a blade's vibration response relative to a bladed disc by considering the bladed disc's rotation, and an effective method for deciding the steady-state of the blade is proposed. The influences of the mass of the damper, the damper's vibrational stiffness, and the external excitation amplitude are discussed. The following conclusions can be drawn:

1. The dynamic model and analysis method presented in this paper are effective to study the influence of a bladed disc's rotation on the dynamic characteristics of a turbine blade. The changes of the convective inertial force and Coriolis inertial force during the rotation of the bladed disc have a significant influence on the hysteretic constructive relationship of friction-relative displacement and the characteristics of the system dynamics.
2. When the friction contact surface is not detached, changing the damper mass, the damper vibration stiffness, and the external excitation amplitude, results in only higher harmonics in the system response and friction force, and in the system's response, bifurcation and chaos cannot be observed.
3. With proper parameters, adding a platform damper will make the turbine blade's vibration reduce obviously. When the steady-state response is periodic, the reduction law of the average power of blade vibration and the maximum absolute value of the steady-state response are basically the same. However, reduction laws of the average power of blade vibration and the maximum absolute value of the transient response are different. With some parameters, the reduction effect of the transient response may be negative. Greater normal pressure would be needed to keep the damper working well when the external excitation amplitude increases.
4. In engineering, the vibrational reduction effects of the steady-state response and the transient response should be analyzed comprehensively in the under platform-damper design.

Author Contributions: S.H. conceived the paper. S.H., W.J., and Z.Y. wrote the paper and clarified the key methods and results. B.H. helped in the programming and data analysis. J.Z. edited the manuscript.

Funding: This research was funded by the National Key R&D Program of China (grant number 2016YFE0125600) and the National Natural Science Foundation of China (grant number 51405452).

Acknowledgments: The authors are grateful to the discussion on compositive motion with Doctor Xu Zhang.

Conflicts of Interest: The authors declare no conflict of interest.

References

1. Qin, J.; Yan, Q.; Huang, W.C. Literature survey of dry friction damper model and analysis method for depressing vibration of rotating blade. *Adv. Aeronaut. Sci.* **2018**, *34*, 25–33.
2. Menq, C.H.; Chidamparam, P. Friction damping of two-dimensional motion and application in vibration control. *J. Sound Vib.* **1991**, *144*, 427–447. [CrossRef]
3. Sanliturk, K.Y.; Ewins, D.J. Modelling two-dimensional friction contact and its application using harmonic balance method. *J. Sound Vib.* **1996**, *193*, 511–523. [CrossRef]
4. Xia, F. Modelling of a two-dimensional Coulomb friction oscillator. *J. Sound Vib.* **2003**, *265*, 1063–1074. [CrossRef]
5. Shan, Y.C. Investigation on the Vibration Control of Turbo-Machine Blade by Dry Friction Damping. Ph.D. Thesis, Beihang University, Beijing, China, 2002.
6. Shan, Y.C.; Zhu, Z.G. Theoretical and numerical methods for solving friction force of circular motion in contact plane. *J. Aerosp. Power* **2006**, *17*, 447–450.
7. Ma, H.; Xie, F.; Nai, H.; Wen, B. Vibration characteristics analysis of rotating shrouded blades with impacts. *J. Sound Vib.* **2016**, *378*, 92–108. [CrossRef]
8. He, B.; Ouyang, H.; Ren, X.; He, S. Dynamic response of a simplified turbine blade model with under-platform dry friction dampers considering normal load variation. *Appl. Sci.* **2017**, *7*, 228. [CrossRef]
9. Botto, D.; Umer, M. A novel test rig to investigate under-platform damper dynamics. *Mech. Syst. Signal Proc.* **2018**, *100*, 344–359. [CrossRef]
10. Botto, D.; Gastadi, C.; Gola, M.M.; Umer, M. An experimental investigation of the dynamics of a blade with two under-platform dampers. *J. Eng. Gas Turbines Power* **2018**, *140*, 032504. [CrossRef]
11. Umer, M.; Botto, D. Measurement of contact parameters on under-platform dampers coupled with blade dynamics. *Int. J. Mech. Sci.* **2019**, *159*, 450–458. [CrossRef]
12. Liao, M.F.; Li, Y.; Song, M.B.; Wang, S.J. Dynamics modeling and numerical analysis of rotor with elastic support/dry friction dampers. *Trans. Nanjing Univ. Aeronaut.* **2018**, *35*, 69–83.
13. Qi, W.K.; Gao, D.P. Study of vibration response analysis method for the dry friction damping systems. *J. Aerosp. Power* **2006**, *21*, 167–173.
14. Qi, W.K.; Gao, D.P. Microslip study on friction damping and its application in design of vibration reductions. *Acta Aeronaut. Astronaut.* **2006**, *27*, 74–78.
15. Stein, G.J.; Zahoranský, R.; Múčka, P. On dry friction modeling and simulation in kinematically excited oscillatory systems. *J. Sound Vib.* **2008**, *311*, 74–96. [CrossRef]
16. Allara, M. A model for the characterization of friction contacts in turbine blades. *J. Sound Vib.* **2009**, *320*, 527–544. [CrossRef]
17. Sayed, B.A.; Chatelet, E.; Baguet, S.; Jacquet-Richardet, G. Dissipated energy and boundary condition effects associated to dry friction on the dynamics of vibrating structures. *Mech. Mach. Theory* **2011**, *46*, 79–491. [CrossRef]
18. He, S.W.; Ren, X.M.; Qin, W.Y. A method for reducing the blade vibration of platform damper using the macro-micro slip model. *J. Northwest. Polytech. Univ.* **2012**, *27*, 282–288.
19. Gastaldi, C.; Gola, M.M. On the relevance of a microslip contact model for under-platform dampers. *Int. J. Mech. Sci.* **2016**, *115*, 145–156. [CrossRef]
20. Lou, Y.; Jiang, X.; Wang, Y. Modeling of microslip friction and its application in the analysis of under-platform damper. *Int. J. Aeronaut. Space* **2018**, *19*, 388–398.
21. Wang, B.L.; Zhang, X.M.; Wei, H.T. Harmonic balance method for nonlinear vibration of dry friction oscillator with Iwan model. *J. Aerosp. Power* **2013**, *28*, 1–9.

22. Chen, L.; Sun, L. Steady-state analysis of cable with nonlinear damper via harmonic balance method for maximizing damping. *J. Struct. Eng.* **2016**, *143*, 04016172. [CrossRef]
23. Zhang, D.; Fu, J.; Zhang, Q.; Hong, J. An effective numerical method for calculating nonlinear dynamics of structures with dry friction: Application to predict the vibration response of blades with under-platform dampers. *Nonlinear Dyn.* **2017**, *88*, 223–237. [CrossRef]
24. He, B.; Ouyang, H.; He, S.; Ren, X. Stick–slip vibration of a friction damper for energy dissipation. *Adv. Mech. Eng.* **2017**, *9*, 1–13. [CrossRef]
25. Yu, H.; Xu, Y.; Sun, X. Analysis of the non-resonance of nonlinear vibration isolation system with dry friction. *J. Mech. Sci. Technol.* **2018**, *32*, 1489–1497. [CrossRef]
26. Shang, G.B.; Xu, Z.L.; Xiao, J.F. A microslip dry friction model with variable normal load. *J. Vib. Eng.* **2016**, *29*, 444–451.
27. Xu, C.; Li, D.W.; Chen, X.Q.; Wang, D. Vibration analysis for a micro-slip frictional system considering variable normal load. *J. Vib. Shock* **2017**, *36*, 122–127.

© 2019 by the authors. Licensee MDPI, Basel, Switzerland. This article is an open access article distributed under the terms and conditions of the Creative Commons Attribution (CC BY) license (http://creativecommons.org/licenses/by/4.0/).

Article

Validating the Model of a No-Till Coulter Assembly Equipped with a Magnetorheological Damping System

Galibjon M. Sharipov *, Dimitrios S. Paraforos and Hans W. Griepentrog

Institute of Agricultural Engineering (440d), University of Hohenheim, Garbenstr. 9, 70599 Stuttgart, Germany; d.paraforos@uni-hohenheim.de (D.S.P.); hw.griepentrog@uni-hohenheim.de (H.W.G.)
* Correspondence: Galibjon.Sharipov@uni-hohenheim.de; Tel.: +49-711-459-24554

Received: 3 July 2019; Accepted: 18 September 2019; Published: 21 September 2019

Abstract: Variability in soil conditions has a significant influence on the performance of a no-till seeder in terms of an inconsistency in the depth of seeding. This occurs due to the inappropriate dynamic responses of the coulter to the variable soil conditions. In this work, the dynamics of a coulter assembly, designed with a magnetorheological (MR) damping system, were simulated, in terms of vertical movement and ground impact. The developed model used measured inputs from previously performed experiments, i.e., surface profiles and vertical forces. Subsequently, the actual coulter was reassembled with an MR damping system. Multiple sensors were attached to the developed coulter in order to capture its motion behavior together with the profiles, which were followed by the packer wheel. With the aim to validate the correctness of the simulation model, the simulation outputs, i.e., pitch angles and damper forces, were compared to the measured ones. The comparison was based on the root-mean-squared error (RMSE) in percentage, the root-mean-squared deviation (RMSD), and the correlation coefficient. The average value of the RMSE for the pitch angle, for all currents applied on the MR damper, was below 10% and 8% for the speeds of 10 km h^{-1} and 12 km h^{-1}, respectively. For the damper force, these figures were 15% and 13%. The RMSD was below 0.5 deg and 1.3 N for the pitch angle and the damper force, respectively. The correlation coefficient for all datasets was above 0.95 and 0.7 for the pitch angle and the damper force, respectively. Since the damper force indicated a comparatively lower correlation in the time domain, its frequency domain and coherence were investigated. The coherence value was above 0.9 for all datasets.

Keywords: damper force; pitch angle; seeder dynamics; simulation model

---

1. Introduction

In no-till seeding, a consistency in seeding depth plays a very important role in achieving proper seed germination and seedling emergence since both ultimately influence the crop growth, and in general the produced yield [1,2]. The inconsistency in the seeding depth mainly occurs due to the inappropriate dynamic behavior of the furrow components. Harsh soil conditions, such as rough soil surface, residue effects, variable soil density, etc., have a considerable effect on the dynamic performance of the seeding machines. The inability to regulate the vertical dynamics of the coulter assembly, in terms of reaction forces and displacements [3], results in variability of their dynamic response. The forces that are resulting from the interaction between the coulter tine and the soil could characterize the motion behavior of the assembly [4,5]. However, this depends on the components of the assembly that are excited by the soil reaction forces. In addition, the assembly might include the extra packing component, where the developed impact forces also need to be considered [6]. By considering the nature of the forces when regulating the vertical position of the coulter, its performance could be significantly improved, which will result in a better seeding depth. In order to do so, the simulation of the dynamics of the seeder (when the latter is equipped with a damping system), would be the first step towards the implementation of a semi-active damping system on a coulter assembly.

The dynamics of farm machinery have been gaining attention by machine developers due to the obvious advantages such as defining a correct model and evaluating its performance for further optimization of the machine performance. The vertical excessive oscillations of the seeding components occur due to the dynamic response of the machine to soil undulations. This has been reported in reference [7] by simulating the dynamic response of a semi-mounted seeding implement to surface undulations. The effect of the soil undulations and forces on the subsoiler motion behavior has also been investigated by simulating its dynamics [8–10]. Furthermore, an active control mechanism for no-till seed drill under residue-free field conditions was also broadly investigated [11,12]. The control systems that are being developed for regulating the vertical movement of the seeding component could also be integrated into the widely used ISO 11783 (commonly designated as ISOBUS) standard. ISOBUS could offer the necessary communication infrastructure in agricultural machinery to control and receive feedback signals in a standardized manner [13,14]. There have been many types of research conducted on advancing the dynamics of no-till sowing machines to reduce the inconsistency in seeding depth [15]. Controlling the utilized downforce on the coulter assembly could be the best option to achieve better performance [16]; however, the non-linear characteristics of the regulating parts, which consisted of spring and damping elements, of the coulter depth control system introduce some difficulties in designing a proper control strategy [17]. This could be overcome by implementing a dynamic downforce control system, where the regulating elements are actively controlled with different control techniques [18].

In spite of all the above-mentioned studies, very limited research has been conducted on applying a magnetorheological (MR) damping system to a coulter assembly and on modelling the developed assembly dynamics. The MR damper is a semi-actively controlled suspension system that can change its state from viscous to semi-solid within a short time [19,20]. Since the damper contains MR fluid, the system can also behave as a passively controlled suspension when it is supplied with a constant input power [21]. Furthermore, the significant dynamic characteristics of the MR damper such as the broad range of damping property, the short response time, the sufficient temperature range, and the higher yield stress with a lower input, have made it widely recognized in different areas of engineering [22,23]. Due to the aforementioned reasons, the application of the MR damper can be vital to deal with the improper dynamics that result from its response to non-till soil conditions.

As this has been reported in a previous work by the authors of reference [24], by modelling and simulating the dynamic motion behavior of the coulter, when the latter is equipped with an MR damper, the dynamic response of the seeder in terms of the amplitudes of the reaction forces and the vertical displacements was significantly improved. This was followed by the application of the MR damper on the coulter assembly, where the dynamics of the assembly were also captured by leveraging the information from the attached sensors [25]. Subsequently, the performance of the coulter with its corresponding dynamics was evaluated [26,27]. The evaluation showed the necessity for further validation of the simulation model by utilizing the measured inputs from the real-life vertical forces and surface profiles. By validating the developed model, the optimal settings for the MR damper could be defined to test the assembly designed with the MR damper in different field conditions. The defined optimal settings could be employed in developing and testing various control techniques for the MR damping system in order to improve its dynamic performance that will result in a more consistent seeding depth.

The aim of the paper is to validate the developed mathematical model of the dynamics of the coulter assembly with the MR damper. This should be performed by utilizing measured vertical forces and profiles from realistic no-till seeding operations. For validation purposes, the simulation outputs, i.e., pitch angle and damper force, should be compared with the acquired ones from real-life operations. The comparison will be based on criteria such as root-mean-squared error (RMSE), root-mean-squared deviation (RMSD), and correlation coefficient. Furthermore, a correlation between the frequency content of the simulated and measured damper forces, which would based on the coherence value, should be carried out to indicate if the simulation model can produce a similar frequency range as the measured one.

2. Materials and Methods

2.1. Instrumentation and Implementation of the MR Damper

A metal frame with 2 wheels that could be attached on the 3-point hitch of the tractor was constructed to carry the coulter assembly (AMAZONEN-Werke H. Dreyer GmbH & Co. KG, Hasbergen, Germany) (Figure 1). The coulter was fixed to a square rod using rubber rollers. The square rod had one degree of freedom and could rotate while a hydraulic cylinder was used that applied downward pressure to the coulter with a value of 6.89 MPa. To absorb the impact of the reaction forces, an RD-8040-1 MR damper (LORD, Baltimore, MD, USA) was added to the assembly using an extra shank and bearing system at the joint. During the field tests, a variable damping ratio was provided using a controlling device (Wonder box, LORD, Baltimore, MD, USA) by regulating the input voltage, which was applied on the magnetic field in the flow path of the MR fluid. The device regulated the output current, which was linearly proportional to the input voltage of the MR damper. More detailed information about the performance of the MR damper can be found in previous work by the authors of reference [27].

Multiple sensors were employed in order to acquire the field profiles that were followed by the packing wheel and the developed assembly dynamics (Figure 1). The in-field absolute geo-referenced position of the mainframe and the vertical movement of the developed coulter were obtained using an SPS930 total station (Trimble, Sunnyvale, CA, USA) and a DT50 range detector (SICK AG, Waldkirch, Germany), respectively. The total station detected the geo-referenced position of the frame by following an MT900 active prism powered by the tractor (Trimble, Sunnyvale, CA, USA), while the range finder recorded the vertical movement from a reflectance plate attached on the coulter assembly. The tilting information of the developed frame and the assembly were acquired using two VN-100 inertial measurement units (IMUs) (VectorNav, Dallas, TX, USA). The reaction and damper forces were extracted from the data recorded by 6 linear 350 Ohm DY41-1.5 strain gauges (HBM GmbH, Darmstadt, Germany) fixed on the coulter.

Figure 1. The developed prototype with the coulter assembly and the utilized sensors for capturing its dynamics.

2.2. Experiments

The experiments to measure the coulter dynamics and the field profiles were performed at the research farm of the University of Hohenheim located in 48°43′27.33″ N, 9°11′07.69″ E. Before performing seeding operation, experiments were conducted to determine the soil properties [28]. When capturing the dynamic performances, the MR damper was excited with different current input levels (0–1 A with an increment of 0.2 A). This was carried out for 2 driving speeds, i.e., 10 km h^{-1} and 12 km h^{-1}.

The spatial sampling frequency of the attached sensors for the speed of 10 km h^{-1} has been reported in reference [26]. The spatial sampling frequency for the speed of 12 km h^{-1} of strain gauges, IMUs and range finder was equal to 11 mm, 66 mm, and 53 mm, respectively. Since the various sampling rates of the sensors resulted in non-concurrent data, a linear interpolation approach was employed to obtain values at concurrent instances.

2.3. Dynamics of the Developed Assembly

The assembly consisted of the coulter, the wheel shank and the packer wheel. The coulter assembly, together with the joint part, was modelled as a forced pendulum. The pendulum was also considered as damped since the point of the packer wheel touching the ground comprised of a material made out of rubber. The motion equation of the coulter was developed for 2 cases. The first case was for the original coulter assembly, where the packer wheel was considered as a passively controlled system (see Figure 2a). Detailed information can be found in previous work by the authors in reference [24]. In the second case, the semi-active MR damper, as a controlled mass-spring-damper system, was designed to be located between the wheel shank and the packer wheel, having an extra shank (Figure 2b). The extra shank was assumed to have freedom of rotation.

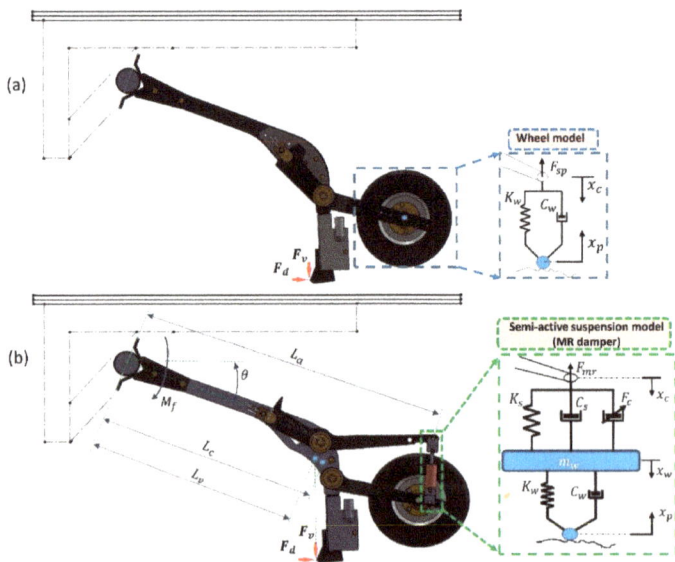

Figure 2. Representative view of (a) the original coulter assembly and (b) the assembly designed with the magnetorheological (MR) damping system.

The weight of the wheel and the coulter were considered as an unsprung mass m_w [kg], and a sprung mass m_c [kg], respectively. The reaction force F_{sp} [N] acting on the wheel was calculated by the deformation and the damping forces of the rubber wheel. The stiffness and the damping forces were introduced by the vertical shift and velocity of the unsprung mass m_w. This was formulated as follows:

$$F_{sp} = K_w(x_p - x_w) + C_w(\dot{x}_p - \dot{x}_w) \qquad (1)$$

where x_w is the shift in the position of the wheel in the vertical plane, in [m]; x_p is the field profile that the wheel followed; K_w is the stiffness coefficient of the rubber tyre [N m^{-1}] and C_w is the damping coefficient [N s m^{-1}]. More detailed information on defining the stiffness and the damping coefficient of the wheel-tire can be found in reference [24].

In this model, the horizontal force F_d was not considered in the modelling due to its minor effect on the vertical motion behavior of the assembly. The vertical movement of the coulter x_c was also expressed with the combination of the pitch angle θ and the distance between the rotating rod and the impact force affecting point L_a. Since the pitch angle of the assembly relative to the main frame was small ($\theta < 5°$), the angle approximation (sin $\theta = \theta$ and cos $\theta = 1$) was used to simplify the equation of motion. The damping force F_{mr} [N] resulting from the MR damping system was the affecting force to the assembly since the MR damper was attached on the fixed shank from one side. Therefore, the profile impact force F_{sp} was substituted by F_{mr}. By taking into account the moment of all the effecting forces and the moment of the constant force, which was applied to the square rod M_f, the equations of motion behavior of the assembly, including the semi-active MR damping system, were the following:

$$\left(m_w + \frac{m_c}{4}\right)L_a^2\ddot{\theta} = F_{mr}L_a - ((m_w + m_c)gL_c + F_vL_v)\theta - M_f \qquad (2)$$

$$m_w\ddot{x}_w + C_s(\dot{x}_w - L_a\dot{\theta}) + K_s(x_w - L_a\theta) + C_w\dot{x}_w + K_w x_w = F_{sp} - F_{mr} + F_c \qquad (3)$$

where K_s and C_s are the stiffness and damping coefficient of the MR damper passive components, in [N m^{-1}] and [N s m^{-1}], respectively; L_c and L_v are the measured lengths of the moments arm for the center of mass and the vertical force [m].

2.4. Bouc-Wen Model

Many different hysteresis models could be applied for the MR damper model to address its hysteresis and nonlinear behavior [29,30]. The Bouc-Wen model (Figure 3) has been proven to be outperforming over the other models [24] when investigating the nonlinear behavior of the MR damper. In the developed model (Figure 2b), F_c is the control force of the MR damper resulting from the pre-yield stress of the damper fluid.

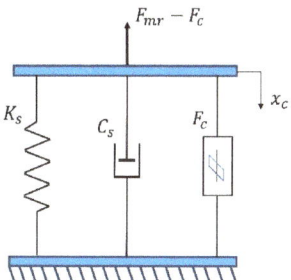

Figure 3. Schematic view of the Bouc-Wen model for the MR damping system.

The resulting force from the Bouc-Wen model was expressed by the following formula:

$$F_{mr} = C_s(u)\dot{x}_c + K_s(u)x_c + \alpha z + F_c \qquad (4)$$

where the evolutionary variable z is the following:

$$\dot{z} = \gamma z |x_c||z|^{n-1} - \beta x_c |z|^n + A x_c \qquad (5)$$

where γ, β, and A are the time-dependent parameters that represent the control of the linear behavior; α is a constant related to the accuracy of the model; n is a response-shape characteristic parameter of the model; $K_s(u)$ and $C_s(u)$ are the voltage-dependent coefficients representing the stiffness and viscous damping, respectively.

The model parameters $K_s(u)$, $C_s(u)$, and α can be defined as a linear function of the applied voltage and are obtained by the following formulas [31]:

$$K_s(u) = K_{s1} + K_{s2} u \qquad (6)$$

$$C_s(u) = C_{s1} + C_{s2} u \qquad (7)$$

$$\alpha = \alpha_1 + \alpha_2 u \qquad (8)$$

where K_{s1}, K_{s2}, C_{s1}, C_{s2}, α_1, α_2 are the necessary parameters to characterize the rheological and magnetic behaviour of the MR damper.

To define the model parameters of the Bouc-Wen model, its response was simulated using the Equations (4)–(8) with the different applied current values on the coil. The parameters K_{s1}, K_{s2}, C_{s1}, C_{s2}, α_1, α_2, were determined by leveraging an empirical formulation that used the experimental data [32,33].

2.5. Evaluation of the Model Performance

All the Equations from (1) to (8) for the coulter dynamics and the hysteresis model of the MR damper were applied and simulated in MATLAB and Simulink. The acquired field surface profiles (Figure 4) and the vertical forces (Figure 5) from the dynamic measurements at the speeds of 10 km h^{-1} and 12 km h^{-1}, when the MR damper was excited with 0.3 A, 0.5 A, and 0.7 A, were used as inputs to the simulation model. The acquired strains were utilized to determine the vertical forces and the damper forces that were acting on the coulter tine and the fixed shank. To define the field profiles, the geo-referenced coordinates of the active prism were transformed to the ground touching point of the wheel. The detailed formulas on calculating all the forces and transformations, in terms of rotations and translations for the profile determination, are given in reference [27].

The outputs of the model were the pitch angle of the assembly movement and the damper force. All parameters and constants of the Bouc-Wen model, as well as the coulter model, can be found in reference [24]. The real-life test of the assembly performance was performed with different current input values loaded on the MR damper (0.0, 0.1, 0.3, 0.5, 0.7, 0.9, 1.0 A). Two different traveling speeds of 10 km h^{-1} and 12 km h^{-1} were set for all performed experiments. The field tests with the current values 0.3 A, 0.5 A, and 0.7 A had the best performance, in terms of coulter dynamics [27], thus they were chosen to be further analyzed in the present work.

In order to validate the correctness of the simulation model, its outputs such as the simulated pitch angles of the assembly vertical motion and the damper forces were compared with the measured ones. The comparison analyses were carried out by calculating the RMSE in percentage, absolute RMSD, and the correlation coefficient between the simulated and the measured data. The RMSE estimated the relative error in percentage between the measured and the simulated datasets. Taking into account the effect of the neglected lateral forces on the tire dynamics, the value of 20% for the RMSE in percentage was considered as a threshold that should not be exceeded [34,35]. The formula of the RMSE in percentage that was used is the following:

$$\varepsilon_{RMS} = \frac{Z^m_{RMS} - Z^s_{RMS}}{Z^m_{RMS}} \times 100\% \qquad (9)$$

where $Z^m_{RMS} = \sqrt{\sum_{i=1}^{N}(z_m(t))^2/N}$; $Z^s_{RMS} = \sqrt{\sum_{i=1}^{N}(z_s(t))^2/N}$; $z_m(t)$ and $z_s(t)$ are the measured and the simulated data, respectively, within the time t that was needed to travel the specified distance. N is the total number of samples.

Subsequently, the average difference between the measured and the simulated outputs was evaluated by determining the RMSD value. This was defined by the following expression:

$$\mu_{RMSD} = \sqrt{\frac{\sum_{i=1}^{N}((z_m(t))^2 - (z_s(t))^2)}{N}} \qquad (10)$$

In addition to the error estimation, the fitting accuracy, in terms of similarity relationship, of the simulated data to the measured data were also addressed with a correlation coefficient. The values higher than 0.6 could be considered as practically acceptable based on the same reasons as above. The correlation coefficient ρ was based on the following equation:

$$\rho = \frac{1}{N-1}\sum_{i=1}^{N}\left(\frac{z_m(t) - \mu_{z_m}}{\sigma_{z_m}}\right)\left(\frac{z_s(t) - \mu_{z_s}}{\sigma_{z_s}}\right) \qquad (11)$$

where μ_{z_m} and σ_{z_m} are the mean value and the standard deviation of $z_m(t)$, respectively, and μ_{z_s} and σ_{z_s} are the mean value and the standard deviation of $z_s(t)$, respectively.

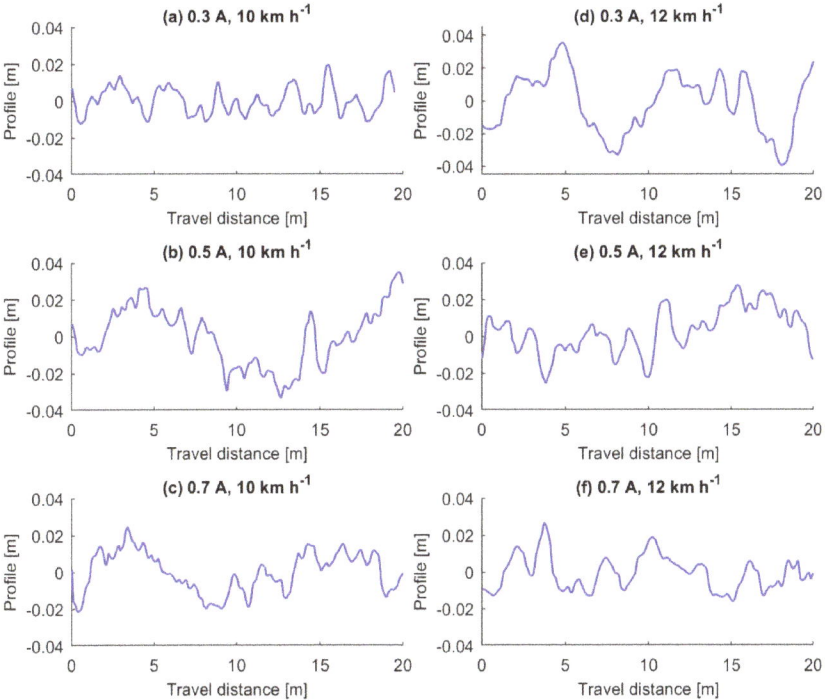

Figure 4. The measured field surface profiles at the speeds of (**a**–**c**) 10 km h^{-1} and (**d**–**f**) 12 km h^{-1}, when the MR damper was supplied with 0.3 A, 0.5 A, and 0.7 A.

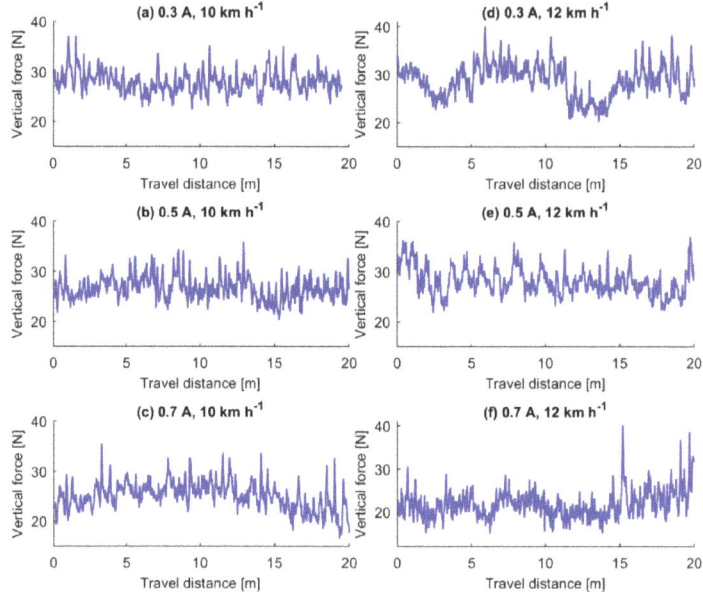

Figure 5. The measured vertical forces at the speed of (**a–c**) 10 km h^{-1} and (**d–f**) 12 km h^{-1}, when the MR damper was supplied with 0.3 A, 0.5 A, and 0.7 A.

Due to the wide range of signal wavelengths of the damper forces, the correlation of the measured and the simulated forces in the frequency domain was carried out based on the coherence analysis. In this case, the relationship between the measured and the simulated forces can be properly evaluated by analyzing the coherence value that was resulted from the cross-spectrum of the wavelengths in the frequency domain. The power spectral density (PSD) of the measured and the simulated datasets were computed to determine the coherence based on the following equation:

$$\alpha_{F_m,F_s}(f) = \frac{|P_{F_m,F_s}(f)|^2}{P_{F_m}(f) \cdot P_{F_s}(f)} \qquad (12)$$

where P_{F_m} and P_{F_s} are the resulted PSD datasets for the measured and the simulated forces, respectively, and P_{F_m,F_s} is the cross-spectrum density of these 2 data sets. A coherence value equal to 1 indicates the maximum possible agreement while a value of 0 denotes no correlation [36].

3. Results and Discussion

Figure 6 presents both measured and simulated pitch angles of the assembly movement for the driving speeds of 10 km h^{-1} and 12 km h^{-1}. Three different cases are highlighted, i.e., when the MR damper was supplied with 0.3 A, 0.5 A, and 0.7 A.

For the same current input levels applied on the MR damper, i.e., 0.3 A, 0.5 A, and 0.7 A, both measured and simulated damper forces for the driving speeds of 10 km h^{-1} and 12 km h^{-1} are shown in Figure 7.

The results of the correlation, in terms of the RMSE, RMSD and correlation coefficient, which were calculated using Equations (9)–(11), respectively, are given in Table 1.

The analyses of the RMSE percentage and RMSD showed good agreement between the simulated and the measured outputs. The average values of the RMSE and the RMSD for the pitch angle at the speed of 10 km h^{-1} were equal to around 10% and 0.4 deg, respectively. These figures were below 8% and around 0.5 deg for the speed of 12 km h^{-1}. Based on the above a high correlation coefficient was determined with an average value of 0.96 and 0.97 for the traveling speed of 10 km h^{-1} and 12 km h^{-1}, respectively. The correlation analyses of the damper forces indicated slightly different results, in terms of a higher RMSE percentage and RMSD value, and smaller correlation coefficient compared to the ones resulted from the pitch angles. The average value of the RMSE percentage was 15% and 13% for the operation speed of 10 km h^{-1} and 12 km h^{-1}, respectively. The average of the RMSD, indicated a value of 1.2 N at the speed of 10 km h^{-1} and 1.3 N at the speed of 12 km h^{-1}. Consequently, the higher error resulted in a lower correlation coefficient with an average value of around 0.7 and 0.69 for these speeds.

Table 1. Root-mean-squared-error (RMSE) in percentage, root-mean-squared-deviation (RMSD) and correlation coefficient.

Speed	10 km h^{-1}						12 km h^{-1}					
Parameter	Pitch			Force			Pitch			Force		
Current [A]	0.3	0.5	0.7	0.3	0.5	0.7	0.3	0.5	0.7	0.3	0.5	0.7
RMSE [%]	11.8	8.6	9.6	12.3	14.8	16.2	6.9	6.7	7.1	11.2	13.6	13.7
RMSD	0.35	0.52	0.34	0.98	1.03	1.69	0.60	0.49	0.35	1.52	0.98	1.38
CorrCoef	0.95	0.987	0.97	0.72	0.67	0.73	0.984	0.97	0.964	0.65	0.72	0.74

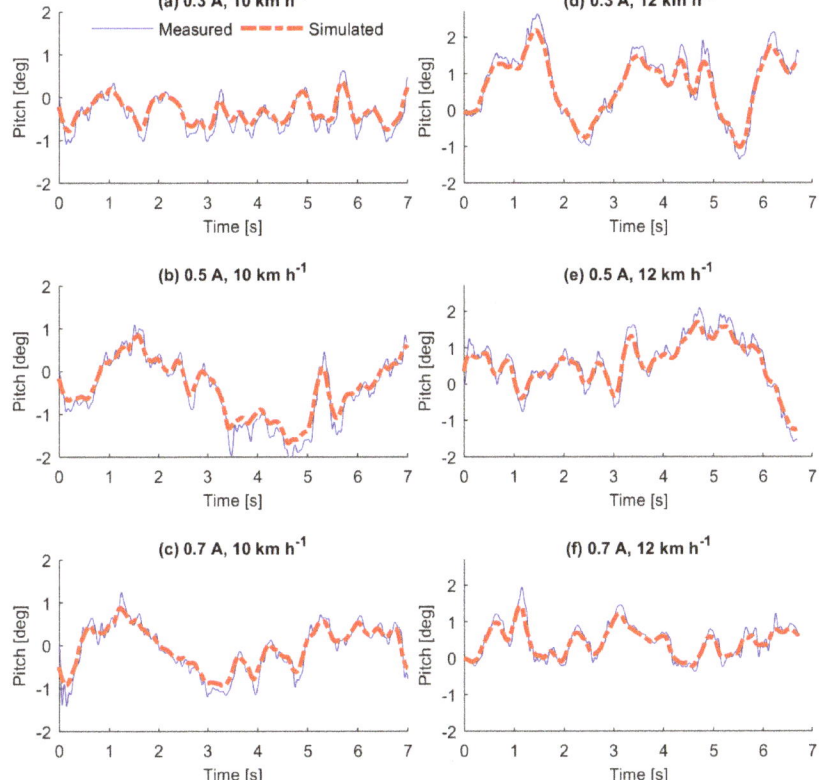

Figure 6. The measured and simulated pitch angle of the assembly movements at the speed of (**a–c**) 10 km h^{-1} and (**d–f**) 12 km h^{-1}, when the MR damper was supplied with 0.3 A, 0.5 A and 0.7 A.

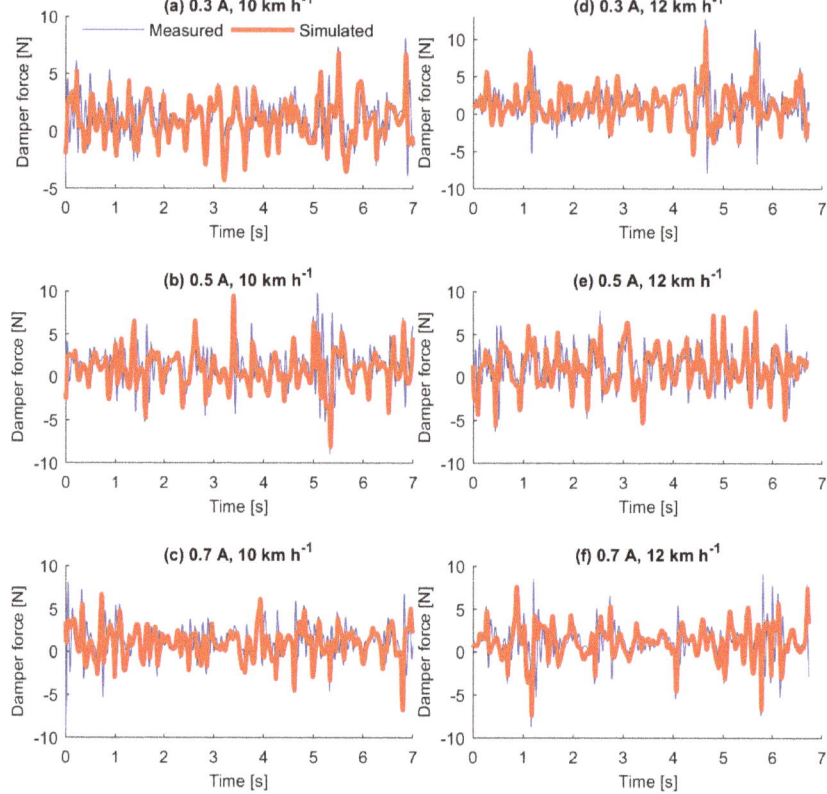

Figure 7. The measured and simulated damper forces resulted from the MR damper at the speed of (**a–c**) 10 km h^{-1} and (**d–f**) 12 km h^{-1}, when the MR damper was supplied with 0.3 A, 0.5 A, and 0.7 A.

Since the correlation analyses in the time domain indicated lower agreement between the measured and the simulated damper forces, a correlation was carried out based on the coherence analyses in the frequency domain (Equation (12)). Figures 8 and 9 present the frequency content of the measured and the simulated damper forces, for the traveling speed of 10 km h^{-1} and 12 km h^{-1}, respectively, together with the corresponding coherence, with current inputs of 0.3 A, 0.5 A, and 0.7 A applied on the MR damper.

Despite the fact that there was a wide range of wavelengths of the damper forces in the time domain, which resulted in lower correlation, the analyses of their frequency content showed good agreement according to the resulted coherence values. The resulting coherence values were above 0.96 at the speed of 10 km h^{-1} for all the current inputs applied on the MR damper. This figure was slightly lower for the speed of 12 km h^{-1} with a coherence value of 0.91.

The analyses in the time domain denoted that all the RMSE percentage, RMSD, and correlation coefficient values were within an acceptable range in terms of the predefined thresholds. In addition, the pattern as the correlation coefficient was high when the RMSE indicated a low error, which confirmed the correctness of the performed correlation. Considering all the above-given results in both time and frequency domain, it could be stated that the simulation model that described the dynamics of the assembly equipped with the MR damper was capable of producing output values within acceptable deviation ranges compared to the observed ones.

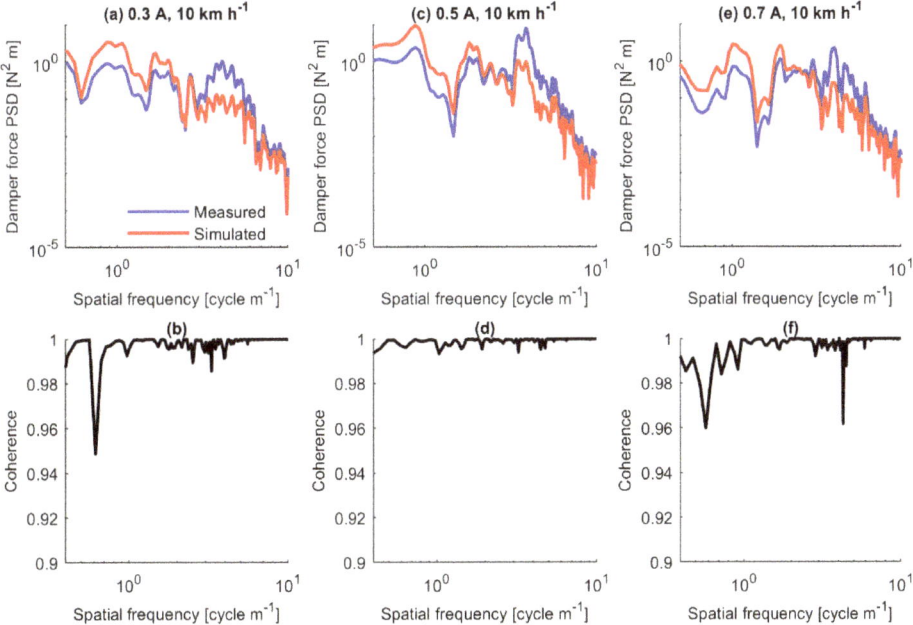

Figure 8. (**a**,**c**,**e**) Power spectral density (PSD) of the measured and the simulated damper forces when the MR damper was supplied with 0.3 A, 0.5 A, and 0.7 A and a traveling speed of 10 km h^{-1}, and (**b**,**d**,**f**) the corresponding coherence.

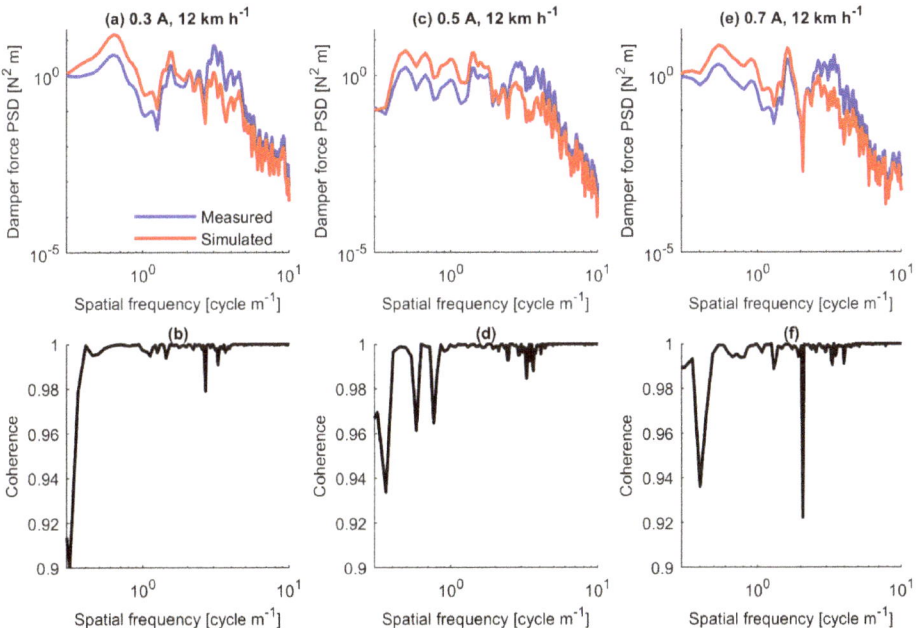

Figure 9. (**a**,**c**,**e**) Power Spectral Density (PSD) of the measured and the simulated damper forces when the MR damper supplied with 0.3 A, 0.5 A, and 0.7 A and a traveling speed of 12 km h^{-1}, and (**b**,**d**,**f**) the corresponding coherence.

4. Conclusions

A mathematical model of the vertical motion dynamics of the coulter assembly with the MR damper was developed. The necessary instrumentation to measure the vertical dynamics of the assembly and the field profiles were presented. The dynamic response, in terms of the damper forces and the pitch angles of the assembly vertical motion, were simulated using the measured surface profiles and vertical forces. The performance of the mathematical model for two different traveling speeds of the seeder (10 km h^{-1} and 12 km h^{-1}) was validated by comparing the RMSE percentage, the RMSD, and the correlation coefficient of the simulated outputs and the measured ones. The comparison analyses depicted a good agreement, with an RMSE percentage below 15% and a correlation coefficient above 0.7, between the simulated and measured datasets for all current inputs on the MR damper. The average difference, in terms of RMSD, also indicated acceptable values for both pitch angle and damper force. This analysis confirmed that the mathematical model is capable of producing similar results compared to the observed ones. Thus, the model could be used as a basis for finding the optimal settings for the MR damper but also to further develop and optimize the entire seeding assembly under various field and technical conditions. Future steps should include the implementation of different control techniques, such as the real-time adaptive PID and fuzzy hybrid controller [37], in order to optimize the dynamic performance of the MR damper and thus improve the assembly performance under different soil conditions.

Author Contributions: Conceptualization, methodology, data acquisition, G.M.S. and D.S.P.; writing—original draft preparation, G.M.S.; writing—review and editing, G.M.S. and D.S.P.; supervision, D.S.P. and H.W.G.

Funding: The authors would like to thank GA nr 213-2723/001–001–EM Action2 TIMUR project that financially covered the study visit for the data acquisition.

Acknowledgments: The authors are very thankful to R. Resch and Ch. Gall from AMAZONEN-WERKE H. Dreyer GmbH & Co. KG (Osnabrück, Germany) for providing the coulter assemblies and the supervisions on the settings.

Conflicts of Interest: The authors declare no conflict of interest.

References

1. Özmerzi, A.; Karayel, D.; Topakci, M. Effect of Sowing Depth on Precision Seeder Uniformity. *Biosyst. Eng.* **2002**, *82*, 227–230. [CrossRef]
2. Collins, B.A.; Fowler, D.B. Effect of soil characteristics, seeding depth, operating speed, and opener design on draft force during direct seeding. *Soil Tillage Res.* **1996**, *39*, 199–211. [CrossRef]
3. Abo Al-Kheer, A.; Eid, M.; Aoues, Y.; El-Hami, A.; Kharmanda, M.G.; Mouazen, A.M. Theoretical analysis of the spatial variability in tillage forces for fatigue analysis of tillage machines. *J. Terramech.* **2011**, *48*, 285–295. [CrossRef]
4. Hasimu, A.; Chen, Y. Soil disturbance and draft force of selected seed openers. *Soil Tillage Res.* **2014**, *140*, 48–54. [CrossRef]
5. Liu, J.; Chen, Y.; Kushwaha, R.L. Effect of tillage speed and straw length on soil and straw movement by a sweep. *Soil Tillage Res.* **2010**, *109*, 9–17. [CrossRef]
6. Baker, C.J.; Saxton, K.E.; Ritchie, W.R.; Chamen, W.C.T.; Reicosky, D.C.; Ribeiro, F.; Justice, S.E.; Hobbs, P.R.; Justice, F.R.S.E. *No-Tillage Seeding in Conservation Agriculture*; CABI: Wallingford, UK, 2006; ISBN 9781845931162.
7. Lawrance, N.S. A method of Analyzing Dynamic Responses of a Semi-Mounted Farm Implement. Ph.D. Thesis, The Ohio State University, Columbus, OH, USA, 1969.
8. Mouazen, A.M.; Anthonis, J.; Saeys, W.; Ramon, H. An Automatic Depth Control System for Online Measurement of Spatial Variation in Soil Compaction, Part 1: Sensor Design for Measurement of Frame Height Variation from Soil Surface. *Biosyst. Eng.* **2004**, *89*, 139–150. [CrossRef]
9. Shahgoli, G.; Fielke, J.; Desbiolles, J.; Saunders, C. Optimising oscillation frequency in oscillatory tillage. *Soil Tillage Res.* **2010**, *106*, 202–210. [CrossRef]
10. Lines, J.A.; Murphy, K. The stiffness of agricultural tractor tyres. *J. Terramech.* **1991**, *28*, 49–64. [CrossRef]

11. Burce, M.E.C.; Kataoka, T.; Okamoto, H. Seeding Depth Regulation Controlled by Independent Furrow Openers for Zero Tillage Systems—Part 1: Appropriate Furrow Opener. *Eng. Agric. Environ. Food* **2013**, *6*, 1–6. [CrossRef]
12. Burce, M.E.C. Active Seed Depth Control for No-tillage Systems Grand Sierra Resort and Casino. In Proceedings of the ASABE Annual International Meeting, Reno, NV, USA, 21–24 June 2009.
13. Suomi, P.; Oksanen, T. Automatic working depth control for seed drill using ISO 11783 remote control messages. *Comput. Electron. Agric.* **2015**, *116*, 30–35. [CrossRef]
14. Paraforos, D.S.; Sharipov, G.M.; Griepentrog, H.W. ISO 11783-compatible industrial sensor and control systems and related research: A review. *Comput. Electron. Agric.* **2019**, *163*, 104863. [CrossRef]
15. Nielsen, S.K.; Munkholm, L.J.; Lamandé, M.; Nørremark, M.; Edwards, G.T.C.; Green, O. Seed drill depth control system for precision seeding. *Comput. Electron. Agric.* **2018**, *144*, 174–180. [CrossRef]
16. Rui, Z.; Tao, C.; Dandan, H.; Dongxing, Z.; Kehong, L.; Xiaowei, Y.; Yunxia, W.; Xiantao, H.; Li, Y.; Wu, Y.X.; et al. Design of depth-control planting unit with single-side gauge wheel for no-till maize precision planter. *Int. J. Agric. Biol. Eng.* **2016**, *9*, 56–64.
17. Gratton, J.; Chen, Y.; Tessier, S. Design of a spring-loaded downforce system for a no-till seed opener. *Can. Biosyst. Eng.* **2003**, *45*, 29–35.
18. Nielsen, S.K.; Norremark, M.; Green, O. Sensor and control for consistent seed drill coulter depth. *Comput. Electron. Agric.* **2016**, *127*, 690–698. [CrossRef]
19. Ahamed, R.; Ferdaus, M.M.; Li, Y. Advancement in energy harvesting magneto-rheological fluid damper: A review. *Korea-Aust. Rheol. J.* **2016**, *28*, 355–379. [CrossRef]
20. Eshkabilov, S.L. Modeling and Simulation of Non-Linear and Hysteresis Behavior of Magneto-Rheological Dampers in the Example of Quarter-Car Model. *Eng. Math.* **2016**, *1*, 19–38.
21. Wang, Z.; Chen, Z.; Gao, H.; Wang, H. Development of a Self-Powered Magnetorheological Damper System for Cable Vibration Control. *Appl. Sci.* **2018**, *8*, 118. [CrossRef]
22. Şahin, İ.; Engin, T.; Çeşmeci, Ş. Comparison of some existing parametric models for magnetorheological fluid dampers. *Smart Mater. Struct.* **2010**, *19*, 035012. [CrossRef]
23. Madhavrao, R.; Mohibb, D.; Jamadar, E.H.; Kumar, H.; Joladarashi, S. Evaluation of a commercial MR damper for application in semi—Active suspension. *SN Appl. Sci.* **2019**, *1*, 993.
24. Sharipov, G.M.; Paraforos, D.S.; Griepentrog, H.W. Modelling and simulation of the dynamic performance of a no-till seeding assembly with a semi-active damper. *Comput. Electron. Agric.* **2017**, *139*, 187–197. [CrossRef]
25. Paraforos, D.S.; Reutemann, M.; Sharipov, G.; Werner, R.; Griepentrog, H.W. Total station data assessment using an industrial robotic arm for dynamic 3D in-field positioning with sub-centimetre accuracy. *Comput. Electron. Agric.* **2017**, *136*, 166–175. [CrossRef]
26. Sharipov, G.M.; Paraforos, D.S.; Pulatov, A.; Griepentrog, H.W. Dynamic performance of a no-till seeding assembly. *Biosyst. Eng.* **2017**, *158*, 64–75. [CrossRef]
27. Sharipov, G.M.; Paraforos, D.S.; Griepentrog, H.W. Implementation of a magnetorheological damper on a no-till seeding assembly for optimising seeding depth. *Comput. Electron. Agric.* **2018**, *150*, 465–475. [CrossRef]
28. Poll, C.; Marhan, S.; Back, F.; Niklaus, P.A.; Kandeler, E. Field-scale manipulation of soil temperature and precipitation change soil CO_2 flux in a temperate agricultural ecosystem. *Agric. Ecosyst. Environ.* **2013**, *165*, 88–97. [CrossRef]
29. Sapiński, B.; Filuś, J. Analysis of parametric models of MR linear damper. *J. Theor. Appl. Mech.* **2003**, *41*, 215–240.
30. Arsava, K.S.; Kim, Y. Modeling of magnetorheological dampers under various impact loads. *Shock Vib.* **2015**, *10*, 20. [CrossRef]
31. Talatahari, S.; Kaveh, A.; Rahbari, N.M. Parameter identification of Bouc-Wen model for MR fluid dampers using adaptive charged system search optimization. *Mech. Sci. Technol.* **2012**, *26*, 2523–2534. [CrossRef]
32. Rashid, M.M.; Rahim, N.A.; Hussain, M.A.; Rahman, M.A. Analysis and experimental study of magnetorheological-based damper for semiactive suspension system using fuzzy hybrids. *IEEE Trans. Ind. Appl.* **2011**, *47*, 1051–1059. [CrossRef]
33. Braz-Cesar, M.T.; Barros, R.C. Semi-active Vibration Control of Buildings using MR Dampers: Numerical and Experimental Verification. In Proceedings of the 14th European Conference on Earthquake Engineering, Ohrid, Macedonia, 30 August–3 September 2010.

34. Ngwangwa, H.M.; Heyns, P.S.; Breytenbach, H.G.A.; Els, P.S. Reconstruction of road defects and road roughness classification using Artificial Neural Networks simulation and vehicle dynamic responses: Application to experimental data. *J. Terramech.* **2014**, *53*, 1–18. [CrossRef]
35. Sharipov, G.M.; Paraforos, D.S.; Griepentrog, H.W. Modelling and simulation of a no-till seeder vertical motion dynamics for precise seeding depth. In Proceedings of the 11th European Conference on Precision Agriculture (ECPA 2017), Edinburgh, UK, 16–17 July 2017; Cambridge University Press: Cambridge, UK, 2017; Volume 8, pp. 455–460.
36. Hawari, H.; Murray, M.H. Correlating track forces and track profile. In Proceedings of the Conference on Railway Engineering CORE 2006, Melbourne, Australia, 1 May 2006; Volume 54, pp. 259–266.
37. Phu, D.X.; An, J.-H.; Choi, S.-B. A Novel Adaptive PID Controller with Application to Vibration Control of a Semi-Active Vehicle Seat Suspension. *Appl. Sci.* **2017**, *7*, 1055. [CrossRef]

© 2019 by the authors. Licensee MDPI, Basel, Switzerland. This article is an open access article distributed under the terms and conditions of the Creative Commons Attribution (CC BY) license (http://creativecommons.org/licenses/by/4.0/).

Article

Control Technology of Ground-Based Laser Communication Servo Turntable via a Novel Digital Sliding Mode Controller

Jianqiang Zhang [1,2], Yongkai Liu [1,2,*], Shijie Gao [1] and Chengshan Han [1]

[1] Changchun Institute of Optics, Fine Mechanics and Physics, Chinese Academy of Sciences, Changchun 130033, China; zhangjq7170@163.com (J.Z.); 13604329504@163.com (S.G.); han_chengshan@163.com (C.H.)
[2] University of Chinese Academy of Sciences, Beijing 100049, China
* Correspondence: liuyk@ciomp.ac.cn; Tel.: +86-0431-86708237

Received: 21 August 2019; Accepted: 18 September 2019; Published: 27 September 2019

Abstract: In this study, a sliding mode control (SMC) algorithm was proposed based on a novel reaching law to solve the nonlinear disturbance problem of a ground-based laser communication turntable. This algorithm is a chatter-free method, in which the coefficient of sliding mode variable structure function is designed as an adaptive function, so the chattering of the sliding mode approaches zero. For any perturbed system, this algorithm can ensure a finite time for the system state to reach the sliding mode surface from any initial state. Additionally, the system will stabilize in the quasi-sliding mode domain (QSMD) with $O(T^3)$ width, where a narrower QSMD width corresponds to stronger robustness toward nonlinear disturbances. Both mathematical calculations and simulations verified the sliding mode and stability of this control algorithm. Experimental results of the velocity closed-loop of pitch axis show that the proposed algorithm effectively improved the anti-nonlinear disturbance ability of the control system compared with the effects of the traditional digital PID and the existing chatter-reduced SMC algorithms, for smooth system operation.

Keywords: reaching law; sliding mode control (SMC); ground-based laser communication turntable; chatter-free; quasi-sliding mode domain (QSMD)

---

1. Introduction

Ground-based laser communication is the communication between two or more terminals on the ground, using the atmosphere as the medium and a laser beam as the carrier. Compared with the traditional microwave communication, laser communication offers advantages of low power consumption, high bandwidth, and strong anti-jamming ability, making development of this technology an important focus in the field of information and communication technology [1–3].

The communication distance between ground-based laser communication terminals can be extensive, the divergence angle of a laser beam is quite small, and the laser can also be affected by atmospheric turbulence, presenting several challenges to this technology [4,5]. Therefore, to meet the power requirements of laser communication, the servo control system requires micro-radian optical axis alignment accuracy. Currently, coarse and fine two-stage tracking technology is used in most ground-based laser communication systems to ensure alignment accuracy [6–8]. Objectively, the coarse tracking control system is the most important part of the control strategy, allowing the isolation of external disturbances to ensure the stability of the optical axis, and the control accuracy determines if the optical axis can be coupled into the fine tracking field of view required for a fine tracking control strategy. A servo turntable is selected here as the actuator, with nonlinear factors such as unbalanced internal torque, friction, torque ripple, model identification error of the control system, and parameter changes, which can lead to unstable operation of the system and restricting system accuracy. Traditional linear control technologies such as use of a digital proportion-integral-differential controller (PID) and lead-lag compensation are unable to meet the accuracy requirements of a coarse tracking system, therefore, a high precision and strong robustness control technology is required to restrain the influence of nonlinear factors on the system and ensure smooth operation of the servo turntable [4].

Several methods have been proposed to suppress the nonlinear disturbance and improve system accuracy, which can be divided into three main approaches. First, the control algorithm structure and the dynamic performance of the control system can be improved by adding a current loop [9], or feedforward compensators [10,11]. Second, the traditional control algorithm can be improved by the incorporation of aspects of modern algorithms, such as fuzzy control [12–14] or adaptive control [15,16], which can be applied to improve the robustness of traditional algorithms. Improved algorithms can restrain the influence of nonlinear disturbance on the system, but this approach cannot overcome all the shortcomings of traditional algorithms. Third, the design of a nonlinear controller based on the modern control algorithms, such as active disturbance rejection [17,18], adaptive [19,20], and predictive [11,21] control strategies can avoid the shortcomings of traditional algorithms, effectively compensate for nonlinear disturbances in the system, and improve the accuracy and stability of the system.

Siding mode control (SMC) is an effective approach for a robust control algorithm, and can effectively suppress the nonlinear disturbance of a system. This anti-disturbance control algorithm can ensure that the system state reaches the sliding surface in a finite time and is stable in the quasi-sliding mode domain (QSMD). Once the system state stabilizes at the QSMD, the system state is invariant to system parameter variations and other nonlinear disturbances such as friction [21]. The narrower the QSMD width, the stronger the system robustness to nonlinear disturbances [22]. However, the direct application of the SMC algorithm to ground-based laser communication servo turntable has been a challenge. Large and high frequency chattering of the controller is caused by the switching function in the traditional SMC algorithm, which affect the overall accuracy and stability of the system, and may lead to damage of experimental equipment [23]. Therefore, improving the traditional SMC algorithm to avoid sliding mode chattering is critical for the effective application of SMC algorithm in specific projects.

Several algorithms have been proposed to improve the sliding mode control. The terminal sliding mode algorithm improves the sliding mode surface function design method, and eliminates sliding mode chattering of the controller output by design of a nonlinear sliding mode surface [24–27]. A higher-order SMC algorithm is applied to extend the design method of traditional SMC output, which eliminates the sliding mode chattering by application of discontinuous control variables to the higher derivatives of sliding mode functions [28]. A robust adaptive second-order SMC was tested to address the tracking problem of uncertain linear systems with both matched and unmatched disturbances, and the results showed that the chattering was removed by applying the sign function for the time-derivative of the control signal [29]. A global SMC was applied to an uncertain chaotic system, and resulted in robustness to multiple delays, parametric uncertainties, and other nonlinear disturbances [30]. A composite nonlinear feedback technique based on a self-tuning integral SMC algorithm was proposed for the robust tracking control of switched systems with uncertainties and input saturation; this technique guaranteed robustness against uncertainties, removed reaching phase, and avoided the chattering problem of sliding mode [31]. A novel adaptive super-twisting-based global-SMC algorithm was proposed to remove the reaching interval and confer robustness and stability to underactuated systems; chatter-free operation was guaranteed by integration of the discontinuous sign function in the control signal [32]. Of these methods, the reaching law function directly defines the approaching movement and sliding mode, making it the most direct, effective, and simple method among the described improved sliding mode algorithms. In this work, a novel chatter-free reaching law algorithm with a disturbance compensation based on the exponential reaching law is proposed for the application in a ground-based laser communication rough tracking control system. This algorithm guarantees that the sliding mode chattering approaches zero and the system state is stabilized in the

QSMD with $O(T^3)$ (T is the sampling period) width, resulting in superior control that is robust to nonlinear disturbances and model parameter variations.

The main findings of this work can be summarized as follows:

(1) The frequency domain characteristic of the turntable pitch axis is tested by the classical sweep method [33]. According to the frequency domain characteristic curve, the system model is obtained by the traditional identification method, which is the precondition for the design of SMC.

(2) A novel reaching law with a disturbance compensator is proposed to solve the chattering problem, which is robust to system model identification error, friction, and other nonlinear disturbances. Both mathematical calculation and simulation support the effectiveness of the algorithm.

(3) The proposed digital SMC algorithm, the traditional digital PID algorithm, and the existing chatter-reduced SMC algorithm were compared. The experimental results show that the proposed algorithm provides higher control accuracy, stronger anti-interference ability, better frequency domain characteristics, and also suppresses chattering for an improved ground-based laser communication servo turntable control system.

2. Model Identification of Ground-Based Laser Communication Servo Turntable

2.1. System Frequency Domain Characteristic Test

The azimuth or pitch axis control structure of ground-based laser communication servo turntable is shown in Figure 1. The system is composed of a velocity loop controller, a power amplifier, a torque motor, a circular grating encoder, and other parts.

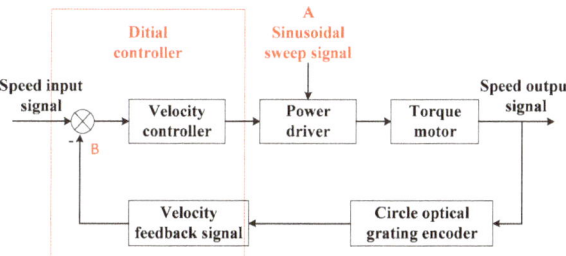

Figure 1. Control structure of the pitch axis servo turntable system.

Traditionally, the sinusoidal sweep method is used to test the frequency domain characteristics of the system. As shown in Figure 1, point A indicates the application of the sinusoidal sweep signal, and the speed record at point B encoder feedback value. The input and output data can be used to obtain the system frequency characteristic curve, and the measured frequency domain characteristics include the power amplifier, motor, mechanical structure, and encoder components.

The sinusoidal sweep signal is input at point A, and its digital sequence is expressed as:

$$\begin{aligned} u(k) &= A(k)\sin[2\pi\omega(k)] \\ \omega(k) &= f_1\big(1 + c(k)\cdot(k\cdot t)^n\big), c(k) = \tfrac{f_2/f_1 - 1}{(n+1)T^n} \end{aligned} \quad (1)$$

where f_1 is the sweep signal starting frequency, f_2 is the sweep signal termination frequency, t is the sampling time, n is the order of the sweep signal, and T is the sweep signal duration.

Considering the pitch axis of ground-based laser communication servo turntable as an example, a frequency sweep experiment was carried out. During the experiment, the frequency range of sinusoidal sweep signal $[f_1, f_2]$ was set to [0.5, 200], the signal amplitude $A(t)$ was set to 10°/s, the sweep duration T was set to 25 s, and the signal sweep order n was set to the third to ensure sufficient low-frequency data. The sweep input signal and system response curve are shown in Figure 2.

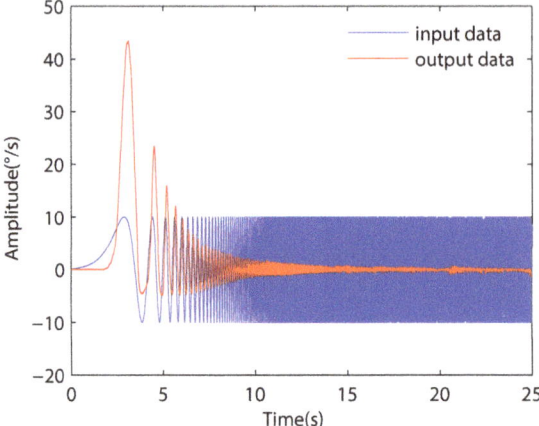

Figure 2. Input and response signal curve of pitch axis open-loop test.

According to the input and output data, the open-loop transmission Bode diagram of the system can be obtained by discrete Fourier transform and power spectrum estimation, as shown in Figure 3. The following conclusions can be drawn explicitly: the turntable pitch axis tends to be linear in the low and middle frequency bands, but there are many nonlinear factors in the system because of the friction of the axis and other factors. There are multi-order resonance links in the high frequency band, and the third-order locked rotor frequency at 105 Hz and the resonant frequency at 113.3 Hz are the main frequencies that affect the system performance. The mechanical structure cannot achieve absolute rigid connection, because of torque imbalance and other nonlinear factors, but the numerical difference between the locked rotor frequency and the resonant frequency is small resulting in strong stiffness.

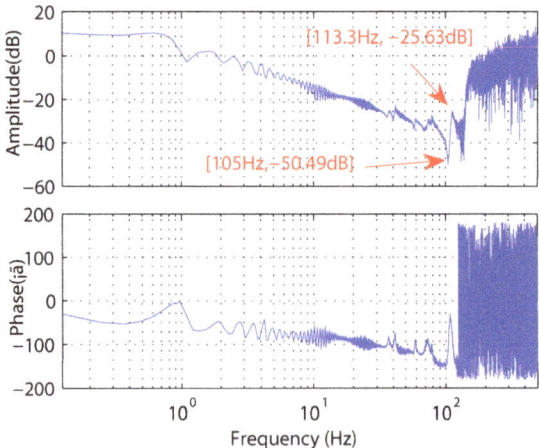

Figure 3. Open-loop frequency characteristic curve of pitch axis velocity.

Remark 1. *It is worth noting that the inverse tangent function is adopted for phase calculation in this method. If the phase difference is less than 90°, the angle value will jump, so ± n180° (n = 1, 2 ...) should be applied according to the phase angle data value.*

2.2. System Model Identification

For practical engineering application, system model identification utilized the mathematical method of medium-low frequency linear fitting and high-frequency two-order model fitting. The essence of this method is to consider the middle and low frequency bands of the system as an ideal linear model and then a polynomial mathematical model is adopted to fit the frequency characteristic curve of the middle and low frequency bands. For the high frequency band, the multi-order resonance frequency in the frequency characteristic curve is fitted by one or more two-order mathematical models. Therefore, the ideal mathematical model of the system can be expressed as follows:

$$G(s) = G_L(s) G_{H_1}(s) G_{H_2}(s) \cdots G_{H_n}(s). \tag{2}$$

In view of Equation (2), $G_L(s)$ represents the transfer function model of ideal linear link in middle and low frequency band, which is composed of inertial link. $G_{Hi}(s)$ ($i = 1, 2, \ldots, n$) represents the transfer function model of the resonance link in high frequency band, which is composed of a two-order mathematical model in series, and its mathematical model is expressed as:

$$G_H(s) = \frac{\frac{1}{\omega_z^2} s^2 + 2 \frac{\zeta_z}{\omega_z} s + 1}{\frac{1}{\omega_p^2} s^2 + 2 \frac{\zeta_p}{\omega_p} s + 1}. \tag{3}$$

The transfer function shown in Equation (3) includes a pair of conjugate complex zero and a pair of complex poles. The complex zeros correspond to systems of high frequency locked rotor frequency ω_z and damping coefficient ζ_z. The double pole corresponds to the resonance frequency ω_p and the damping coefficient ζ_p.

According to the frequency characteristic curve of the pitch axis as shown in Figure 3, the linear transfer function of the middle and low frequency band system can be obtained by the polynomial fitting method as follows:

$$G_L(s) = \frac{3.5}{0.264s + 1}. \tag{4}$$

Shown in the frequency characteristic curve, the third-order resonance mode can have the greatest influence on the high frequency band, and the other modes can be ignored. Therefore, only a second-order mathematical model is sufficient to fit the high-frequency characteristic curve. The high frequency mathematical model is expressed as:

$$G_{H_1}(s) = \frac{(0.0095s)^2 + 0.00054s + 1}{(0.0088s)^2 + 0.00066s + 1}. \tag{5}$$

In conclusion, the system model is expressed by the transfer function as:

$$G(s) = G_L(s) \cdot G_{H_1}(s). \tag{6}$$

The continuous time equation of state corresponds to the transfer function as shown in Equation (6) and can be expressed as:

$$\begin{cases} \dot{x}(t) = Ax(t) + Bu(t) \\ y(t) = Cx(t) \end{cases}, \tag{7}$$

where $x(t) \in R^{n \times 1}$ is the state vector, $y(t) \in R^{n \times 1}$ is the output vector, and $A \in R^{n \times n}$, $B \in R^{n \times n}$, and $C \in R^{1 \times n}$ are system model parameter matrices. Therefore, according to Equations (5) and (6), the parameter matrixes in Equation (7) can be specifically expressed as follows:

$$A = 10^4 \times \begin{bmatrix} -0.0012 & -1.295 & -4.89 \\ 0.0001 & 0 & 0 \\ 0 & 0.0001 & 0 \end{bmatrix}, B = \begin{bmatrix} 1 \\ 0 \\ 0 \end{bmatrix}.$$

$$C = 10^5 \times \begin{bmatrix} 0.0002 & 0.0009 & 1.7123 \end{bmatrix}$$

The characteristic of the open loop frequency domain of the mathematical model shown in Equation (7) has been tested. Given the data presented in Figure 3, the frequency characteristic fitting curve of model identification was constructed and is shown in Figure 4 and the model identification error frequency characteristic curve is presented in Figure 5.

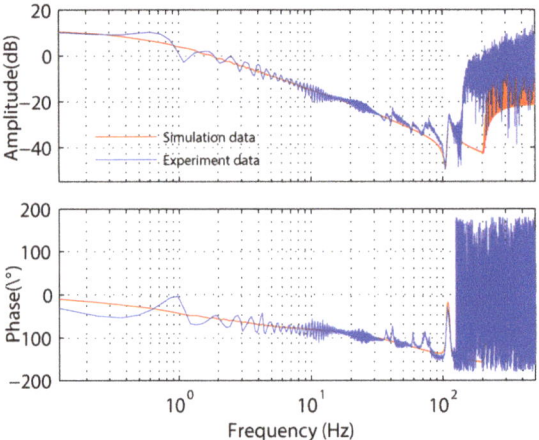

Figure 4. Frequency characteristic fitting curve of model identification.

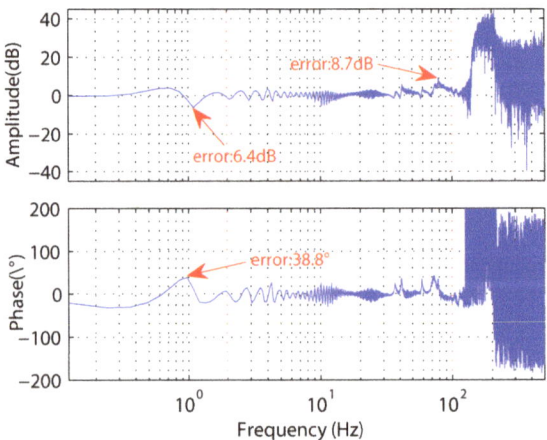

Figure 5. Model identification error frequency characteristic curve.

According to the comparison curve of frequency characteristics, the following conclusions can be drawn. The frequency is less than 2 Hz, and the motor is in the running start stage. There is an effect of the static friction force, and the running state is unstable, with obvious amplitude-frequency characteristic curve fluctuation and phase-frequency hysteresis characteristics. The frequency is greater than 2 Hz, and the model frequency characteristic curve can better fit the trend of the actual frequency characteristic curve, however nonlinear disturbances such as friction force and torque imbalance in the system can significantly affect the identification accuracy. The absolute value of the amplitude frequency characteristic curve identification error is less than 8.7 dB. For engineering practice, model identification error and nonlinear disturbance restrict dramatic improvement of the control performance, but a traditional PID controller cannot effectively suppress low and medium frequency nonlinear disturbance, so cannot meet the requirements of high control accuracy of ground-based laser communication. Therefore, it is important to develop a high-precision control algorithm that can better suppress the nonlinear disturbance.

3. Design and Simulation of a Velocity Loop Controller Based on Discrete SMC

This section presents a novel discrete SMC algorithm with a disturbance compensator for low-frequency nonlinear disturbances and model identification error.

3.1. Basic Theory of Discrete SMC

3.1.1. System Discrete Time Ideal Equation of State

According to the identification model equation of state shown in Equation (7), the ideal state equation of the system can be expressed as follows:

$$\dot{x}(t) = Ax(t) + Bu(t) + \zeta(t). \tag{8}$$

For the digital control system, the sampling frequency is 1 KHz, and the discretized equation of state is expressed as:

$$x(k+1) = A_k x(k) + B_k u(k) + \zeta(k), \tag{9}$$

where $A_k = A_o + \Delta A(k)$, $B_k = B_o + \Delta B(k)$ are respectively represented as the real state matrix and output matrix of the system. A_o, B_o are nominal parts of A_k and B_k respectively, which can be obtained by Matlab as follows:

$$A_o = \begin{bmatrix} 0.98 & -12.87 & -48.52 \\ 0.001 & 0.99 & -0.024 \\ 0 & 0.001 & 1 \end{bmatrix}, B_o = 10^3 \cdot \begin{bmatrix} 0.992 \\ 0.0005 \\ 0 \end{bmatrix}.$$

The modeling error and the parameter variation $\Delta A(k)$, $\Delta B(k)$ are assumed to be differentiable with respect to temporal series k, and $\zeta(k)$ represents nonlinear factors of the system.

Assumption 1. *The uncertainties $\Delta A(k)$, $\Delta B(k)$, and $\zeta(k)$ are bounded and satisfy the "matching" condition that $\Delta A(k)$, $\Delta B(k)$, and $\zeta(k) \in span\{B_k\}$.*

Therefore, $\Delta A(k)$, $\Delta B(k)$, and $\zeta(k)$ must be able to be unified as nonlinear factors for theoretical research [34–38], as follows:

$$\varepsilon(k) = \Delta A(k)x(k) + \Delta B(k)u(k) + \zeta(k), \tag{10}$$

where $\zeta(k)$ satisfies $\|\zeta(k)\| < \zeta_{max}$ with $\|\cdot\|_\infty$ being the vector infinity norm and ζ_{max} being an unknown constant.

Therefore, the discrete equation of state shown in Equation (9) can be rewritten as:

$$x(k+1) = A_k x(k) + B_k u(k) + \varepsilon(k). \tag{11}$$

3.1.2. Discrete Sliding Mode Function and the Sliding Mode Surface

In this study, the classical sliding mode function was adopted, which is defined as follows:

$$s(k) = C_e e(k) = C_e(x(k) - R_n(k)), \tag{12}$$

where $C_e = [C_{e(1)}, C_{e(2)}, \ldots, C_{e(n-1)}, 1]$ is the sliding mode coefficient matrix, and $e(k)$ represents the error matrix between the ideal state and the actual state. Therefore, the sliding mode surface can be defined as:

$$S = \{e(k)|C_e e(k) = 0\}. \tag{13}$$

Lemma 1. *The sliding mode coefficient matrix parameters $C_{e(1)}, C_{e(2)}, \ldots, C_{e(n-1)}$ satisfy the polynomials $p^{n-1} + C_{e(n-1)} p^{n-2} + \ldots + C_{e(2)} + C_{e(1)}$, and must be Hurwitz polynomials; p is the Laplace operator [35].*

3.1.3. The Quasi Sliding Mode Domain

Definition 1. *For the perturbed system, the system will be in a quasi-sliding-mode (QSM) in the Δ vicinity of the sliding surface, not at the sliding surface. This specified domain where the QSM occurs is called the quasi-sliding-mode domain (QSMD) s_Δ, the positive constant Δ is the QSMD width, and s_Δ is the QSMD [39–41],*

$$s_\Delta = \{s(k)| |s(k)| \leq \Delta\} \tag{14}$$

The condition in which the system state is stable in the QSMD is defined as:

$$\begin{cases} -\Delta < s(k+1) < s(k), s(k) > \Delta \\ s(k) < s(k+1) < \Delta, s(k) < -\Delta \\ |s(k+1)| \leq \Delta, |s(k)| < \Delta \end{cases}$$

3.2. A Novel Chatter-Free Approach Law Sliding Mode Control Based on Disturbance Compensator

The SMC algorithm is widely applied in control systems because of its simple structure and good dynamic performance. After stabilization of the system states in the QSMD, the SMC is consistent with the nonlinear disturbance, and the smaller the width of the QSMD, the stronger the robustness of the system to the nonlinear factors. However, chattering is the main problem of SMC, which can lead to high frequency noise in the system, with adverse effects on the stability and control accuracy of the system. In this section, a novel chatter-free SMC algorithm is proposed to reduce the effects of chattering on the system and to suppress nonlinear disturbances in the low and medium frequency bands of the system.

The novel chatter-free sliding mode approach law based on exponential approach law is shown as follows:

$$\begin{cases} s(k+1) = (1 - \lambda T)s(k) - \kappa T \cdot \eta(k) \cdot \text{sgn}[s(k)] + \omega(k) \\ \eta(k) = \dfrac{|e_1(k)|^\alpha}{\delta + \left(1 + |e_1(k)|^{\alpha-1} - \delta\right) e^{-\mu \cdot |s(k)|^\gamma}} \end{cases}, \tag{15}$$

in this formula, $\kappa > 0$ is the coefficient of variable structure function; T is the sampling time of the system; $\eta(k)$ is the error-related adaptive function, where $0 < \delta < 1$, $\gamma > 1$, $\mu > 0$; $|e_1(k)|$ represents the first state error of the state error vector, if $|e_1(k)| > 1$, $0 < \alpha < 1$, if $0 < |e_1(k)| \leq 1$, $\alpha > 1$; $\omega(k)$ represents the disturbance compensator, which can be expressed as:

$$\omega(k) = C_e(\varepsilon(k) - 2\varepsilon(k-1) + \varepsilon(k-2)). \tag{16}$$

Remark 2. *If the system state is far away from the sliding mode surface, according to formula (12), s(k) and $|e_1(k)|$ tend to be great, so the coefficient of the switching function sgn[s(k)] tends to be $\kappa T \cdot |e_1|^\alpha/\delta$, which is greater than κT. The approaching movement is the state of the system that gradually approaches the sliding mode surface driven by the reaching law. If the system state is near the sliding surface, $s(k) \approx 0$, and the coefficient of the switching function sgn[s(k)] tends to $\kappa T \cdot |e_1|^\alpha/(1+|e_1|^\alpha)$, which is far smaller than κT, so the system state is stable in the QSMD for sliding mode motion. Thus, chattering on the sliding surface is reduced.*

Lemma 2. *According to [22,42], $\varepsilon(k) = O(T)$, $\varepsilon(k) - \varepsilon(k-1) = O(T^2)$, (T is the sampling period), where O(T), and $O(T^2)$ represent the disturbance estimation error and are the first-order O(T) and the second-order O(T), respectively; $O(T) > O(T^2)$. Therefore, the magnitude of the disturbance estimation error shown in Equation (16) is the third-order O(T), $\varepsilon(k) - 2\varepsilon(k-1) + \varepsilon(k-2) = O(T^3)$, $O(T^3) < O(T^2)$.*

According to Equation (11), (12), (15), and (16), the sliding mode controller can be deduced as follows:

$$u(k) = (C_e B_k)^{-1} [C_e(R_n(k+1) - A_k x(k)) + (1 - \lambda T)s(k) \\ - \kappa T \cdot \eta(k) \text{sgn}[s(k)] - C_e(2\varepsilon(k-1) - \varepsilon(k-2))] \tag{17}$$

Remark 3. *The disturbance function $\varepsilon(k-1)$ of the sliding mode controller (17) is usually deduced or calculated by the "time delay estimation method" [22,39,42–44]:*

$$\varepsilon(k-1) = x(k) - A_k x(k-1) - B_k u(k-1). \tag{18}$$

Remark 4. *According to Lemma 2, the width of the QSMD of the proposed SMC is related to the disturbance compensator, and its width is $O(T^3)$ order, which is smaller than the $O(T^2)$ order width described in [23,42,45]. Therefore, the proposed algorithm attains stronger robustness and higher control accuracy.*

3.3. Proof of Robustness and Stability of SMC

Theorem 1. *The absolute value of the system's nonlinear function equation as shown in Equation (16) has an upper bound, and this upper bound is assumed as ω. The trajectories of the system from any initial state must arrive at the sliding mode surface driven by the proposed algorithm.*

Proof. There must be an initial state such that the sign of the sliding mode switching functions s(0), s(1), ... s(n) does not change, where n is a positive integer. The following proofs are discussed for two cases: $s(0) < 0$ and $s(0) \geq 0$.

1. If $s(k) \geq 0$ ($k = 0, 1, \ldots n$)

Assuming that the system does not cross the sliding surface within the n step, recursive formulas can be obtained according to formula (15).

$$\begin{aligned} s(1) &= (1 - \lambda T)s(0) - (\kappa T \cdot \eta(0) + \omega(0)) \\ s(2) &= (1 - \lambda T)^2 s(0) - (1 - \lambda T)(\kappa T \cdot \eta(0) - \omega(0)) - \kappa T \cdot \eta(1) - \omega(1) \\ &\vdots \\ s(n) &= (1 - \lambda T)^n s(0) - \sum_{i=0}^{n-1} (1 - \lambda T)^{n-1-i} (\kappa T \cdot \eta(i) - \omega(i)) \end{aligned} \tag{19}$$

There must be a positive number δ so that the following formula is workable:

$$\sum_{i=0}^{n-1} (1 - \lambda T)^{n-1-i} (\kappa T \cdot \eta(i) - \omega(i)) = \sum_{i=0}^{n-1} (1 - \lambda T)^{n-1-i} \delta. \tag{20}$$

Assuming that the system trajectory reaches the sliding mode surface at time m, then $s(m) = 0$, and according to Equations (19) and (20), the following equation can be obtained,

$$s(m) = (1-\lambda T)^m s(0) - \delta \cdot \sum_{i=0}^{m-1}(1-\lambda T)^{m-1-i} = (1-\lambda T)^m s(0) - \delta \frac{1-(1-\lambda T)^m}{\lambda T} \qquad (21)$$

Therefore, the arrival time m can be expressed as:

$$m = \log_{1-\lambda T} \frac{1}{\lambda T \cdot s(0)/\delta + 1}. \qquad (22)$$

2. Similarly, if $s(k) < 0$ ($k = 0, 1, \ldots n$), the moment that the system state reaches the sliding mode surface can be expressed as:

$$m = \log_{1-\lambda T} \frac{1}{-\lambda T \cdot s(0)/\delta + 1}. \qquad (23)$$

Above all, driven by the sliding mode controller shown in formula (17), the state trajectory of the system from any initial position can reach the sliding mode surface in a limited time, and the arrival time is expressed as follows:

$$m = \log_{1-\lambda T} \frac{1}{\lambda T \cdot |s(0)|/\delta + 1}. \qquad (24)$$

Theorem 1 has been proved. □

Theorem 2. *Driven by the sliding mode controller, once the system state reaches the sliding mode surface, it will be stable in the QSMD and cannot escape, the control system is bounded. The QSMD can be expressed as:*

$$\Phi = \{s(k)||s(k)| \leq \Delta = \kappa T \cdot \eta + \omega\}, \qquad (25)$$

where $\omega/\kappa T \leq \eta < 1$, ω is the upper of the disturbance.

Proof. The proof process can be divided into two cases: $s(0) < 0$ and $s(k) \geq 0$ ($k = 0, 1, \ldots n$).
1. If $s(k) \geq 0$ ($k = 0, 1, \ldots n$), it can be obtained that

$$\begin{aligned} s(k+1) &= (1-\lambda T)s(k) - \kappa T \cdot \eta(k)\text{sgn}[s(k)] + \omega(k) \\ &< s(k) - \kappa T \cdot \eta + \omega < s(k) \end{aligned} \qquad (26)$$

Therefore, when $s(k) \geq 0$, the value of $s(k)$ decreases successively. Assuming that the system state is in the QSMD at time n, the system state at time $n + 1$ must be in the QSMD.
2. If $s(k) < 0$ ($k = 0, 1, \ldots n$), similar conclusions can be drawn that

$$\begin{aligned} s(k+1) &= (1-\lambda T)s(k) - \kappa T \cdot \eta(k)\text{sgn}[s(k)] + \omega(k) \\ &> s(k) + \kappa T \cdot \eta + \omega > s(k) \end{aligned} \qquad (27)$$

Therefore, when $s(k) < 0$, the value of $s(k)$ increases successively. When the system state is in the QSMD at time n, the system state at time $n + 1$ must also be in the QSMD.

Above all, $|s(k)|$ deceases with time. Once the trajectory of the system reaches the sliding surface, the system state will stabilize in the QSMD. At this moment, the system is strongly robust to nonlinear disturbances. Therefore, the control algorithm is stable, and Theorem 2 is proved. □

Remark 5. *In view of [36,44,46–49], the overestimation and underestimation problems may exist when the state is stable in the QSMD, we will study it in the future work.*

3.4. Sliding Mode Simulation

In this section, using the pitch axis system model of the ground-based laser communication servo turntable shown in Equation (11) as the simulation model, the stability of the proposed sliding mode algorithm was analyzed as follows. Assuming that the nonlinear disturbance of the system is $\varepsilon(k)$ = [0;0;2.5sin(2$k\pi$) + 0.5], the initial state of the system is [−1;−0.8;0], and the target state is [0;0;0].

According to the discrete state equation of the system, the sliding mode coefficient shown in Equation (12) is a three-order matrix such that $C_e = [C_{e1}, C_{e2}, 1]$, the parameters should be chosen as [1,2,1] to satisfy lemma 1. The parameters of the sliding mode controller (17) were selected as: $\lambda = 5$, $\delta = 0.15$, $\alpha = 2$, $\mu = 10$, and $\gamma = 2$. To assess the proposed algorithm, it was compared to the classical exponential reaching law (28) with a $O(T^2)$ disturbance compensator and the chatter-reduced reaching law (29) with the $O(T^3)$ disturbance compensator described in Ref. [22]. The exponential function coefficients are the same, with values of $\lambda_c = 5$, $\lambda_{[22]} = 5$, and the variable structure function coefficients κ, κ_c, $\kappa_{[22]}$ are determined by the disturbance estimation error in the next section. Other coefficients in the algorithm of Ref. [22] were set to $\delta_{[22]} = 0.15$, $\mu_{[22]} = 10$, and $\gamma_{[22]} = 2$.

$$s(k+1) = (1 - \lambda_c T)s(k) - \kappa_c T \cdot \text{sgn}[s(k)] + C_e \varepsilon(k-1). \tag{28}$$

$$\begin{cases} s(k+1) = \left(1 - \lambda_{[22]} T\right)\phi_{[22]}(k) \cdot s(k) - \frac{\kappa_{[22]} T}{\phi_{[22]}(k)} \text{sgn}[s(k)] + C_e[\varepsilon(k) - 2\varepsilon(k-1) + \varepsilon(k-2)] \\ \phi_{[22]}(k) = \delta_{[22]} + \left(1 - \delta_{[22]}\right) e^{-\mu_{[22]}|s(k)|^{\gamma_{[22]}}} \end{cases} \tag{29}$$

The comparison curves of disturbance estimation errors between the three algorithms were determined and are shown in Figure 6. The maximum $O(T^3)$ amplitude was 0.01 for the algorithm disturbance estimation error curve for both the proposed algorithm and the Ref. [22] algorithm. This value is much smaller than that of the classical exponential reaching law algorithm, which exhibits a maximum $O(T^2)$ amplitude of 0.15. Therefore, the variable structure function coefficients κ and $\kappa_{[22]}$ of both algorithms are set as 0.012, and the variable structure function coefficient κ_c of the classical algorithm is set as 0.17. According to Theorem 2, utilization of this parameter, larger than the maximum perturbation estimation, can ensure the stability of the sliding mode trajectory in the QSMD.

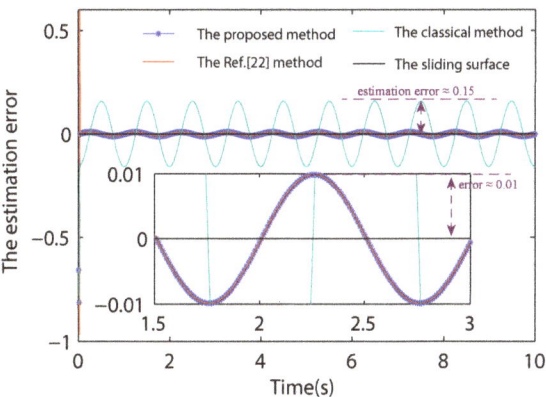

Figure 6. Disturbance estimation error contrast curve.

Figure 7 shows the sliding mode trajectory contrast curves. Three reaching law algorithms can ensure that the system state arrives at the sliding mode surface in a limited time and is stable in the QSMD with a fixed width. Theorems 1 and 2 are proved here. Compared with the traditional exponential reaching law algorithm, the QSMD width of the proposed reaching law algorithm is narrower, $\Delta_p \approx 0.038$, compared to the QSMD width of the traditional algorithm of $\Delta_c \approx 0.31$. It should be noted that a smaller QSMD width correlates to stronger robustness of the system to nonlinear disturbances. Compared with the chatter-reduced reaching law algorithm of Ref. [22], chattering in sliding mode can change to be continuous around the sliding surface, and the widths of both algorithms are almost equal. The proposed algorithm obviously eliminates sliding mode chattering and improves the stability of the system without reducing the robustness of the system.

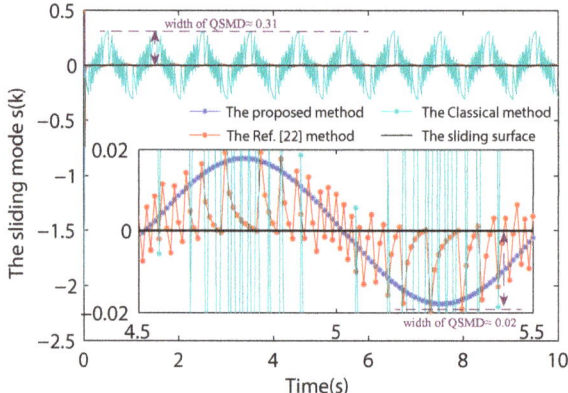

Figure 7. Sliding mode trajectory contrast curves.

4. Experimental Verification of Control Algorithm

In this study, a velocity closed-loop experiment was performed, based on the servo turntable pitch axis system for ground-based laser communication. The experimental setup is shown in Figure 8. The experimental devices include the pitch axis motor, which is a DC torque motor (J175LYX04, Chengdu Precision Motor Factory, China), and a C01 motor drive module as the power amplifier (Chengdu Precision Motor Factory, China). The angle value of the pitch motor is set using an encoder (RA26BEA115B05F, Renishaw, UK). The digital controller is composed of a DSP (TMS320F2812) and FPGA (Altera EP1C12Q240). The DSP was used to store the control algorithm, which calculated the control and attitude signals. The FPGA was used to receive and transmit command signals to achieve logic control of the circuit.

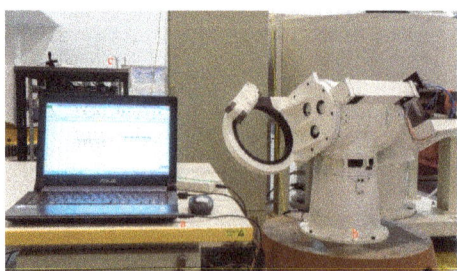

Figure 8. Ground-based laser communication servo turntable experimental platform: (a) master computer; (b) ground-based laser communication servo turntable; (c) digital controller (power amplifier).

The experimental structure is shown in Figure 1, and the input speed signals are fixed, sinusoidal, or closed-loop sweep frequency signals for the speed closed-loop performance test. The classical incremental digital PID controller, the chatter-reduced sliding mode controller of Ref. [22], and the proposed discrete sliding mode controller with disturbance compensation algorithm were tested for application as the speed loop controller. The parameters of the incremental PID controller were set as $k_p = 0.8$ and $k_i = 0.16$. The variable structure function coefficients of the proposed and Ref. [22] controllers were set to $\kappa T = 0.56$, and the other parameters were set in accordance with the sliding mode simulation setting.

4.1. Closed-Loop Experiment of Fixed Speed Signal

For the closed-loop experiment of fixed speed signal, the fixed speed signal of 0.1 mrad/s was input into the pitch axis control system of the ground-based laser communication servo turntable. Under the control of the proposed chatter-free sliding mode controller, the Ref. [22] chatter-reduced sliding mode controller, or the incremental digital PID controller, the motor operation was compared, as shown in Figure 9.

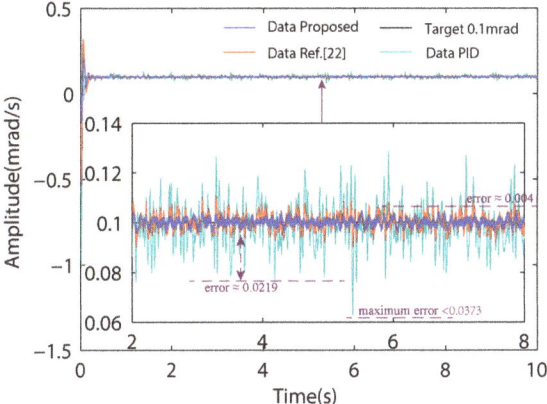

Figure 9. Contrast curve of rated speed 0.1 mrad/s experiment.

Comparison reveals that all three algorithms can guarantee operation of the motor at the target speed of 0.1 mrad/s. However, there are obvious nonlinear noises in the operation curve of the digital PID controller, the maximum absolute speed error of the PID control accuracy is 37.3 μrad/s, with an absolute error root of mean square (RMS) of 8.3 μrad/s and an average absolute error of 6.5 μrad/s. The nonlinear noise was obviously suppressed for both the proposed method and the method described in Ref. [22]. The maximum absolute speed error of the proposed SMC is 6.9 μrad/s, the absolute error RMS is 1.5 μrad/s and the average absolute error is 1.2 μrad/s. These values are much higher than those obtained for classical PID algorithm, and are slightly smaller than that obtained for the Ref. [22] algorithm, which exhibited a maximum error of 8.3 μrad/s, an average error of 2.3 μrad/s, and RMS of 2.9 μrad/s. Therefore, highest control accuracy was obtained for the proposed chattering-free algorithm, with slightly reduced the performance of the chatter-reduced algorithm described in Ref. [22] because of chattering.

4.2. Sinusoidal Guidance Experiments

For the sinusoidal guidance experiments, the input signal was set to sinusoidal 0.1 mrad/s·sin(2 Tk), for a sampling time $T = 0.001$. The velocity sinusoidal guidance contrast curve and sinusoidal guidance error curve are shown as follows.

According to the data presented in Figure 10, the pitch axis motor exhibits smooth sinusoidal operation driven by three controllers, but there are large torque ripples because of nonlinear factors such as friction and nonuniform torque. The data presented in Figure 11 show the absolute value of error. When the servo turntable stably tracks the sinusoidal guidance signal, the absolute error RMS of the PID algorithm is 13.7 μrad/s, the average absolute error is 9.9 μrad/s, and the maximum absolute error is 46 μrad/s. The absolute error RMS of the chatter-reduced algorithm described in Ref. [22] is 2.8 μrad/s, the average absolute error is 2.4 μrad/s, and the maximum absolute error is 15.6 μrad/s. The absolute error RMS of the proposed chatter-free algorithm is 2.3 μrad/s, the average absolute error is 2.1 μrad/s, and the maximum absolute error is 8.1 μrad/s. There is obvious high frequency noise caused by chattering using the algorithm from [22], but this noise is not present when using the proposed chatter-free algorithm. Therefore, to more clearly analyze the frequency distribution of noise, the sinusoidal guidance absolute error was next analyzed by single-sided Fourier transform in the frequency domain, as shown in Figure 12.

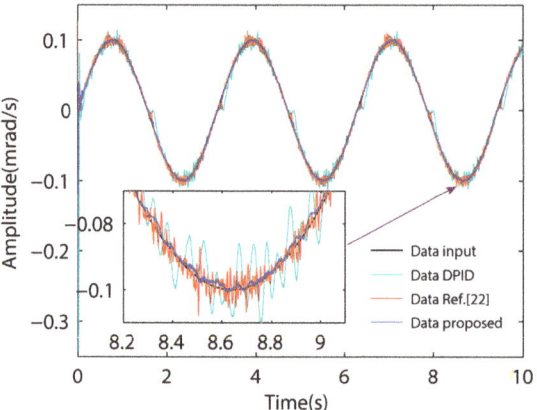

Figure 10. Velocity sinusoidal guidance contrast curve.

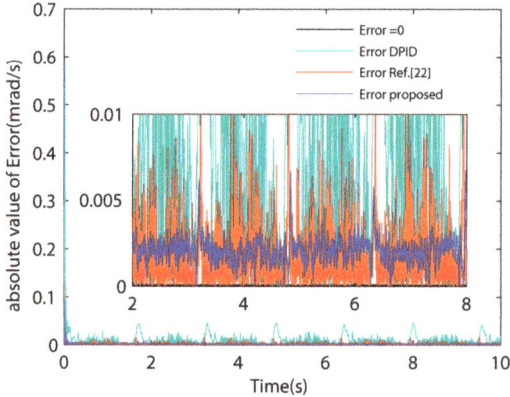

Figure 11. Absolute value of sinusoidal guidance error curve.

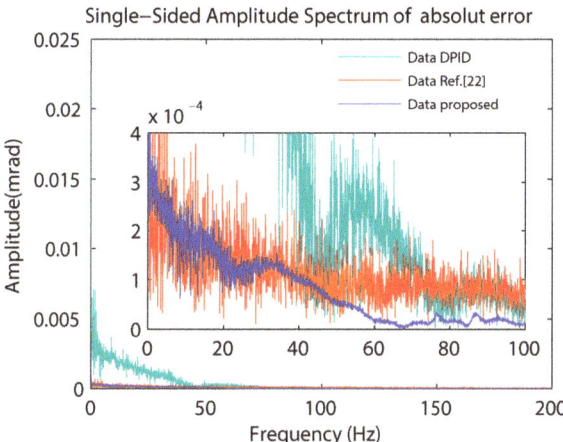

Figure 12. Frequency domain analysis of guidance absolute error curve.

According to the results presented in Figure 12, nonlinear disturbance can be effectively suppressed by both the proposed and the algorithm described in [22]. The PID linear controller exhibits limited ability to suppress medium and low frequency noise. There is obvious high frequency noise in the error curve caused by the sliding mode chattering of the Ref. [22] chatter-reduced algorithm, reducing the accuracy and the stability of the system. The high frequency noise is significantly reduced by the proposed chatter-free algorithm.

Meanwhile, it is noted that the biased error generated by the proposed algorithm means that the switching gain is not the optimal, but it is sufficient to prove the advancement of the proposed algorithm. If the switching gain is set to be larger, $\kappa T = 0.7$, the biased error can be eliminated as shown in Figure 13.

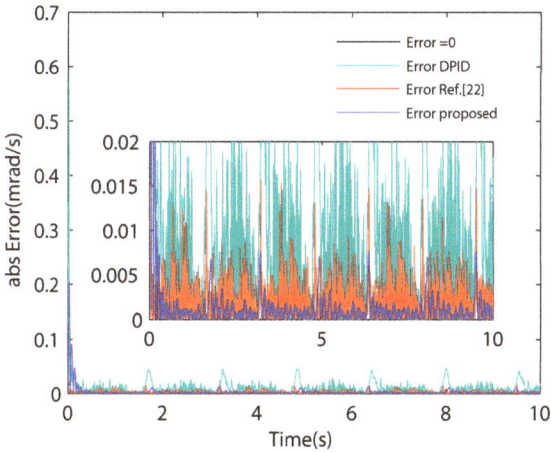

Figure 13. Absolute value of sinusoidal guidance error curve.

4.3. Closed-Loop Sweep Experiment

In the closed-loop sweep experiment, the input sweep signal was set as shown in Equation (1), the amplitude was set to 1 mrad/s, and the frequency range was set to 0.01 Hz–200 Hz. The closed-loop sweep curve of pitch axis and the frequency characteristic curve of control system are shown in Figures 14 and 15.

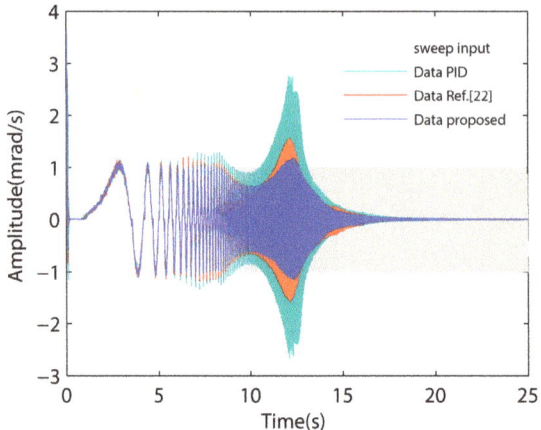

Figure 14. Closed-loop sweep curve of pitch axis.

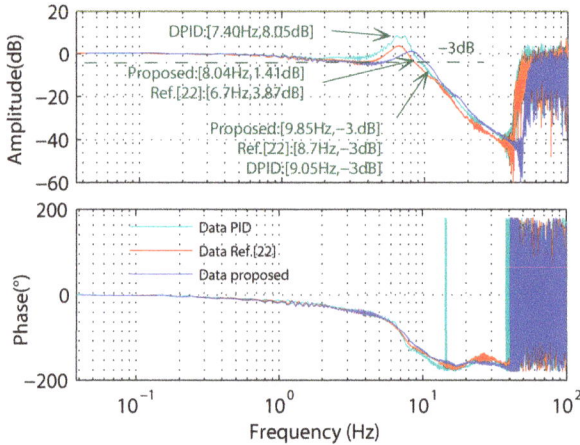

Figure 15. Contrast curve of the frequency characteristics of the control system.

According to the frequency characteristic curves shown in Figures 14 and 15, the closed-loop bandwidth of the proposed algorithm is ~9.85′Hz, slightly higher than that of the digital PID, ~9.05 Hz. The closed-loop bandwidth using the Ref. [22] algorithm is the smallest, ~8.7 Hz. Therefore, the dynamic performance of the proposed chatter-free algorithm is much better than that of the previously described chatter-reduced algorithm [22]. The resonance peak of the proposed algorithm is ~1.41 dB, which is much smaller than that of the previous algorithm [22] of ~3.87 dB. The resonance peak of the digital PID is the largest, ~8.05 dB. The smaller the resonance peak is, the better the stability of the system.

4.4. Summary

All results of the three experiments are shown in Table 1.

Table 1. Summary of experimental results.

Value	Fixed Speed Signal (μrad/s)			Sinusoidal Guidance (μrad/s)			Closed-loop Sweep	
Algorithms	RMS	Average	Maximum	RMS	Average	Maximum	Closed-loop Bandwidth	Resonance Peak
PID	8.3	6.5	37.3	13.7	9.9	4.6	9.05Hz	8.05dB
Ref. [22]	2.9	2.3	10.5	2.8	2.4	15.6	8.7Hz	3.87dB
Proposed	1.5	1.2	6.9	2.3	2.1	8.1	9.85Hz	1.41dB

5. Conclusion

In this study the pitch axis of the ground-based laser communication servo turntable was considered as the research object, and the frequency domain characteristic of the pitch axis was tested by the sweeping frequency method. Based on the frequency domain characteristic curve, the pitch axis system model was established using a classical model identification method and analyzing the influence of nonlinear disturbance on the system. To address the problems of model identification error and nonlinear disturbance, a novel chatter-free SMC algorithm with a disturbance compensator is proposed. The new algorithm is robust to the system model identification error, friction, and other nonlinear disturbances, and shows good stability by both theoretical calculation and simulation techniques. Finally, the digital PID controller, the chatter-reduced sliding mode controller [22], and the proposed chatter-free sliding mode controller were tested in closed-loop control experiments of pitch axis speed with rated, sinusoidal, or sweep input. The experimental results show that the proposed chatter-free algorithm exhibits higher control accuracy, stronger anti-interference ability, better frequency domain characteristics, and also suppresses chattering for an effective ground-based laser communication servo turntable control system.

The study of the overestimation and underestimation problems of the switching gain when the states are stable in the QSMD will be explored in our future work.

Author Contributions: Project administration, C.H.; resources, S.G.; writing—original draft, J.Z.; writing—review and editing, Y.L.

Funding: This research was funded by Research project of scientific research equipment of Chinese Academy of Sciences (2017); funded by CIOMP-Fudan University Joint Fund, grant number: Y8O732E.

Conflicts of Interest: The authors declare no conflict of interest.

References

1. Jiao, D.D.; Jing, G.; Jie, L.; Xue, D.; Xu, G.J.; Chen, J.P.; Dong, R.F.; Tao, L.; Zhang, S.G. Development and application of communication band narrow linewidth lasers. *Acta Phys. Sin.* **2015**, *64*. [CrossRef]
2. Mita, T.; Ohashi, J.; Venkatesan, M.; Marma, A.S.P.; Nakamura, M.; Plowe, C.V.; Tanabe, K. Acquisition and tracking control of satellite-borne laser communication systems and simulation of downlink fluctuations. *BMC Syst. Biol.* **2006**, *45*, 1–20.
3. Smith, R.J.; Casey, W.L.; Begley, D.L. Wideband laser diode transmitter for free-space communication. *Opt. Eng.* **1988**, *27*, 344–351. [CrossRef]
4. Jian, G.; Yong, A. Design of active disturbance rejection controller for space optical communication coarse tracking system. In Proceedings of the AOPC: Advances in Laser Technology & Applications, Beijing, China, 5–7 May 2015.
5. Shimizu, M.; Yamashita, T.; Eto, D.; Toyoshima, M.; Takayama, Y. Cooperative Control Algorithm of the Fine/Coarse Tracking System for 40Gbps Free-Space Optical Communication. In Proceedings of the AIAA International Communications Satellite System Conference, Ottawa, ON, Canada, 24–27 September 2012.
6. Kingsbury, R.; Cahoy, K. Development of a pointing, acquisition, and tracking system for a CubeSat optical communication module. In Proceedings of the SPIE LASE, San Francisco, CA, USA, 1–6 February 2015.

7. Li, M.Q.; Jiang, S.H. Design of Optimum Controller of APT Fine Tracking Control. System. *Adv. Mat. Res.* **2013**, *712–715*, 2738–2741. [CrossRef]
8. Min, Z.; Liang, Y. Compound tracking in ATP system for free space optical communication. In Proceedings of the International Conference on Mechatronic Science, Jilin, China, 19–22 August 2011.
9. Fei, L.; Yan, Z.; Duan, S.; Yin, J.; Liu, B.; Liu, F. Parameter Design of a Two-Current-Loop Controller Used in a Grid-Connected Inverter System With LCL Filter. *IEEE Trans. Ind. Electron.* **2009**, *56*, 4483–4491.
10. Grzesiak, L.M.; Tarczewski, T. PMSM servo-drive control system with a state feedback and a load torque feedforward compensation. *Int. J. Comput. Math. Electric. Electron. Eng.* **2013**, *32*, 364–382. [CrossRef]
11. Shen, W.H.; Chen, X.Q.; Pons, M.N.; Corriou, J.P. Model predictive control for wastewater treatment process with feedforward compensation. *Chem. Eng. J.* **2009**, *155*, 161–174. [CrossRef]
12. Carvajal, J.; Chen, G.; Ogmen, H. Fuzzy PID controller: Design, performance evaluation, and stability analysis. *Inf. Sci.* **2000**, *123*, 249–270. [CrossRef]
13. Li, H.X.; Zhang, L.; Cai, K.Y.; Chen, G. An improved robust fuzzy-PID controller with optimal fuzzy reasoning. *IEEE Trans. Syst. Man Cybern. Part B Cybern.* **2005**, *35*, 1283–1294. [CrossRef]
14. Mann, G.I.; Hu, B.G.; Gosine, R.G. Analysis of direct action fuzzy PID controller structures. *I IEEE Trans. Syst. Man Cybern. Part B Cybern.* **1999**, *29*, 371–388. [CrossRef]
15. Jung, J.W.; Leu, V.Q.; Do, T.D.; Kim, E.K.; Han, H.C. Adaptive PID Speed Control. Design for Permanent Magnet Synchronous Motor Drives. *IEEE Trans. Power Electron.* **2014**, *30*, 900–908. [CrossRef]
16. Kuc, T.Y.; Han, W.G. An adaptive PID learning control of robot manipulators. *Automatica* **2000**, *36*, 717–725. [CrossRef]
17. Li, S.; Xu, Y.; Di, Y. Active disturbance rejection control for high pointing accuracy and rotation speed. *Automatica* **2009**, *45*, 1854–1860. [CrossRef]
18. Zhu, E.; Pang, J.; Na, S.; Gao, H.; Sun, Q.; Chen, Z. Airship horizontal trajectory tracking control based on Active Disturbance Rejection Control. (ADRC). *Nonlinear Dyn.* **2012**, *75*, 725–734. [CrossRef]
19. Bo, E. Adaptive control stability, convergence, and robustness: Shankar Sastry and Marc Bodson. *Automatica* **1993**, *29*, 802–803.
20. Liu, Z.G.; Wu, Y.Q. Universal strategies to explicit adaptive control of nonlinear time-delay systems with different structures. *Automatica* **2018**, *89*, 151–159. [CrossRef]
21. Sun, T.; Pan, Y.; Zhang, J.; Yu, H. Robust model predictive control for constrained continuous-time nonlinear systems. *Int. J. Control* **2017**, *91*, 1–16. [CrossRef]
22. Ma, H.; Wu, J.; Xiong, Z. A Novel Exponential Reaching Law of Discrete-Time Sliding-Mode Control. *IEEE Trans. Ind. Electron.* **2017**, *64*, 3840–3850. [CrossRef]
23. Du, H.; Yu, X.; Chen, M.Z.Q.; Li, S. Chattering-free discrete-time sliding mode control. *Automatica* **2016**, *68*, 87–91. [CrossRef]
24. Abidi, K.; Xu, J.X.; She, J.H. A Discrete-Time Terminal Sliding-Mode Control. Approach Applied to a Motion Control. Problem. *IEEE Trans. Ind. Electron.* **2009**, *56*, 3619–3627. [CrossRef]
25. Li, S.; Du, H.; Yu, X. Discrete-Time Terminal Sliding Mode Control. Systems Based on Euler's Discretization. *IEEE Trans. Autom. Control* **2014**, *59*, 546–552. [CrossRef]
26. Xu, Q. Digital Integral Terminal Sliding Mode Predictive Control. of Piezoelectric-Driven Motion System. *IEEE Trans. Ind. Electron.* **2016**, *63*, 3976–3984. [CrossRef]
27. Mobayen, S. Adaptive Global Terminal Sliding Mode Control. Scheme with Improved Dynamic Surface for Uncertain Nonlinear Systems. *Int. J. Control Autom. Syst.* **2018**, *16*, 1692–1700. [CrossRef]
28. Levant, A. Homogeneity approach to high-order sliding mode design. *Automatica* **2005**, *41*, 823–830. [CrossRef]
29. Mobayen, S.; Tchier, F. A novel robust adaptive second-order sliding mode tracking control technique for uncertain dynamical systems with matched and unmatched disturbances. *Int. J. Control Autom. Syst.* **2017**, *15*, 1097–1106. [CrossRef]
30. Afshari, M.; Mobayen, S.; Hajmohammadi, R.; Baleanu, D. Global Sliding Mode Control. Via Linear Matrix Inequality Approach for Uncertain Chaotic Systems With Input Nonlinearities and Multiple Delays. *J. Comput. Nonlinear Dyn.* **2018**, *13*, 3. [CrossRef]
31. Mobayen, S.; Tchier, F. Composite nonlinear feedback integral sliding mode tracker design for uncertain switched systems with input saturation. *Commun. Nonlinear Sci. Numer. Simul.* **2018**, *65*, 173–184. [CrossRef]

32. Mobayen, S. Adaptive global sliding mode control of underactuated systems using a super-twisting scheme: an experimental study. *J. Vib. Control* **2019**, *25*, 2215–2224. [CrossRef]
33. Yongting, D.; Hongwen, L.I.; Tao, C. Dynamic analysis of two meters telescope mount control system. *Opt. Prec. Eng.* **2018**, *26*, 654–661.
34. Kwan, C. Further results on variable output feedback controllers. *IEEE Trans. Autom. Control* **2001**, *46*, 1505–1508. [CrossRef]
35. Cong, B.L.; Chen, Z.; Liu, X.D. On adaptive sliding mode control without switching gain overestimation. *Int. J. Robust Nonlinear Control* **2014**, *24*, 515–531. [CrossRef]
36. Roy, S.; Roy, S.B.; Kar, I.N. Adaptive–Robust Control. of Euler–Lagrange Systems With Linearly Parametrizable Uncertainty Bound. *IEEE Trans. Control Syst. Technol.* **2018**, *26*, 1842–1850. [CrossRef]
37. Huang, Y.J.; Kuo, T.C.; Chang, S.H. Adaptive sliding-mode control for nonlinear systems with uncertain parameters. *IEEE Trans. Syst. Man Cybern. B Cybern.* **2008**, *38*, 534–539. [CrossRef] [PubMed]
38. Hung, J.Y.; Gao, W.; Hung, J.C. Variable Structure Control: A Survey. *IEEE Trans. Ind. Electron.* **1993**, *40*, 21. [CrossRef]
39. Ma, H.; Wu, J.; Xiong, Z. Discrete-Time Sliding-Mode Control. with Improved Quasi-Sliding-Mode Domain. *IEEE Trans. Ind. Electron.* **2016**, *63*, 6292–6304. [CrossRef]
40. Zhang, X.; Sun, L.; Ke, Z.; Li, S. Nonlinear Speed Control. for PMSM System Using Sliding-Mode Control. and Disturbance Compensation Techniques. *IEEE Trans. Power Electron.* **2013**, *28*, 1358–1365. [CrossRef]
41. Niu, Y.; Ho, D.W.C.; Wang, Z. Improved sliding mode control for discrete-time systems via reaching law. *Control Theory Applic.* **2010**, *4*, 2245–2251. [CrossRef]
42. Su, W.C.; Drakunov, S.V.; Özgüner, Ü. An O(T2) boundary layer in sliding mode for sampled-data systems. *IEEE Trans. Autom. Control* **2000**, *45*, 482–485.
43. Morgan, R.Ü.Ö. A decentralized variable structure control algorithm for robotic manipulators. *IEEE J. Robot. Autom.* **1985**, *1*, 8. [CrossRef]
44. Roy, S.; Lee, J.; Baldi, S. A New Continuous-Time Stability Perspective of Time-Delay Control: Introducing a State-Dependent Upper Bound. Structure. *IEEE Control Syst. Lett.* **2019**, *3*, 475–480. [CrossRef]
45. Qu, S.; Xia, X.; Zhang, J. Dynamics of Discrete-Time Sliding-Mode-Control. Uncertain Systems With a Disturbance Compensator. *IEEE Trans. Ind. Electron.* **2014**, *61*, 3502–3510. [CrossRef]
46. Roy, S.; Roy, S.B.; Lee, J.; Baldi, S. Overcoming the Underestimation and Overestimation Problems in Adaptive Sliding Mode Control. *IEEE/ASME Trans. Mechatron.* **2019**. [CrossRef]
47. Roy, S.; Roy, S.B.; Kar, I.N. A New Design Methodology of Adaptive Sliding Mode Control for a Class of Nonlinear Systems with State Dependent Uncertainty Bound. In Proceedings of the 15th International Workshop on Variable Structure Systems (VSS), Graz, Austria, 9–11 July 2018.
48. Sharma, N.K.; Roy, S.; Janardhanan, S. New design methodology for adaptive switching gain based discrete-time sliding mode control. *Int. J. Control* **2019**. [CrossRef]
49. Sharma, N.K.; Roy, S.; Janardhanan, S.; Kar, I.N. Adaptive Discrete-Time Higher Order Sliding Mode. *IEEE Trans. Circuits Syst. II Express Br.* **2019**, *66*, 4.

© 2019 by the authors. Licensee MDPI, Basel, Switzerland. This article is an open access article distributed under the terms and conditions of the Creative Commons Attribution (CC BY) license (http://creativecommons.org/licenses/by/4.0/).

Article

Numerical Investigation on Unsteady Separation Flow Control in an Axial Compressor Using Detached-Eddy Simulation

Mingming Zhang [1] and Anping Hou [2],*

[1] College of Applied Sciences, Beijing University of Technology, Beijing 100124, China
[2] School of Energy and Power, Beihang University, Beijing 100191, China
* Correspondence: houap@buaa.edu.cn; Tel.: +86-010-8231-6624

Received: 2 July 2019; Accepted: 8 August 2019; Published: 12 August 2019

Abstract: Unsteady excitation has proved its effectiveness in separation flow control and has been extensively studied. It is observed that disordered shedding vortices in compressors can be controlled by unsteady excitation, especially when the excitation frequency coincides with the frequency of the shedding vortex. Furthermore, former experimental results indicated that unsteady excitation at other frequencies also had an impact on the structure of shedding vortices. To investigate the impact of excitation frequency on vortex shedding structure, the Detached-Eddy Simulation (DES) method was applied in the simulation of shedding vortex structure under unsteady excitations at different frequencies in an axial compressor. Effectiveness of the DES method was proved by comparison with URANS results. The simulation results showed a good agreement with the former experiment. The numerical results indicated that the separation flow can be partly controlled when the excitation frequency coincided with the unsteady flow inherent frequency. It showed an increase in stage performance under the less-studied separation flow control by excitation at a certain frequency of pressure side shedding vortex. Compared with other frequencies of shedding vortices, the frequency of pressure side shedding vortex was less sensitive to mass-flow variation. Therefore, it has potential for easier application on flow control in industrial compressors.

Keywords: unsteady flow control; vortex dynamics; separation flow; Detached-Eddy Simulation

1. Introduction

The flow separation at the compressor trailing edge is unavoidable due to the high adverse pressure gradient, especially in modern gas turbines which have an increased loading than in the past. Investigations in turbomachinery have shown that the separation vortex is one of the main sources of loss at near stall point, and it can reduce the stage efficiency as well as the stall range. Therefore, methods aiming to control and decrease the separation have been extensively studied.

Analysis of separation flow started in the 1960s with experiments in airfoils with a large attack angle [1]. In the primary stage, passive control methods were analyzed, and then the active controls of separation flow were considered. Active control methods such as suction blowing on the blade surface were confirmed to have a better capacity than passive control [2]. Further development of the active control method was unsteady excitation control. Dynamic excitation control methods can provide more performance improvement with less injected mass flow than the constant excitation control [3–6], because the separation flow structure at large attack angles is inherently unsteady. To be precise, the structure of vortex shedding is physically similar to the Karman vortex street structure [7,8].

Under different types of unsteady excitation control methods, the separated flow can be controlled to be in order, including unsteady suction blowing, upstream wake excitation, oscillating guide vanes, and total pressure fluctuation in the incoming flow by trumpets [9–11]. In an axial compressor stage, it had been increased to a maximum of 40% of the stall margin by dynamic air injectors in the experiment. The mass-flow rate in these air injectors was less than 1% of the compressor flow rate [10]. Koc applied plasma actuators to provide an active separation control on a bluff body, and a "locked-on" effect was shown when the excitation frequency approached the natural vortex shedding frequency [11]. Experimental [12] and numerical analysis [13] in an axial compressor test rig demonstrated that the stall boundary, pressure rise, and near-stall efficiency were all increased with unsteady excitation by an appropriate frequency. The maximum enhancements under separation control were 5.4% increasement in total pressure rise, 5.5% increasement in the efficiency of the compressor, and 30.7% increasement in the relative stall margin. In these studies, the maximum enhancement on stage performance as well as stall range could be acquired when the dynamic excitation frequency was in coincidence with the vortex shedding frequency.

However, there are few investigations that focus on the mechanism of non-vortex-shedding frequency excitation, which was also proved to have a remarkable effect on cascade performance [12]. For the sake of explanation on the experimental phenomenon and improvement of the current unsteady separation control theory, the Detached-Eddy Simulation (DES) method was applied in this paper as a high-fidelity numerical simulation tool. The same low-speed axial compressor test rig as used by Li [12] was applied. In order to analyze the impact of excitation on vortex shedding structure, this study analyzed the performances under unsteady excitations at different frequencies in the compressor model.

Numerical results that were confirmed by previous experimental works were composed of three classifications of separation flow control under unsteady excitations: Vortex-shedding control (VSC), suction-side separation vortex control (SSVC), and pressure-side separation vortex control (PSVC). It could be concluded that these separation control methods were effective when the excitation frequency coincided with the inherently unsteady flow frequency. In this test case, the enhancement of stage performance can be achieved at the whole working range by excitation at a certain frequency of pressure side shedding vortex. Compared with traditional unsteady separation control, the new unsteady excitation method may be easy to apply in industrial compressors and has a great potential.

2. Numerical Approach

The DES method based on the SST k-ω turbulence model [14–16] was used in this study. Detached-Eddy Simulation (DES) is a hybrid large eddy simulation (LES)/RANS method. The concept of DES is to calculate the boundary layer by a RANS turbulence model and to switch the turbulence model into a LES mode in detached regions. Compared with using LES in the global computational domain, the DES method not only ensures the calculation accuracy, but also improves the calculation efficiency and saves the calculation cost. Because of the balance of computation resources and fidelity, the DES method had been widely applied in capturing vortex-shedding structures in turbomachines [17–19]. In the DES calculation, the LES model is activated in the region where the turbulence length predicted by the RANS model is larger than the local mesh scale. The turbulence length scale in an SST k-ω model and in the DES model are shown in Equations (1) and (2), respectively:

$$l_{k-\omega} = k^{0.5}/(\beta \times \omega), \quad (1)$$

$$l_{DES} = \min(l_{k-\omega}, C_{DES}\Delta). \quad (2)$$

The dissipative term of the k-transport equation, the only modified term in the DES model, is transferred to Equation (3):

$$D^k_{DES} = \rho k^{2/3}/l_{DES}. \quad (3)$$

The commercial CFD software ANSYS CFX was used for the DES calculation. The advection term was discretized by the High Resolution Scheme, and the Second Order Backward Euler method was applied in the discretization of transient terms. According to the experimental results, the highest frequency of interest in this unsteady phenomenon of this stage would be less than 2500 Hz, so there must be more than 40 time steps in a blade computing pitch. In the unsteady simulation, every blade passing pitch was divided into 105 time steps with each time step including a maximum of 50 inner interaction steps. In this paper, total performance parameters and separation positions were dealt with time-averaged results recorded after a stable quasi-periodic flow state achieved. Fast Fourier Transformation analysis was based on the data collected in over 200 blade-passing pitch cycles.

3. Compressor Test Model

The rotor stage of a low-speed axial compressor test rig shown in Figure 1 in Beihang University [12] was applied in this analysis. The test rig consists of a 13-blade stator row and a 19-blade rotor row with the C4 profile (Figure 2). At the mid-span, the rotor blade has a chord length of 52 mm, a solidity of 0.605, a stagger angle of 34.49 degree, and an outlet angle of 41.99 degree. The design speed of the test rig is 3000 rpm, and the mass flow is 2.40 m^3/s at the design point with a total pressure rise equal to 1500 Pa. A detailed introduction of the test rig is presented in Li [12] and Zhang [20]. In this study, a blade cascade extended from the rotor's mid-span profile was applied as the compressor test model. The length in the vertical direction was set as 20% span according to the literature.

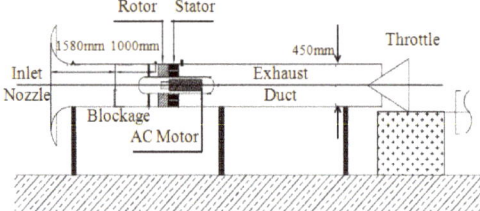

Figure 1. Compressor test rig.

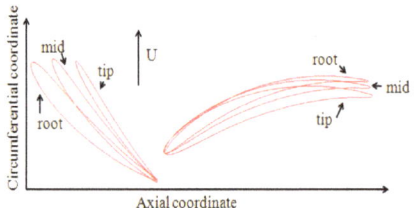

Figure 2. Blade sections at three spanwise locations.

An axial total pressure fluctuation in the inlet boundary was added as the source of unsteady excitation, which was set to physically imitate the periodic sound excitation generated by pneumatic speakers. The schematic plot of the sound generator and the figure in the experiment by Li [12] are shown in Figure 3.

To imitate the inlet unsteady excitation, the total pressure at inlet condition was set as Equation (4), where P_{inlet} was the total pressure on the inlet boundary, t was the physical time during the unsteady simulation, A_m was the maximum excitation amplitude of pressure, and f was the frequency of the excitation. Total temperature and velocity at inlet directions were kept constant in the computations.

$$P_{inlet} = 101325 + A_m \times \sin(f \times 2\pi \times t). \tag{4}$$

The maximum excitation amplitude of pressure was set to be 600 Pa in the experiment, and the maximum flow control effect under unsteady excitation could be obtained at that amplitude. In the computations, the maximum amplitude was extended to 1000 Pa. The computation boundary conditions were set up based on the parameters of a near-stall working state with a total pressure rise over 1600 Pa in the calculation. Periodic boundary conditions were set between the up and down side of the computation domain as well as between the left and right side. The total pressure boundary condition was used at the inlet and the averaged static pressure boundary condition was set to the outlet.

 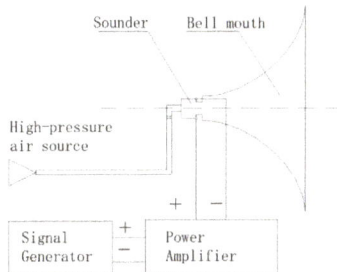

Figure 3. The schematic plot of the sound generator in the experiment.

Since DES results can be sensitive to grid resolution, four sets of mesh solutions with the same topology were studied to guarantee the accuracy of results. The Y-plus of the meshes near the blade surface were kept smaller than one in all four cases. The separation positions were chosen as the reference of shedding vortex structure (identification of separation position would be described in Section 4), and the results with the four meshes were shown in Figure 4. When the amount of mesh grids exceeded 7.36 million, the simulated separation positions were beginning to converge. Therefore, the mesh solution with 7.36 million mesh grids was chosen as the basic solution, which contained 347 nodes in the axial direction, 134 nodes in the pitchwise direction, and 125 nodes in the spanwise direction near the blade region.

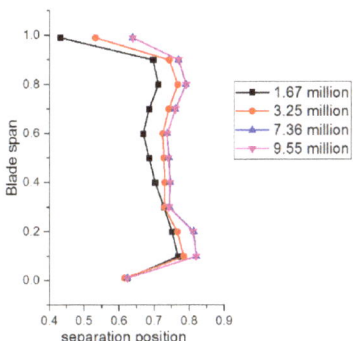

Figure 4. Separation positions with different grids.

From the numerical results, it could be concluded that the separation positions were varying less than 5% between 30% and 70% of the blade span, which indicated that the vortex shedding flow could be possibly considered as a quasi-3D phenomenon in circumferential direction at the middle span of this blade. In order to exclude the influence of hub/shroud secondary flow and focus on the separation flow, a 1 mm-thick curved slice at 50% span of the rotor blade was applied as a quasi-3D mode. Reliability of the quasi-3D model on the simulation of shedding vortex structure by high-fidelity simulation was verified by Zhao [21]. The quasi-3D model shared the same mesh grid distribution in the B2B section with the 3D mesh grid. The free slip boundary conditions were applied on the top and bottom surfaces of the quasi-3D model.

4. Results

4.1. Analysis on the Inherent Unsteady Flow Structure

The natural type of unsteady separation flow structure was analyzed as a reference state in this section. The DES simulations have been performed at different mass-flow coefficients. Characteristic lines of the quasi-3D model were shown in Figure 5. The mass-flow coefficient was defined as the ratio of averaged axial velocity to averaged total velocity. The pressure coefficient ΔCp was defined as Equation (5):

$$\Delta Cp = \Delta p / \left(\frac{1}{2}\rho U_m^2\right) \tag{5}$$

where Δp was pressure raise, U_m was the rotational speed at midspan, and ρ was the density.

In continuum mechanics, the vorticity $\vec{\omega}$ is a pseudovector field defined as the curl of the flow velocity $\vec{u}$ vector. It describes the local spinning motion of a continuum near some point. The definition can be expressed by the vector analysis formula:

$$\vec{\omega} \equiv \nabla \times \vec{u} \tag{6}$$

where ∇ is the del operator. The vorticity of a two-dimensional flow is always perpendicular to the plane of the flow, and therefore can be considered a scalar field. In this paper the working point A near the stall was selected as the research working point to investigate unsteady separation flow structure. The instantaneous vorticity contour at the working point A was presented in Figure 6. As the graph showed, the separation on the suction side started near the middle length of chord, and then the separation vortex spread downstream and induced a shedding vortex at the trailing edge of blade.

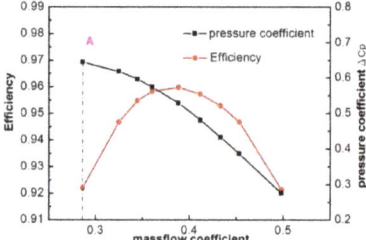

Figure 5. Characteristic lines of the quasi-3D model.

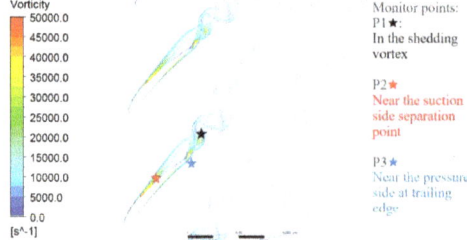

Figure 6. Instantaneous vorticity contour at working point A.

The vortex shedding phenomenon at the blade trailing edge, which was physically similar to the Karman vortex street, could be described as that separation vortices on the two sides of the blade combined with each other at the trailing edge with different vorticity directions, and then shed off in pairs when the balance of vorticity was obtained. Analyzed from this physical mechanism, three different types of unsteady flows existed in the flow field: Shedding vortex, suction-side separation vortex, and pressure-side separation vortex. To obtain the characteristic frequencies of the three inherently unsteady flows, the analysis on vorticity fluctuations on monitor points P1, P2, and P3 (marked in Figure 6) was conducted. The vorticity fluctuations in the time domain and corresponding spectrum analysis at the three monitor points at working point A are shown in Figure 7.

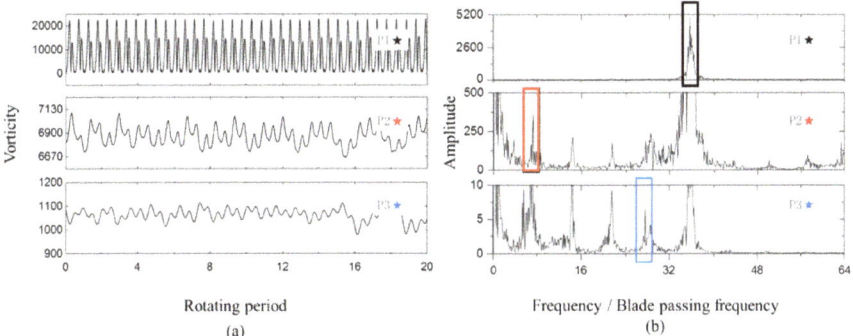

Figure 7. Vorticity fluctuations in time domain (**a**) and frequency domain (**b**) at monitor points.

A dominant frequency can be identified in all three locations, which was 36.5 times the rotation frequency (RF). The maximum amplitude at this frequency was located in the region of shedding vortices, which signified that it was the frequency of shedding vortex F_{shed}. The dominant frequency marked by the red square reached its maximum amplitude near the separation point on the suction side, which indicated the suction-side vortex separated at 7.12 times the rotation frequency (F_{ss}, frequency of suction-side vortex shedding). The dominant frequency at 27.6 times the rotation frequency (marked by the blue square) reached its maximum amplitude near the trailing edge in the pressure side. Based on the physical mechanism, this frequency was the inherent frequency of the pressure-side separation vortex named as F_{ps}.

Variations of the frequencies to mass-flow coefficients are shown in Figure 8 for the three types of unsteady separation flows. It could be concluded that the vortex shedding frequency F_{shed} varied more than 100% with mass-flow coefficients, while the frequencies F_{ps} and F_{ss} were insensitive to the variation of working conditions.

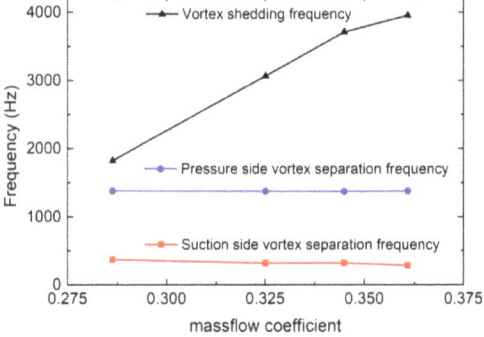

Figure 8. Variation of inherent frequencies with mass flow.

It had to be mentioned that the vortex shedding frequency measured by experiments [12] at the near stall point was 36.8 times the rotation frequency (36.8 RF). The agreement in F_{shed} between numerical and experimental approaches indicated that the method used in this research was sure in capturing the vortex shedding structure of separation flow. Thus, it was chosen as working point A to investigate separation flow characteristics in this paper. And the experimental results at the near stall point in Reference [12] are shown in the following section to verify the phenomenon shown in numerical studies.

4.2. Identification of Separation Position

Time-averaged wall shear stress distribution on the rotor blade surface at working point A is shown in Figure 9. The separation position on the suction side was identified where the wall shear stress on the suction side switched from positive to negative. At working point A, the flow on the suction side separated at 43.5% of the chord length from the leading edge. Figure 10 shows the variation of separation positions to the mass-flow coefficients. The *y*-axis label of the separation position was the chordwise of the blade. The vortices separated almost at the trailing edge of the blade in the large flow rate. The separation on the suction-side separation emerged distinctly when the mass-flow coefficient was less than 0.42. And the separation position moved towards the leading edge with the decrease of the mass-flow coefficient because of the increase in the blade attack angle.

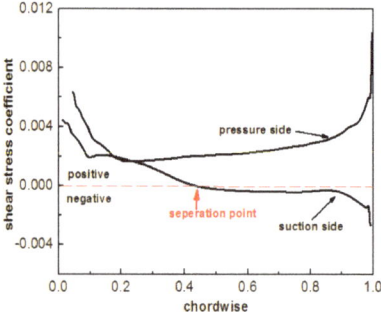

Figure 9. Wall shear stress distribution on the rotor blade surface.

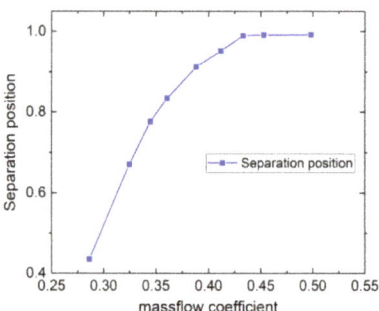

Figure 10. Variation of the separation position with mass flow.

To investigate the loss generated by separation flow quantitatively, 21 monitor points of total pressure were distributed uniformly in the circumferential direction at the middle span of rotor exit. With the decrease of the mass-flow coefficient, the area of total pressure loss was in expansion with the increase of the minimum amplitude of the pressure coefficient, as shown in Figure 11. The phenomenon was consistent with conclusions from previous studies that the loss generated by separation flow increased with the decrease of the mass-flow coefficient. In the next section, the changes of loss in distribution under unsteady excitations will be presented.

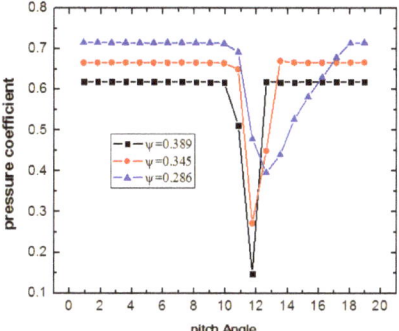

Figure 11. Pressure distribution at pitchwise to different mass-flow coefficients.

4.3. Selection of Parameters of Excitation

Unsteady excitations can cause an influence on mass-flow rate as well as pressure raise and efficiency, which makes it difficult to make a comparison between the effects under different excitation frequencies. To make a quantitative comparison, separation position on the suction side was chosen as the characteristic parameter for evaluating the effectiveness of separation control. Excitation, which had positive effects on stage performance, would translocate the separation position to the trailing edge. It meant that the separated flow was restructured, and delayed. Total pressure distributions of the blade downstream and vorticity contours were also presented to analyze the transformation of pressure loss regions. Investigations in this section were all carried at working point A, and all excitation frequencies were set as integral multiples of RF.

According to previous studies [9–13], excitation at the vortex shedding frequency had a remarkable effect on control of the flow field. Therefore, excitations with different amplitudes at the same frequency of 36 times the RF were first investigated to analyze the effect of excitation amplitude A_m on the control of the flow field.

As indicated in Figure 12, the separation position kept moving to the trailing edge with the increase in excitation amplitude A_m. And the efficiency under different excitations were all increased. Based on the results, the positive influence on the separation control and efficiency enhancement can be obtained at excitations with computed amplitudes. But there existed a peak efficiency point that indicated the maximum benefit was gained for the performance by the excitation condition. In the following research, 400 Pa was chosen as the amplitude of unsteady excitation because of the maximum efficiency improvement in this condition.

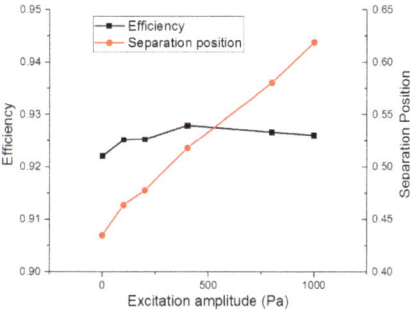

Figure 12. Influence of excitation amplitude on efficiency and separation position.

Then the frequency of excitation was varied from 0 to 50 times the RF, which contained 27 quantities. And the effects of these frequencies on the control of separation vortex structure are shown in Figure 13. From the figure, it was concluded that the positive effects on the vortex shedding were obtained when the frequencies of excitation were closed to harmonic frequencies of F_{shed} (marked by the black line) and harmonic frequencies of F_{ps} (marked by the blue line). Excitations with the frequencies near F_{ss} and its harmonic frequencies (marked by the red lines) could cause a dominant negative effect on the separation control. These phenomena can be verified by experimental results on the variation of relative efficiency at the near stall point [12]. For the difference of frequencies and effects between the experimental and computed results, there were two possible causes. Because of the limitation of CFD in simulating the turbulent flow, the predicted structure of vortices still had a little discrepancy in the experiment. Under excitation, the response of the flow field had also made a certain change. Corresponding to the relative efficiency involved in the experiment, the separation position change was used to represent the positive and negative effects in numerical simulation. Compared with the description of efficiency gain, this method may be better in characterizing the changes of flow-field structure. Despite these inaccuracies, the curves calculated were overall a close match to the shape of the experimental data.

Corresponding to excitations with frequencies of three harmonics, the influences of unsteady excitations on vortex shedding flow were divided into three different types: Vortex-shedding control (SVS), suction-side separation vortex control (SSVS), and pressure-side separation vortex control (PSVS). The three types of separation flow control methods would be discussed separately in the following section.

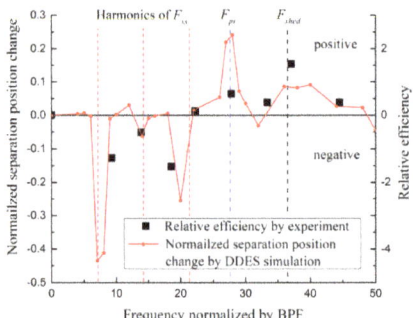

Figure 13. Effect of unsteady excitation at near stall point.

4.4. Response of Separation Flow under Unsteady Excitations

For the traditional method to control the separation of flow field, it was common to excite the separated flow with the frequency of shedding vortices. Under this condition the working range and total performance of the stage also had been improved, which had been widely investigated by researchers [9–13]. In this section, the excitation by the frequency of shedding vortices on the separation flow was also conducted. As Figure 13 indicated, positive effects were indeed obtained on the control of separation flow at a range of ±12.5% around F_{shed}.

As shown in Figure 14, the structure of the vortex under excitation was similar to the initial state, which was presented in Figure 6 without excitation. From the total pressure distribution in Figure 15, the separation zone and total pressure loss were all reduced with the unsteady excitation at the frequency of 36 times the RF. The mechanism of this phenomenon was that periodic excitation rectified the separation flow and injected energy into the vortex shedding flow. While increasing the strength of shedding vortices, the separation flow on the suction side of blade was suppressed. Then the pressure loss was also reduced. Inhabitations of the suction-side separation and rectification of the vortex shedding structure under SVC had been optimized. But along with the variation of flow rate, the characteristic frequency of wake shedding vortices also changed. It was very difficult to accurately excite the flow field with the frequency of wake shedding vortices at all operating points.

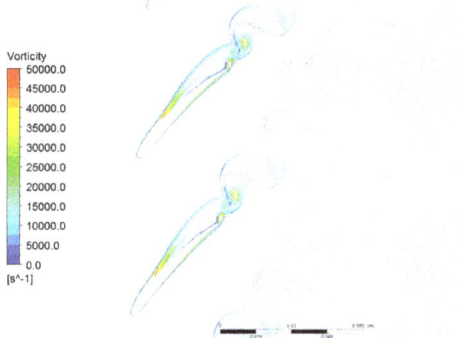

Figure 14. Instantaneous vorticity contour with excitation at 36 times the RF.

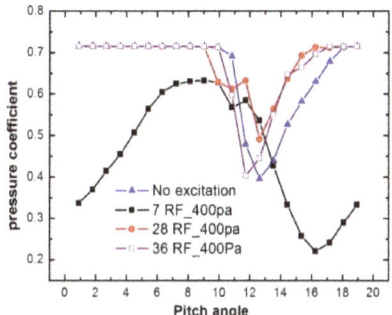

Figure 15. Pressure loss comparison with and without excitations.

For the suppression of suction-side separated flow, the effectiveness of the excitations with the frequency of suction-side separation vortices was checked as the next step. The vortex frequency of the suction surface was not the dominant frequency of the flow field, as indicated in Figure 7. The vorticity of the suction-side separation vortex was about an order magnitude lower than the vorticity of shedding vortices. However, after the excitation at ±5% of the range around the harmonics of F_{ss}, it was shown as a remarkable negative separation control effect, as expressed in Figure 13. In order to investigate the mechanism for the negative effect, the computation example with the excitation at the frequency of seven times of the RF was analyzed here. The instantaneous vorticity contour under excitation was presented in Figure 16. With the excitation, the vortex on the suction side separated forward to the leading edge of blade. And the structure of the vortices became disordered, and the flow field deteriorated markedly. The separation flow increased the flow instability and could even cause stall flow in the cascade. Loss distribution in Figure 15 showed that the pressure coefficient at the cascade exit dropped sharply, and the average pressure loss downstream of the blade increased significantly. Excitation that coincided with the eigenfrequency of suction-side separation transfused energy to the separation flow and thus led the separation position forward. This could cause the reduction of the working range and increase the loss in the flow field.

For the analysis above, it was concluded that the method with activating the flow field may not always have a positive effect. The effectiveness depended on the response of the separation flow field to excitations. Next, the excitations with the frequency of pressure side separation vortices (28 times the RF) were conducted to check whether it worked or not. From Figure 13, it could be seen that unsteady excitations at frequencies around the F_{ps} ± 7.5% range had a magnificent effect on the separation control. Instantaneous separation vortex structure under PSVC (Figure 17) was very different from its inherent structure (Figure 6). The separation region of suction side vortices had been decreased, and the strength of the shedding vortices also had been weakened. The comparisons of total pressure distribution with and without excitations in Figure 15 indicated that the core area of pressure loss moved to the pressure side of the blade, while the loss zone on the suction side decreases significantly with the unsteady excitation. With input of the excitation energy, the strength of the pressure-side separation vortices increased under the excitation. And because of the weak influence of the suction separation flow on the vortex shedding, the flow field deviated the vortex shedding to the pressure side and accelerated the shedding procedure under the increasing vorticity in pressure-side separation. As a result, the suction-side separation flow was restrained with a reduction of pressure loss. Even with the relative increase of pressure loss in the pressure side, the stage performance was still enhanced.

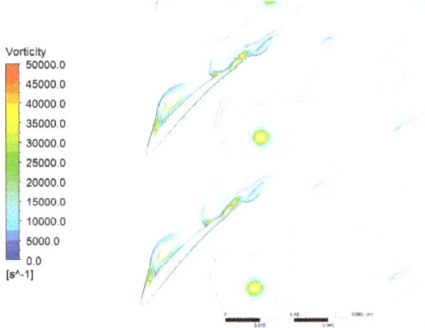

Figure 16. Instantaneous vorticity contour with excitation at 7 times the RF.

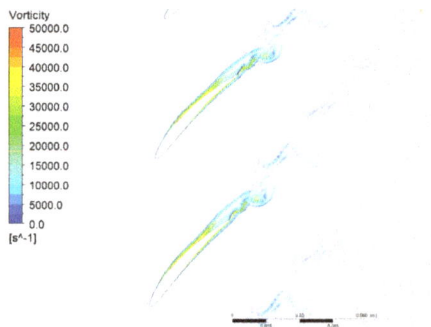

Figure 17. Instantaneous vorticity contour with excitation at 28 times the RF.

5. Discussion on the Application of PSVC

It could be pointed out that the maximum efficiency enhancement was obtained under the excitation with the frequency of shedding vortex from Figure 13. However, the frequency of unsteady excitation must be altered along with mass-flow coefficients to achieve the positive effects under vortex shedding control. As indicated in Figure 8, the vortex shedding frequency was sensitive to the mass-flow rate. The traditional separation control method, which was based on VSC, demanded the variation of frequency at excitation, altering with the working points. The effect of separation control was mainly dependent on the frequency of excitation. So, it was very difficult to apply in an industry application, which required alternating the frequency of wake shedding vortices needed at all operating points.

Meanwhile, the variations of the other two vortex separation flow frequencies F_{ps} and F_{ss} were relatively much smaller than the variation of the vortex shedding frequency to the change of working conditions, as shown in Figure 6. Though the effective range of PSVC was less than that of VSC, it was highly possible to obtain a performance enhancement at the whole working range under a certain excitation frequency that equaled the effect under excitation of frequency F_{ps}.

To verify this conjecture, the computations with the excitation at the frequency of 28 times the RF were conducted with DES. And the effects on the stage performance under excitation were presented in Figure 18. It was shown that under the type of separation control of PSVC, separation positions on the suction side of blade were delayed almost in the whole working range. And the cascade performance was also enhanced. As a less investigated kind of separation control method, the PSVC was indeed able to obtain stage performance improvement at the whole working range with a certain frequency. This result turned out to be a case that it was easier to apply in industries by the PSVC than that of the traditional unsteady separation control method.

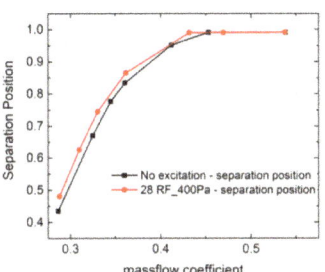
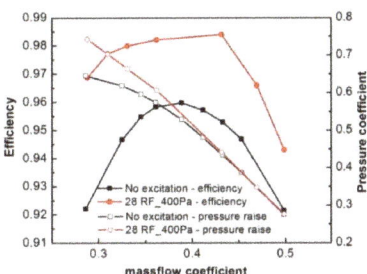

Figure 18. Variation of separation position on the suction side (**left**) and total performance (**right**) under pressure-side vortex control (PSVC).

6. Conclusions

In this paper, the control of the vortex shedding structure in a low-speed axial compressor model was analyzed using the DES method. Three different types of vortex control methods under unsteady excitations were classified by numerical results and previous experimental data.

The transformation of separation vortices in this test case indicated that, besides the traditional unsteady separation control method, excitation at other inherent separation frequencies could also have remarkable effects on the control of separation flow. The classifications of SSVC and PSVC effectively had been complementary to the separation control theory and demonstrated the potential capability on control of an unsteady vortex structure with neglected eigen frequencies. Specifically, unsteady excitations at frequencies around F_{ps} had a significant separation control effect, while excitations at frequencies around F_{ss} reduced the stability of separation flow.

The performance improvement obtained by the traditional unsteady control required the alteration of excitation frequencies at different working points, because the vortex shedding frequency was sensitive to the mass-flow rate. Meanwhile, the stage performance enhancement by PSVC, which had not been widely studied before, could be achieved at the whole working range by excitation at a certain frequency in the compressor test case.

Author Contributions: M.Z. conceived and designed the numerical simulation; M.Z. performed the simulation and analyzed the data; A.H. supervision and administration; M.Z. writing the original draft; A.H. reviewed and edited the final draft.

Funding: This research was funded by the National Science and Technology Major Project, grant number 2017-II-0009-0023.

Conflicts of Interest: The authors declare no conflict of interest.

Abbreviations

The following abbreviations are used in this manuscript:

A_m	amplitude of unsteady excitation
$C_{DES}\Delta$	local mesh spacing
D^k_{DES}	dissipation rate in the equation for the turbulent kinetic energy in DES model
F	frequency of unsteady excitation
F_{ps}	frequency of pressure side vortex shedding
F_{shed}	frequency of trailing edge vortex shedding
F_{ss}	frequency of suction side vortex shedding
P_{inlet}	total pressure at the stage inlet boundary
$l_{k-\omega}$	local turbulent length scale in the SST $k-\omega$ turbulence model
l_{DES}	local turbulent length scale in DES model
k	turbulent kinetic energy
t	physical time during the unsteady simulation
β	SST Closure Constant
ρ	density
Ψ	mass-flow coefficient
ω	specific rate of dissipation
Δp	pressure raise
U_m	rotational speed at midspan
$\Delta C p$	pressure coefficient, $\Delta p / (\frac{1}{2}\rho U_m^2)$
Δ	averaged wall shear stress
C_δ	shear stress coefficient, $\delta / (\frac{1}{2}\rho U_m^2)$

References

1. Ringleb, F.O. Separation control by trapped vortices. *Bound. Layer Flow Control* **1961**, *1*, 265–294.
2. Sinha, S.K. Active flexible walls for efficient aerodynamic flow separation control. In Proceedings of the AIAA-99-3123, 17th Applied Aerodynamics Conference, Las Vegas, NV, USA, 28 June–1 July 1999.
3. Goldstein, M.E.; Hultgren, L.S. Boundary-Layer Receptivity to Long-Wave Free-Stream Disturbances. *Annu. Rev. Fluid Mech.* **1989**, *21*, 137–166. [CrossRef]
4. You, D.; Moin, P. Active control of flow separation over an airfoil using synthetic jets. *J. Fluids Struct.* **2008**, *24*, 1349–1357. [CrossRef]
5. Wang, C.; Tang, H.; Duan, F.; Yu, S.C. Control of wakes and vortex-induced vibrations of a single circular cylinder using synthetic jets. *J. Fluids Struct.* **2016**, *60*, 160–179. [CrossRef]
6. Braza, M.; Hourigan, K. Unsteady separated flows and their control. *J. Fluids Struct.* **2008**, *24*, 1151–1155. [CrossRef]
7. Ericsson, L.G.J. Karman Vortex Shedding and the Effect of Body Motion. *AIAA J.* **1980**, *18*, 935–944. [CrossRef]
8. Ning, W.; He, L. Some modelling issues on trailing edge vortex shedding. In Proceedings of the 99-GT-183, ASME 1999 International Gas Turbine and Aeroengine Congress and Exhibition, Indianapolis, Indiana, 7–10 June 1999; American Society of Mechanical Engineers: New York, NY, USA, 1999.
9. Greenblatt, D.; Wygnanski, I.J. The control of flow separation by periodic excitation. *Prog. Aerosp. Sci.* **2000**, *36*, 487–545. [CrossRef]
10. Koc, I.; Britcher, C.; Wilkinson, S. Investigation and active control of bluff body vortex shedding using plasma actuators. In Proceedings of the AIAA 2012-2958, 6th AIAA Flow Control Conference, New Orleans, LA, USA, 25–28 June 2012.
11. Day, I.J. Active Suppression of Rotating Stall and Surge in Axial Compressors. *J. Turbomach.* **1993**, *115*, 40–47. [CrossRef]
12. Li, Z.-P.; Li, Q.-S.; Yuan, W.; Hou, A.-P.; Lu, Y.-J.; Wu, Y.-L. Experimental study on unsteady wake impacting effect in axial-flow compressors. *J. Sound Vib.* **2009**, *325*, 106–121. [CrossRef]
13. Zheng, X.-Q.; Zhou, X.-B.; Zhou, S. Investigation on a Type of Flow Control to Weaken Unsteady Separated Flows by Unsteady Excitation in Axial Flow Compressors. *J. Turbomach.* **2005**, *127*, 489–496. [CrossRef]
14. Spalart, P.R. Comments on the feasibility of LES for wings, and on a hybrid RANS/LES approach. *Adv. DNS/LES* **1997**, *1*, 4–8.

15. Spalart, P.R.; Deck, S.; Shur, M.L.; Squires, K.D.; Strelets, M.K.; Travin, A.; Shur, M. A New Version of Detached-eddy Simulation, Resistant to Ambiguous Grid Densities. *Theor. Comput. Fluid Dyn.* **2006**, *20*, 181–195. [CrossRef]
16. Spalart, P.R. Detached-eddy simulation. *Annu. Rev. Fluid Mech.* **2009**, *41*, 181–202. [CrossRef]
17. Strelets, M. Detached eddy simulation of massively separated flows. In Proceedings of the AIAA-2001-879, 39th Aerospace Sciences Meeting and Exhibit, Reno, NV, USA, 8–11 January 2001.
18. Menter, F.R.; Kuntz, M. *Kuntz. Adaptation of Eddy-Viscosity Turbulence Models to Unsteady Separated Flow Behind Vehicles. The Aerodynamics of Heavy Vehicles: Trucks, Buses, and Trains*; Springer: Berlin/Heidelberg, Germany, 2004; pp. 339–352.
19. Liu, R.Y. Unsteady Numerical Investigation on Cascade Separation Vortex Flow Based on Delayed Detached-Eddy Simulation. *J. Aerosp. Power* **2017**, *1*, 16–26.
20. Zhang, M.; Hou, A. Investigation on stall inception of axial compressor under inlet rotating distortion. *J. Mech. Eng. Sci.* **2017**, *231*, 1859–1870. [CrossRef]
21. Zhao, L. Large eddy simulation of flow field unsteady oscillation in a two-dimensional moving compressor cascade. *J. Aerosp. Power* **2015**, *30*, 248–256.

© 2019 by the authors. Licensee MDPI, Basel, Switzerland. This article is an open access article distributed under the terms and conditions of the Creative Commons Attribution (CC BY) license (http://creativecommons.org/licenses/by/4.0/).

Article

Application of Multiple-Scales Method for the Dynamic Modelling of a Gear Coupling

Enrico Pipitone, Christian Maria Firrone * and Stefano Zucca

Dipartimento di Ingegneria Meccanica e Aerospaziale (DIMEAS), Politecnico di Torino,
Corso Duca degli Abbruzzi, 24, 10129 Torino, Italy; enrico.pipitone@polito.it (E.P.); stefano.zucca@polito.it (S.Z.)
* Correspondence: christian.firrone@polito.it

Received: 13 February 2019; Accepted: 12 March 2019; Published: 23 March 2019

Abstract: Thin-walled gears, designed for aeronautical applications, have shown very rich dynamics that must be investigated in advance of the design phase. One of the signatures of their dynamics is coupling due to the meshing teeth which stand-alone gear models cannot capture. This paper aims to investigate the dynamics of thin-walled gears considering time-varying coupling due to the gear meshing. Each gear is modelled with lumped parameters according to a local rotating reference system and the coupling is modelled by a traveling meshing stiffness. The set of equations of motion is solved by the non-linear Method of Multiple-Time-Scales (MMTS). MMTS is a very powerful technique that is widely used to solve perturbation problems in many fields of mathematic and physics. In the analyzed numerical test case, the relevance of gear coupling is demonstrated as well as the capability of the MMTS to capture the fundamental features of the system dynamics. In this study the analytical methodology, which uses MMTS, allows for the calculation of the forced response of the system made of two meshing gears despite the presence of a parametric quantity, i.e., the mesh stiffness. The calculation is performed in the frequency domain using modal coordinates, which ensures a fast computation. The result is compared with time domain analysis for validation purposes.

Keywords: dynamics; mesh stiffness; forced response; time-variant parameters; Method of Multiple Time-Scales; cyclic-symmetric systems dynamics; lumped parameters model

1. Introduction

The need to identify a non-linear methodology for a dynamic study of two meshing gears moves from the evidence of some critical resonances occurring during operations, which cannot be investigated by analyzing a single gear considered as a stand-alone component, but it requires the analysis of the overall system which can be made of two or more than two meshing gears (planetary system), where time-varying parameters and non-linearities appear in the equations of motion. In practice, it is experimentally verified that one gear can influence the dynamics of the other meshing gears, under certain conditions, causing unexpected resonances, which are dangerous for the overall system. Then, a dynamic coupling is established between the meshing gears. This phenomenon is mainly due to the fluctuation of the mesh stiffness at the meshing teeth, which varies because of the different mesh conditions and contact points during meshing. Fluctuations of the mesh stiffness can induce severe instability conditions and affect also the resonances of the system. The phenomenon of dynamic coupling can be experimentally verified in industrial applications, in particular for aeronautical applications where the gears, having specific mechanical characteristics and working at critical speed regimes, show mutual interactions, which largely affect the forced response of the system. In more detail, the dynamic coupling causes critical resonances on a gear, which are induced by the excitation of the mode shapes of the meshing gear. The presence of these mutual interactions among the components leads to a different study of the system, which must include all the gears involved in the interactions.

The dynamic coupling is a direct consequence of the variations of tooth flexibility (mesh stiffness) because of the different contact conditions during rotation, but also because of the variations of the contact ratio. Thus, time-varying mesh stiffness causes the system to be non-linear. It is worth remembering that here the term "non-linear" is adopted to highlight the fact that the system is not the usual Linear Time-Invariant System (LTIS), but a Linear Time-Variant System (LTVS). Being the system of the type LTVS, it is not possible to compute the forced response by inverting the dynamic stiffness matrix as for a LTIS, since the latter is a time-variant parameter inside the equation of motion. Nevertheless, the dependent variable of the equation of motion does not show elevation to powers or other non-linearities (e.g., Duffing equation). Then, the superimposition effect principle is valid for such a system and it will be used in the following discussion. As a consequence, the adoption of a "non-linear" method is needed to compute the forced response of the system. The dynamics of gear systems have been extensively studied by researchers for decades and still represent an important matter of interest for the understanding of phenomena affecting the dynamics of such systems. The evaluation of the mesh stiffness variations and the related non-linear aspects have a primary importance for researchers who provide several modelling solutions of the phenomenon [1] for mathematical definition. According to the different types of modelling of the mesh stiffness, many works provided different methodologies for the computation of the response of the system, according to different levels of complexity. Most of the works focused on the combined effect of mesh stiffness variation and backlash between the meshing teeth, which affects largely the response of the meshing gears, developing non-linear methodologies for the iterative and numerical computation of the response [2–9]. Other studies focused on the analysis of the instability conditions, which can be caused by the fluctuations of the mesh stiffness involving sometimes wide operational speed ranges of the gears and can lead to failures. Works on instability provide an analytical solution, using perturbation methods (e.g., method of multiple time-scales, MMTS) to establish relations between the analyzed instability conditions and the entity of the mesh stiffness fluctuations [10]. Recently, instability analyses and forced response studies were extended to more complex systems like planetary gear systems [11–14]. Most of the cited works analyze the dynamics of a gear system by considering the gears as rigid bodies connected by the mesh stiffness and introducing the transmission error between two meshing gears, which consider the fluctuations of an equivalent tooth compliance that excites the system.

In this paper, the aim is to consider the gears as compliant bodies and compute analytically the forced response of the system excited both parametrically and externally. The backlash phenomenon is not considered at this stage in order to focus the attention on the phenomenon of the dynamic coupling and on the method to be developed to study the phenomenon without the nonlinearity introduced by intermitting contacts during meshing. Here, transmission error cannot be used anymore since the gear bodies are considered as compliant. The gears, which constitute the overall system, are linked together by means of a time-variant mesh stiffness, which acts on the nodes of the teeth, where the contact takes place. In other words, the system sees both a parametric excitation and an external force exciting the system. The methodology developed here applies the Method of Multiple-Time-Scale (MMTS) to compute the frequency response of a single mesh gear pair, modelled with lumped parameters, and investigate the dynamic coupling, which is established between the gears, verifying the mutual interactions and resonances induced by the phenomenon. MMTS allows a good approximation to the solution of the problem by introducing "scales" variables, which will substitute the independent variable of the problem. The solution of the problem passes through the elimination of the so-called "secular terms". This procedure represents a necessary solvability conditions for the solution of the problem. In this paper, numerical examples of forced response are reported, based on test cases. Upon these test-case analyses, the methodology is finally validated by means of direct time integration (DTI) of the non-linear equations of motion.

2. Model of the System

The system under analysis is made of two meshing gears (Gear-1 or G_1, and Gear-2 or G_2, Figure 1). For each gear, a local reference system rotating with the gear itself is defined. Each gear is divided into sectors (Z_1 for Gear-1, and Z_2 for Gear-2), one per each tooth (Figure 2). A gear ratio η can be defined for the system under analysis as the ratio between Z_1 and Z_2. Each sector is modelled as a lumped parameter model with two degrees of freedom (dof), or nodes, one for the tooth and one for the gear sector wheel (Figure 3). The rotation of each gear around its own axis is allowed (no radial or axial displacement are allowed, only tangential displacement is allowed). The latter assumption is reasonable for the case of thin walled spur gears where radial and axial displacements can be assumed as negligible. The sectors are then coupled together. The periodic coupling between the teeth of the two gears is modelled by a time-variant mesh stiffness $K_M(t)$, described in more detail in Section 3.

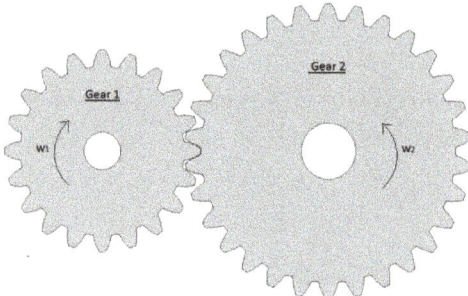

Figure 1. System of two meshing gears.

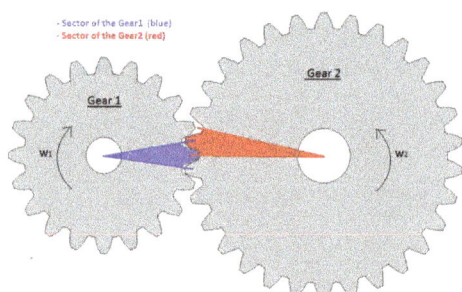

Figure 2. Representation of the sectors of the gears.

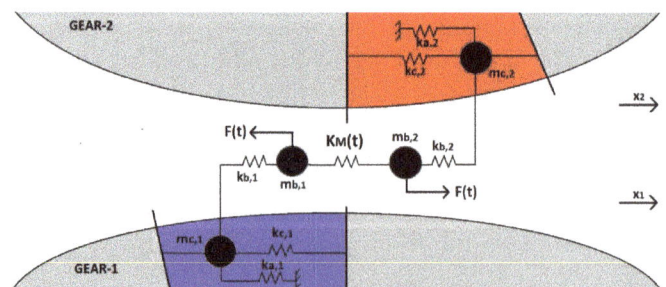

Figure 3. Lumped parameters model of the sectors of the gears Gear-1 and Gear-2.

As shown in Figure 3, the two gears are constrained to ground by means of the stiffness elements $k_{a,1}$ and $k_{a,2}$ respectively. The mechanical characteristics of mass $m_{b,1}$ and $m_{b,2}$, and stiffness $k_{b,1}$ and $k_{b,2}$ are associated to the teeth of the gears, whose displacement coordinates are $x_{G_1,t}$ and $x_{G_2,t}$ (respectively for the teeth of G_1 and the teeth of G_2). The mechanical characteristics of mass $m_{c,1}$ and $m_{c,2}$, and stiffness $k_{c,1}$ and $k_{c,2}$ are associated to the gear sector wheel, whose displacement coordinates are $x_{G_1,c}$ and $x_{G_2,c}$ (respectively for the sector wheel of G_1 and the sector wheel of G_2). In Figure 3, a force acting on the meshing teeth dof (equal but with opposite direction for the tooth of G_1 and the tooth of G_2) is shown, which represents the force establishing between them when the meshing couple is in contact. Obviously, when the couple is not meshing, no forces are exchanged. Then, the force travels along the circumference of each gear as the system rotates. The excitation force will be discussed more in detail in Section 5.

The physical displacement vector (with the corresponding size) is:

$$\{x\} = \left\{ \{x_{G_1}\}^T, \{x_{G_2}\}^T \right\}^T_{1 \times N}, \tag{1}$$

where,

$$\{x_{G_1}\}^T = \left\{ \{x_{G_1,c}\}^T, \{x_{G_1,t}\}^T \right\}^T_{1 \times 2Z_1}, \tag{2}$$

$$\{x_{G_2}\}^T = \left\{ \{x_{G_2,c}\}^T, \{x_{G_2,t}\}^T \right\}^T_{1 \times 2Z_2}, \tag{3}$$

with $N = 2Z_1 + 2Z_2$. $\{x\}$ including Gear-1 displacement coordinates, x_{G_1}, subdivided into $x_{G_1,c}$ which indicates the displacement of the nodes of the gear wheel, and $x_{G_1,t}$ indicating the displacement of the teeth. The same holds for x_{G_2} of Gear-2. The equation of motion, in matrix form, can be written in general as:

$$M\ddot{x} + \hat{C}\dot{x} + \hat{K}(t)x = \hat{F}(t), \tag{4}$$

where M is the mass matrix; $\hat{C}$ is the damping matrix; $\hat{K}(t)$ is the stiffness matrix which includes time-variant parameters, corresponding to the mesh stiffness $K_M(t)$ used to couple the two gears; $\hat{F}(t)$ is the force vector containing non-zero values for the teeth dof. As anticipated before, $\hat{F}(t)$ is a time-variant vector since the mesh force passes from one tooth to another as the system rotates. Then, each tooth is periodically subjected to a force excitation due to the meshing, where the period is equal to the rotation period of the gear. In the next section, the mesh stiffness will be discussed, then the assembly of the matrices will be presented in Section 4.

3. Definition of Mesh Stiffness

During meshing, many factors can induce fluctuations of the stiffness characteristics of the teeth. As explained before, the fluctuations of the mesh stiffness can be due to different contact conditions given by different contact ratios and contact positions along the tooth face. Hertzian contact phenomena can also influence the stiffness of the teeth. The combined effect of all these fluctuation sources produces a time history of the mesh stiffness acting on a single tooth. In this paper the time history of the mesh stiffness is not investigated in detail and it is approximated to a rectangular waveform traveling from one tooth to another one. In more detail, the mesh stiffness, which couples the nth tooth pair, is assumed to have a constant value k_t when the nth tooth pair is in contact and a null value when the contact is missing. Within the meshing time interval, the constant value, k_t, assumed by the mesh stiffness can represent an equivalent mean value of a real trend during meshing. Since rotating reference systems are used, in each gear the mesh stiffness rotates with the same speed as the gear but in the opposite direction. In Figure 4 the time history of a generic mesh stiffness $K_M(t)$ is shown with equivalent value k_t and unitary contact ratio, acting on a single tooth of a gear, with Z sectors (teeth) rotating at certain speed with revolution period T. In this qualitative example of mesh stiffness, one can distinguish between the meshing time interval, when the tooth is in contact, from the rest of the time history when the tooth is not in contact and the mesh stiffness assumes a null value. Once a full revolution is performed, so after a period T, the tooth experiences again the mesh stiffness.

The rectangular waveform can be translated into a sum of harmonics by developing the Fourier Series of time trend:

$$K_M(t) = K_c + \sum_{s=1}^{\infty} [K_{Va}^s \cos(s\Omega t) + K_{Vb}^s \sin(s\Omega t)], \tag{5}$$

where K_c is the mean value of the function, s is the harmonic index, Ω is the speed of the gear ($T = 2\pi/\Omega$), while K_{Va}^s and K_{Vb}^s are the coefficients of the "cosine" harmonics and the "sine" harmonics, respectively. The expression of the Fourier series can be further manipulated by means of the Euler formula, to redefine Equation (5) as the real-valued form of the complex notation of the Fourier Series which will be used in the following discussion (Equation (6)).

$$K_M(t) = K_c + K_V(t) = K_c + \sum_{s=1}^{\infty} \left[K_V^s e^{i s \Omega t} + \overline{K_V^s} e^{-i s \Omega t} \right], \tag{6}$$

where:

$$K_V^s = \frac{1}{2}(K_{Va}^s - i K_{Vb}^s); \overline{K_V^s}, \text{ complex conjugate of } K_V^s. \tag{7}$$

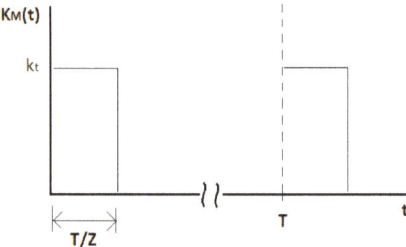

Figure 4. Qualitative time history of the mesh stiffness over one revolution period.

In Figure 5a,b a numerical example is shown (with $Z_1 = 10$, $Z_2 = 20$ and $k_t = 10^6$ N/m) where T_1 and T_2 are the revolution periods of Gear-1 and Gear-2 respectively, and they are different from one another since $Z_1 \neq Z_2$.

By looking at the previous example, it is easy to note that the periodic time history of the mesh stiffness of the two gears is different. As a matter of fact, the rectangular waveform is the same, but its period differs in the two gears, being T_1 for Gear-1 and T_2 for Gear-2. Here, this concept is clarified with an example. Let us consider the previous system with $Z_1 = 10$ teeth and $Z_2 = 20$ teeth. In Figure 6a the couple 1-1 (tooth-1 of G_1—tooth-1 of G_2) starts meshing for an angle $\theta_1 = 0°$ of G_1. After a full revolution of G_1 (Figure 6e) tooth-1 of G_1 meshes again but now with tooth-11 of G_2. The same is for the other couples 2-2 and 2-12 (Figure 6b–f) and so on. Thus, a certain couple (i-j) meshes with a base period that is twice the base period T_1. Of course, the latter relation changes for systems with different number of teeth.

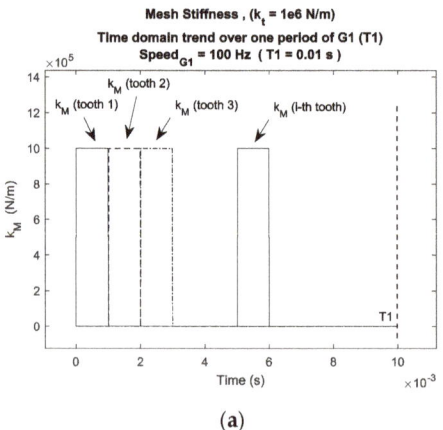

(**a**)

Figure 5. *Cont.*

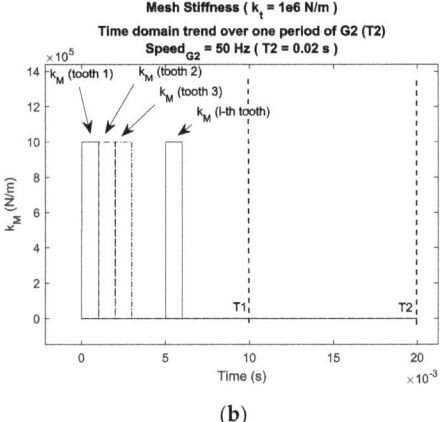

(b)

Figure 5. (a) Mesh Stiffness traveling on G_1. (b) Mesh Stiffness traveling on G_2.

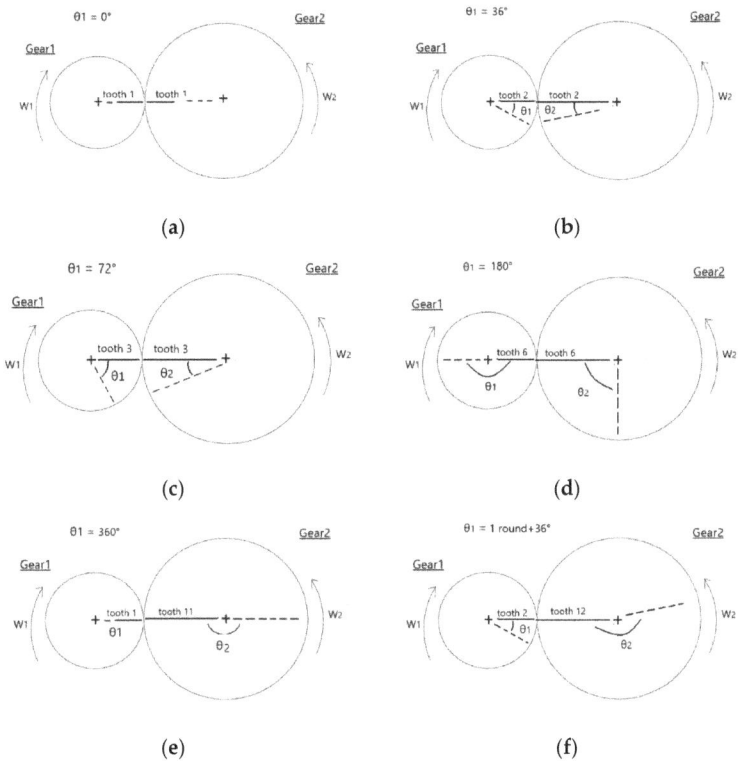

Figure 6. (a) Mesh couple for $\theta_1 = 0°$; (b) Mesh couple for $\theta_1 = 36°$; (c) Mesh couple for $\theta_1 = 72°$; (d) Mesh couple for $\theta_1 = 180°$; (e) Mesh couple for $\theta_1 = 360°$; (f) Mesh couple for $\theta_1 = 1$ round+36°.

4. Equations of Motion and Construction of the Matrices

Having adopted a lumped parameters model, it is convenient to write, for clarity's sake, an equation of motion (considering at this stage the un-damped unforced system) of one dof connected to the i_{th} tooth of Gear-1. Then, let us consider the previous example of a system constituted by one gear (Gear-1) having 10 teeth (Z_1) and a second gear (Gear-2) having 20 teeth (Z_2). Let us write the equation of motion of tooth-1 of Gear-1. By looking at Figure 7 below, neglecting at this stage the presence of damping and excitation force, the resulting equation of motion is:

$$m_b \ddot{x}_{t_1,1} + k_b(x_{t_1,1} - x_{c_1,1}) + K_{M_{1,1}}(t)\left(x_{t_1,1} - x_{t_2,1}\right) + K_{M_{1,11}}(t)\left(x_{t_1,1} - x_{t_2,11}\right) = 0, \qquad (8)$$

where m_b refers to mass of the teeth; k_b, nominal stiffness of the teeth; $K_{M_{1,1}}(t)$ and $K_{M_{1,11}}(t)$ mesh stiffness coupling the 1-1 teeth pair and 1-11 teeth pair respectively.

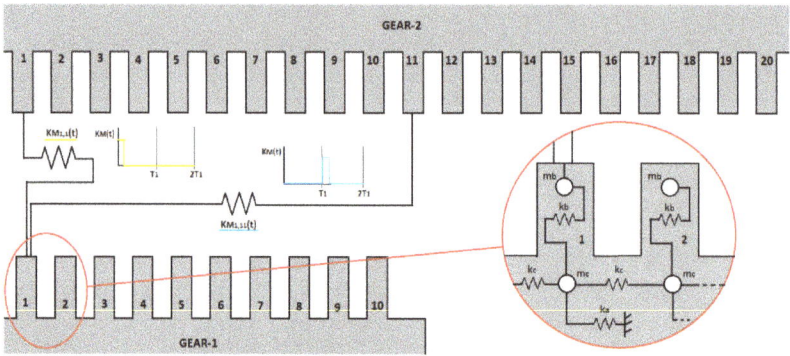

Figure 7. Linearized system showing meshing stiffnesses involving tooth-1 of G_1. The gears are constrained to the ground by means of the stiffness k_a.

It is possible to write the latter equation in a clearer form as follows:

$$m_b \ddot{x}_{t_1,1} + k_b x_{t_1,1} + \left(K_{M_{1,1}}(t) + K_{M_{1,11}}(t)\right) x_{t_1,1} - K_{M_{1,1}}(t) x_{t_2,1} - K_{M_{1,11}}(t) x_{t_2,11} - k_b x_{c_1,1} = 0. \qquad (9)$$

In Equation (9), two types of time-variant stiffness can be highlighted. The first type includes $K_{M_{1,1}}(t)$ and $K_{M_{1,11}}(t)$. The two terms refer to a specific teeth pair that meshes during operation. They are characterized by a rectangular waveform having a period T_{pair} that is strictly dependent on the number of teeth of the two gears. It is easy to derive T_{pair} and to relate it to the revolution periods of the two gears, T_1 and T_2 respectively. In general, T_{pair} is a multiple of both the periods T_1 and T_2. It is convenient to write the latter relation in a mathematical form:

$$T_{pair} = n_{T_1} T_1 = n_{T_2} T_2 \qquad (10)$$

where the coefficients n_{T_1} and n_{T_2} define the multiple of the respective revolution periods. In the example analyzed here it is easy to note that $n_{T_1} = 2$ and $n_{T_2} = 1$. In other words, a specific pair meshes after two revolution of the Gear-1 as well as after one revolution of Gear-2. The other type of variable stiffness term which appears into Equation (9) is the term $K_{M_{1,1}}(t) + K_{M_{1,11}}(t)$. This sum creates a rectangular waveform different from the previous one. As a matter of fact, $K_{M_{1,1}}(t) + K_{M_{1,11}}(t)$ is a waveform with the same constant value k_t but with a period exactly equal to the revolution period T_1, as shown in Figure 8.

Then, for all the equations of motion of the dof belonging to Gear-1, two different sets of harmonics appear into the equation. The fundamental frequencies, which describe the two sets, are respectively:

$$\Omega_1 = \frac{2\pi}{T_1}, \qquad (11)$$

$$\Omega_3 = \frac{2\pi}{T_{pair}} \tag{12}$$

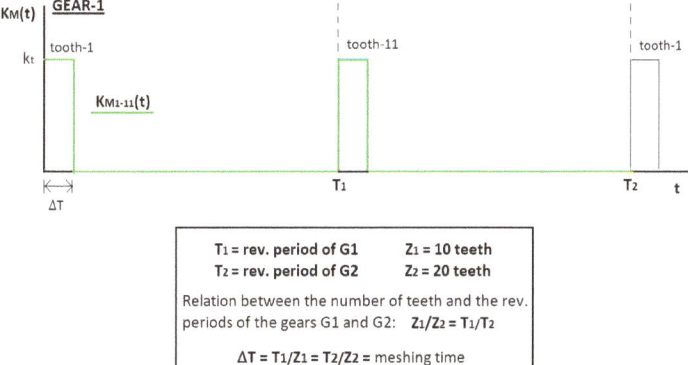

Figure 8. Time history of the time-variant terms that appear into the equation of motion Equation (9).

Following the same approach, it is possible to write the equations of motion for the dof belonging to Gear-2. For those equations, still two types of variable stiffness can be distinguished into the resulting equation (here the equation of motion is not reported because is similar to Equation (9)). The first type is again the rectangular waveform of the single pair of teeth described by the period T_{pair}, so by the fundamental frequency Ω_3. The second type of time-variant stiffness term is the sum of all the other *pair* waveforms involved into the equation, but now the resulting rectangular waveform has a base period equal to the revolution period of Gear-2, i.e., T_2. Therefore, another set of harmonics appears in the equations of motion of the system and it is described by the fundamental frequency Ω_2, which is the speed of Gear-2

$$\Omega_2 = \frac{2\pi}{T_2} \tag{13}$$

Finally, the construction of the stiffness matrix $\hat{K}(t)$ allows to write the full stiffness matrix in the next form (Equation (14)), by properly distinguishing between the terms referred to the three different sets of harmonics, having fundamental frequencies Ω_1, Ω_2 and Ω_3:

$$\hat{K}(t) = \left(K_c + \hat{K}_{V1}(t) + \hat{K}_{V2}(t) + \hat{K}_{V3}(t)\right), \tag{14}$$

$$\hat{K}_{V1}(t) = \sum_{s_1=1}^{\infty} [K_{V1}^{s_1} e^{is_1\Omega_1 t} + \overline{K_{V1}^{s_1}} e^{-is_1\Omega_1 t}], \tag{15}$$

$$\hat{K}_{V2}(t) = \sum_{s_2=1}^{\infty} [K_{V2}^{s_2} e^{is_2\Omega_2 t} + \overline{K_{V2}^{s_2}} e^{-is_2\Omega_2 t}], \tag{16}$$

$$\hat{K}_{V3}(t) = \sum_{s_3=1}^{\infty} [K_{V3}^{s_3} e^{is_3\Omega_3 t} + \overline{K_{V3}^{s_3}} e^{-is_3\Omega_3 t}]. \tag{17}$$

The mass matrix M and the stiffness matrix $\hat{K}(t)$ are obtained by writing the equation of motion for each dof of the system as shown in Equation (9) according to the matrix notation. The damping matrix $\hat{C}$ will be defined later in Section 6.1 assuming a modal damping ratio ς to the mode shapes of the system.

5. Excitation Force

A mesh force applied to two meshing teeth excites the system. The mesh force $F(t)$ of Figure 3 corresponds to a torque (for example applied on G_1) and a resistant torque (applied on G_2) acting on the system. Since the mesh force rotates along the gears as the system rotates, each tooth of a gear undergoes periodically the same mesh force, with a period that is equal to the rotation period of the gear itself (T_1 for teeth of G_1 and T_2 for the teeth of G_2) each tooth is subjected to the same force with a time delay. During meshing, the value of the mesh force is assumed to be constant. Then, considering the generic i_{th} tooth of a gear, the mesh force $F_{Gi}(t)$ acting on it will have the same trend of the mesh stiffness in the time domain (rectangular waveform, Figure 4). As a consequence, the force vector $\hat{F}(t)$ in the equation of motion Equation (4) contains, in correspondence to the teeth dof, the mesh force trends $F_{Gi}(t)$, properly phased in the time domain according to the position of the tooth which is considered (e.g., if tooth-1 of G_1 undergoes $F_{1_1}(t)$, tooth-2 of G_1 undergoes $F_{1_2}(t) = F_{1_1}(t + \Delta T)$, being ΔT the meshing time interval). Finally, the force vector $\hat{F}(t)$ can be represented as

$$\hat{F}(t) = \left\{ \begin{array}{c} \{0\}_{Z_1 \times 1} \\ \left\{ \begin{array}{c} F_{1_1} \\ \vdots \\ F_{1_{2Z_1}} \end{array} \right\}_{Z_1 \times 1} \\ \{0\}_{Z_2 \times 1} \\ \left\{ \begin{array}{c} F_{2_1} \\ \vdots \\ F_{2_{2Z_2}} \end{array} \right\}_{Z_2 \times 1} \end{array} \right\}_{N \times 1} \quad (18)$$

As for the mesh stiffness, it is convenient to express the mesh force as a Fourier series, since it is periodic in the time domain. This will be advantageous for the computation of the forced response (Section 6), since each harmonic will be considered separately, computing the overall response as a superimposition of the effects due to each harmonic. As a matter of fact, the system is a linear time-variant system whereby the superimposition principle is valid. Then, let us write the expression of the force function, in the Fourier series. It is convenient to distinguish between the force acting on the teeth nodes of Gear-1 from the force acting on the teeth nodes of Gear-2. The two force sets, expressed in Fourier Series, have fundamental frequencies that are the speed of the Gear-1 (Ω_1) for the excitation force terms acting on the teeth nodes of Gear-1 and the speed of Gear-2 (Ω_2) for the excitation force terms acting on the teeth nodes of Gear-2. As a consequence, let us write the general expression of the force acting on the i_{th} tooth of G_1 and the force acting on the j_{th} tooth of G_2 respectively in Equations (19) and (20):

$$F_{1_i}(t) = F_{1_i,0} + \sum_{k_1=1}^{\infty} [F_{1_i a}^{k_1} \cos(k_1 \Omega_1 t) + F_{1_i b}^{k_1} \sin(k_1 \Omega_1 t)]; \quad (19)$$

$$F_{2_j}(t) = F_{2_j,0} + \sum_{k_2=1}^{\infty} [F_{2_j a}^{k_2} \cos(k_2 \Omega_2 t) + F_{2_j b}^{k_2} \sin(k_2 \Omega_2 t)]. \quad (20)$$

As for the mathematical expression of the mesh stiffness (Section 3), The Fourier Series of the mesh force can be written by means of the exponential notation. Equations (19) and (20) become:

$$F_{1_i}(t) = F_{1_i,0} + \sum_{k_1=1}^{\infty} [F_{1_i}^{k_1} e^{ik_1 \Omega_1 t} + \overline{F_{1_i}^{k_1}} e^{-ik_1 \Omega_1 t}], \quad (21)$$

$$F_{2_i}(t) = F_{2_i,0} + \sum_{k_1=1}^{\infty} [F_{2_i}^{k_2} e^{ik_1\Omega_1 t} + \overline{F_{2_i}^{k_2}} e^{-ik_2\Omega_2 t}], \qquad (22)$$

With

$$F_{G_i}^{k_G} = \frac{1}{2}\left(F_{G_i a}^{k_G} - i F_{G_i b}^{k_G}\right), \qquad (23)$$

$$\overline{F_{G_i}^{k_G}} = \frac{1}{2}\left(F_{G_i a}^{k_G} + i F_{G_i b}^{k_G}\right). \qquad (24)$$

Being the forcing functions $F_{1_i}(t)$ and $F_{2_j}(t)$ expressed as a series of harmonics with frequencies $k_1\Omega_1$ and $k_2\Omega_2$ respectively, it is convenient to separate the force vector $F(t)$ into a sum of two vectors whose components can be expressed as harmonics having frequencies $k_1\Omega_1$ and $k_2\Omega_2$, respectively $\hat{F}_1(t)$ (Equation (25)) and $\hat{F}_2(t)$ (Equation (26)). Thus, let us represent the two vectors as follows:

$$\hat{F}_1(t) = \left\{ \begin{array}{c} \{0\}_{Z_1 \times 1} \\ \left\{ \begin{array}{c} F_{1_1} \\ \vdots \\ F_{1_{2Z_1}} \end{array} \right\}_{Z_1 \times 1} \\ \{0\}_{2Z_2 \times 1} \end{array} \right\}_{N \times 1}, \qquad (25)$$

$$\hat{F}_2(t) = \left\{ \begin{array}{c} \{0\}_{2Z_1 \times 1} \\ \{0\}_{Z_2 \times 1} \\ \left\{ \begin{array}{c} F_{2_1} \\ \vdots \\ F_{2_{2Z_2}} \end{array} \right\}_{Z_2 \times 1} \end{array} \right\}_{N \times 1}. \qquad (26)$$

The overall forcing function vector $F(t)$ is

$$\hat{F}(t) = \hat{F}_1(t) + \hat{F}_2(t). \qquad (27)$$

6. Forced Response Computation with MMTS

Section 6 deals with the development of a general analytical solution of the forced response of the system under exam using MMTS. MMTS is a very used technique able to obtain approximations of solutions to non-linear problems. It works by substituting different "scales" variables (according to the level of approximation the user desires) to the independent variable of the equation, treating them as independent variables. Being a frequency-based method, it allows the study of the response of a complex system in a very flexible way, reducing considerably the computational time, with respect to other time-based methods which provide a numerical and iterative solution to the problem. Before going through the discussion of MMTS, it is advantageous to rewrite the equation of motion Equation (4) in terms of modal coordinates. The use of modal coordinates leads to a great simplification of the problem by decoupling the equation of motions. Moreover, it allows the direct evaluation of the effect of a harmonic component of the excitation force on the respective mode shape that is excited by that component. Thus, the following Paragraph 6.1 is dedicated to the transformation of the equation of motion in modal coordinates while the mathematical development of MMTS is discussed in Paragraph 6.2.

6.1. Modal Analysis and Transformation of the Equation of Motion in Modal Coordinates

Since the stiffness matrix $\hat{K}(t)$ contains time-variant elements, the computation of modal analysis cannot be performed in general. For this reason, the computation of the modal analysis is performed on the mean part of the stiffness matrix K_C (Equation (14)) (K_C is a symmetric matrix). Thus, let us consider the time-invariant (unforced and undamped) part of the equation of motion Equation (4) (Equation (28)) and perform the modal analysis (Equation (29)).

$$M\ddot{x} + K_C x = 0; \tag{28}$$

$$\det\left(K_C - \omega^2 M\right) = 0, \text{ eigenvalue problem} \rightarrow \omega_n^2, \psi_n, n = 1 \div N; \tag{29}$$

$$\Psi = [\psi_1, \ldots, \psi_n, \ldots, \psi_N]_{N \times N}, \text{ modal matrix}$$
$$\omega_n, \text{ natural frequency}$$

From modal analysis, the natural frequencies ω_n and mode shapes ψ_n of the (mean) overall system are obtained. Then, let us apply Direct Modal Transformation (DMT, Equation (30)) to the equation of motion using the modal matrix Ψ and multiply the latter by the transpose of the modal matrix Ψ^T. The resulting equation written in modal coordinates (u, vector of modal coordinates) is reported in Equation (31).

$$x = \Psi u. \tag{30}$$

$$M_{mod}\,\ddot{u} + \hat{C}_{mod}\,\dot{u} + K_{c,mod}\,u + \hat{D}(t)\,u = \hat{P}(t), \tag{31}$$

with: M_{mod}, modal mass matrix; $\hat{C}_{mod}$, modal damping matrix; $K_{c,mod}$, modal mean stiffness matrix;

$$\hat{D}(t) = \Psi^T \hat{K}_V(t) \Psi; \tag{32}$$

$$\hat{P}(t) = \Psi^T \hat{F}(t) \tag{33}$$

If the mode shapes are normalized with respect to the unitary modal masses, M_{mod} is equal to the identity matrix and $K_{c,mod}$ is a diagonal matrix having on its main diagonal the eigenvalues of the system ω_n^2. As anticipated in Section 2, damping is not modelled physically into the model. For the sake of simplicity, damping is introduced by means of the modal damping ratio ς_n to be associated to each n_{th} mode shape. It is possible to obtain the modal damping $\hat{c}_n$ as:

$$\hat{c}_n = 2\frac{\varsigma_n}{\sqrt{k_{modn} \times m_{modn}}} = 2\frac{\varsigma_n}{\sqrt{\omega_n^2}}, = 1 \div N; \tag{34}$$

with m_{modn} modal mass of the n_{th} mode shape ($m_{modn} = 1$, if the mode shapes are normalized with respect to the unitary modal masses) and k_{modn} modal stiffness of the n_{th} mode shape ($k_{modn} = \omega_n^2$, if the mode shapes are normalized with respect to the unitary modal masses). Then, the modal damping matrix $\hat{C}_{mod}$ is a diagonal matrix having on its main diagonal the modal damping $\hat{c}_n$, satisfying the following relation that allows computation of the damping matrix in physical coordinates.

$$C = \Psi^{T-1} \hat{C}_{mod} \Psi^{-1} \tag{35}$$

The diagonalization of most of the matrices inside the equation of motion allows us to write singularly the equations of motion in modal coordinates (Equation (36)). The only term that is not diagonalized is $\hat{D}(t)$ being a non-symmetric matrix that was not involved in the modal analysis. As a consequence, in the n_{th} equation of motion in modal coordinates $\hat{D}(t)$ must be expressed as sum of the products of the elements $\hat{D}_{nr}(t)$ (elements of the matrix $\hat{D}(t)$ at n_{th} row and r_{th} column) times the r_{th} modal displacement u_r.

$$\ddot{u}_n + \hat{c}_n \dot{u}_n + \omega_n^2 u_n + \sum_{r=1}^{N} \{\hat{D}_{nr}(t)\,u_r\} = \hat{P}_n(t), \quad n = 1 \div N \tag{36}$$

The Equation (36) represents the useful equation for the development of MMTS by means of which it is possible to compute the modal response u_n, in the frequency domain, of the generic n_{th}

mode shape subjected to a modal force $\hat{P}_n(t)$. The development of MMTS will be presented in the next Paragraph 6.2 starting from the single n_{th} equation of motion in modal coordinates (Equation (35)).

6.2. Forced Response Computation Using MMTS

MMTS operates by substituting the independent time variable t with the time scales t_0, t_1, t_2, ..., where $t_0 = \varepsilon^0 t$, $t_1 = \varepsilon^1 t$ and $t_2 = \varepsilon^2 t$ and ε is the scale factor which describes the time scale. The relation between the original time variable and the new time scales is expressed in Equation (37). Here, the series approximating the old variable t is truncated at the second order (power ε^2):

$$t = t_0 + t_1 + o(\varepsilon^2) = t + \varepsilon t + o(\varepsilon^2) = t + \tau + o(\varepsilon^2). \tag{37}$$

As a consequence, the dependent variable $u_n(t)$ as well as the derivative operators need to be written (Equations (38)–(40)), in the new time scale variables:

$$u_n = u_{n0}(t, \tau) + \varepsilon\, u_{n1}(t, \tau) + o(\varepsilon^2), \quad n = 1 \div N; \tag{38}$$

$$\frac{d}{dt} \Rightarrow \frac{\partial}{\partial t} + \varepsilon \frac{\partial}{\partial \tau} + o(\varepsilon^2), \tag{39}$$

$$\frac{d^2}{dt^2} \Rightarrow \frac{\partial^2}{\partial t^2} + 2\varepsilon \frac{\partial^2}{\partial t \partial \tau} + o(\varepsilon^2). \tag{40}$$

The solution to the problem requires the proper manipulation of the equations of motion. Recalling Equation (36), the manipulation consists of associating some terms of the equation to the coefficient ε (Equations (41)–(43)). Let us associate the time-variant part of the stiffness matrix, $\hat{D}(t)$, the damping matrix $\hat{c}_n$ and the modal force $\hat{P}_n(t)$ to the scale factor ε:

$$\hat{D}(t) = \varepsilon\, D(t), \tag{41}$$

$$\hat{c}_n = \varepsilon\, c_n; \tag{42}$$

$$\hat{P}_n(t) = \varepsilon\, P_n(t). \tag{43}$$

Substituting Equations (41)–(43) into Equation (36), the latter becomes:

$$\ddot{u}_n + \varepsilon\, c_n \dot{u}_n + \omega_n^2\, u_n + \sum_{r=1}^{N} \{\varepsilon\, D_{nr}(t)\, u_r\} = \varepsilon\, P_n(t), \quad n = 1 : N. \tag{44}$$

Once the new equation of motion Equation (44) is defined, the MMTS operates by substituting the new expression of the modal response $u_n(t)$ (Equation (38)) and the new derivative operators (Equations (39) and (40)) into Equation (44). The resulting equation of motion will be an equation where all the terms are characterized by a multiplying coefficient that is in general a power of the scale factor ε (ε^n). Then, a separation of the terms according to the power of ε is performed, by creating n different equations. The new equations are reported below (Equations (45) and (46)).

- Equation corresponding to the power ε^0:

$$\ddot{u}_{n0} + \omega_n^2 u_{n0} = 0. \tag{45}$$

- Equation corresponding to the power ε^1:

$$\ddot{u}_{n1} + \omega_n^2 u_{n1} = -2 \frac{\partial^2 u_{n0}}{\partial t \partial \tau} - \sum_{r=1}^{N} \{D_{nr}(t)\, u_{r0}\} - c_n \frac{\partial u_{n0}}{\partial t} + P_n(t). \tag{46}$$

The solution of Equation (45) can be written in the general form

$$u_{n0} = A_n(\tau)e^{i\omega_n t} + \overline{A_n}(\tau)e^{-i\omega_n t} = A_n(\tau)e^{i\omega_n t} + CC, \quad (47)$$

where $A_n(\tau)$ is a function of the time variable τ. Substituting Equation (47) into Equation (46) and adopting the notation in Equation (48), one obtains Equation (49):

$$(\dot{\ldots}) = \frac{\partial(\ldots)}{\partial t}; \quad (\ldots)' = \frac{\partial(\ldots)}{\partial \tau}; \quad (48)$$

$$\ddot{u}_{n1} + \omega_n^2 u_{n1} = -2\left[i\omega_n A'_n e^{i\omega_n t} + CC\right] - \sum_{r=1}^{N}\left\{D_{nr}(t)\left[A_r e^{i\omega_r t} + CC\right]\right\} - c_n\left[i\omega_n A_n e^{i\omega_n t} + CC\right] + P_n(t), \quad (49)$$

where the quantity CC represents the complex conjugate of the previous term.

In order to solve the N equations of motion (Equation (49)) in the modal coordinate u it is necessary to develop the terms $D_{nr}(t)$ and $P_n(t)$ in their harmonic series:

$$D(t) = D_1(t) + D_2(t) + D_3(t), \quad (50)$$

$$D(t) = \sum_{s_1=1}^{\infty}[D_1^{s_1} e^{i s_1 \Omega_1 t} + \overline{D_1^{s_1}} e^{-i s_1 \Omega_1 t}] + \sum_{s_2=1}^{\infty}[D_2^{s_2} e^{i s_2 \Omega_2 t} + \overline{D_2^{s_2}} e^{-i s_2 \Omega_2 t}] \\ + \sum_{s_3=1}^{\infty}[D_3^{s_3} e^{i s_3 \Omega_3 t} + \overline{D_3^{s_3}} e^{-i s_3 \Omega_3 t}] \quad (51)$$

$$P(t) = P_1(t) + P_2(t), \quad (52)$$

$$P(t) = P_{1,0} + P_{2,0} + \sum_{k_1=1}^{\infty}[P_1^{k_1} e^{i k_1 \Omega_1 t} + \overline{P_1^{k_1}} e^{-i k_1 \Omega_1 t}] \\ + \sum_{k_2=1}^{\infty}[P_2^{k_2} e^{i k_2 \Omega_2 t} + \overline{P_2^{k_2}} e^{-i k_2 \Omega_2 t}]. \quad (53)$$

It is worth to remind that the fundamental frequency Ω_1 is the speed of G_1, the fundamental frequency Ω_2 is the speed of G_2, linked to Ω_1 through the gear ratio $\eta = \frac{Z1}{Z2}$ and the fundamental frequency Ω_3 is the frequency whereby a generic pair of teeth meshes (see Equations (11)–(13)). Substituting the expressions Equation (51) and Equation (53) into Equation (49), one obtains the extended p_{th} equation of motion in modal coordinates with all the time-variant parameters developed inside (see Appendix A, Equation (A2)). This equation contains all the harmonics of the excitation force, but they can be treated singularly by computing the forced response to each harmonic component of the force. Finally, a sum of all the response contributions can be performed, thanks to the superimposition effect principle, so to compute the overall multi-harmonic response. Thus, the following discussion analyzes the forced response to the generic k_{th} harmonic component of $P_1(t)$ acting on the teeth of G_1. The same approach is used to compute the forced response to the second set of harmonics $P_2(t)$ acting on the nodes of G_2, but here they are not treated for sake of clarity. Let us consider the following equation (Equation (54)) of the p_{th} modal equation of the system, where only the generic k_{th} harmonic component of the force function $P_1(t)$ is considered:

$$\ddot{u}_{p1} + \omega_p^2 u_{p1} = -2i\omega_p A'_p e^{i\omega_p t} \\ - \sum_{r=1}^{N}\sum_{s_1=1}^{\infty}\left[D_{1pr}^{s_1} A_r e^{i(s_1\Omega_1 + \omega_r)t} + D_{1pr}^{s_1} \overline{A_r} e^{i(s_1\Omega_1 - \omega_r)t}\right] \\ - \sum_{r=1}^{N}\sum_{s_2=1}^{\infty}\left[D_{2pr}^{s_2} A_r e^{i(s_2\eta\Omega_1 + \omega_r)t} + D_{2pr}^{s_2} \overline{A_r} e^{i(s_2\eta\Omega_1 - \omega_r)t}\right] \\ - \sum_{r=1}^{N}\sum_{s_3=1}^{\infty}\left[D_{3pr}^{s_3} A_r e^{i(s_3 \frac{\Omega_1}{\eta T_1}+\omega_r)t} + D_{3pr}^{s_3} \overline{A_r} e^{i(s_3\frac{\Omega_1}{\eta T_1}-\omega_r)t}\right] \\ -ic_p\omega_p A_p e^{i\omega_p t} + P_{1p}^{k_1} e^{i k_1 \Omega_1 t} + CC. \quad (54)$$

It is now possible to investigate and remove the unwanted secular terms, inside the latter equation. The elimination of secular terms represents a solvability condition for the solution of the problem, because of the additional freedom introduced with the new independent variables. In order to eliminate secular terms, the resonant terms of each equation need to be forced to zero. The discussion upon secular terms research and elimination is faced more in detail in the Appendix B. Two types of resonant terms can be distinguished inside the equation Equation (54) which can give secular terms: the first type gives "exact" secular terms and they are reported in Equations (55) and (56); the second type can gives *nearly* secular terms when the excitation frequency $k_1\Omega_1$ approaches to ω_p, and they are reported in Equations (57)–(59).

$$-2i\omega_p A'_p e^{i\omega_p t}, \tag{55}$$

$$-ic_p\omega_p A_p e^{i\omega_p t}, \tag{56}$$

$$D_{1pr}^{s_1} \overline{A_r} e^{i(s_1\Omega_1 - \omega_r)t}, \tag{57}$$

$$D_{2pr}^{s_2} \overline{A_r} e^{i(s_2 \eta \Omega_1 - \omega_r)t}, \tag{58}$$

$$D_{3pr}^{s_3} \overline{A_r} e^{i(s_3 \frac{\Omega_1}{n_T 1} - \omega_r)t}. \tag{59}$$

Since we are interested in the computation of the p_{th} modal response, it is convenient to introduce an auxiliary frequency variable σ to express the neighborhood of the excitation frequency $k_1\Omega_1$ to the p_{th} natural frequency ω_p:

$$k_1\Omega_1 = \omega_p + \varepsilon\sigma; \tag{60}$$

$$\Omega_1 = \frac{\omega_p}{k_1} + \varepsilon\frac{\sigma}{k_1}. \tag{61}$$

By properly substituting the frequency variable σ and performing secular terms elimination by equating to zero the sum of all the possible secular terms, in their critical conditions (see Appendix B for a detailed development), one obtains the following equation in the unknown A_p, which is the amplitude of the p_{th} modal response u_{p0}:

$$\begin{aligned}-2i\omega_p A'_p - D_{1pp}^{(2k_1)} \overline{A_p} e^{i 2\sigma\tau} - D_{2pp}^{(2\eta k_1)} \overline{A_p} e^{i 2\sigma\tau} - D_{3pp}^{(2 n_{T_1} k_1)} \overline{A_p} e^{i 2\sigma\tau} \\ -ic_p\omega_p A_p + P_{1p}^{k_1} e^{i\sigma\tau} = 0.\end{aligned} \tag{62}$$

Now, let

$$A_p = a_p e^{i\sigma\tau}. \tag{63}$$

Substituting Equation (63) into Equation (62), it follows

$$-2i\omega_p\left(a'_p + i\sigma a_p\right)e^{i\sigma\tau} - \mathcal{D} \overline{a_p} e^{i\sigma\tau} - ic_p\omega_p a_p e^{i\sigma\tau} + P_{1p}^{k_1} e^{i\sigma\tau} = 0, \tag{64}$$

where

$$\mathcal{D} = D_{1pp}^{(2 k_1)} + D_{2pp}^{(2\eta k_1)} + D_{3pp}^{(2 n_{T_1} k_1)}. \tag{65}$$

In order to have a steady-state solution $a'_p = \frac{\partial a_p}{\partial \tau}$ has to be null. By eliminating the common term $e^{i\sigma\tau}$, the equation Equation (64) becomes:

$$2\sigma\omega_p a_p - \mathcal{D} \overline{a_p} - ic_p\omega_p a_p + P_{1p}^{k_1} = 0. \tag{66}$$

From Equation (66) it is possible to derive analytically an expression of a_p as a function of the frequency variable σ. As a consequence, the analytical solution of the modal response u_{p0} (Equation (67)) due to the k_{th} harmonic of the excitation force $P_1(t)$ is derived.

$$u_{p0} = A_p(\tau)e^{i\omega_p t} + CC = a_p(\sigma) e^{i(\omega_p + \sigma\varepsilon)t} + CC. \tag{67}$$

Since a_p is a complex quantity, it is convenient to express a_p according to its real and imaginary parts (Equations (68)–(70)). The analytical expression of the real and imaginary parts is derived as a function of σ:

$$a_p = a_R + i\, a_I; \tag{68}$$

$$a_R = \frac{\mathcal{P}_I(\mathcal{D}_I - c_p \omega_p) + \mathcal{P}_R(2\sigma \omega_p + \mathcal{D}_R)}{\omega_p^2 (c_p^2 + 4\sigma^2) - \mathcal{D}_R^2 - \mathcal{D}_I^2}, \tag{69}$$

$$a_I = \frac{a_R(\mathcal{D}_I + c_p \omega_p) - \mathcal{P}_I}{2\sigma \omega_p + \mathcal{D}_R}, \tag{70}$$

where:

$$P_{1p}^{k_1} = \mathcal{P}_R + i\, \mathcal{P}_I, \tag{71}$$

$$\mathcal{D} = \mathcal{D}_R + i\, \mathcal{D}_I, \tag{72}$$

The latter expressions (Equations (68)–(72)) allow the computation of the modal response of the p_{th} mode shape in the frequency domain. Each mode shape, connected to a specific nodal diameter of the system, is excited in resonance by some *EO* of the mesh force, according to the law reported in the following equation Equation (73), from gear dynamics theory [9]:

$$EO = mZ \pm ND, \quad m = 1, 2, \ldots \tag{73}$$

Thus, the construction of the multi-harmonic forced response is computed by considering, one by one, each *EO* of the mesh force, associating it to the mode shapes, which are described by the relative nodal diameter *ND*, excited by the selected *EO* and computing the modal responses. Once the modal responses in the frequency domain are computed for all the *EO*, they are transformed through DMT (Equation (30)), passing from modal coordinates to physical coordinates, and the forced response of a given node is developed in the time domain (in our case the nodes of the teeth of the gears) as the sum of all the mode shapes contributions (i.e., DMT). The validity of the superimposition effect principle (as explained in the Introduction) allows the summation of all the responses due to the different *EO* in the time domain. The result will be a multi-harmonic response, developed in the time domain. Since the response is not described by a single harmonic it is not possible to acquire the amplitude of the time domain response. Thus, the latter will be expressed by acquiring the peak-to-peak measure of the time domain trend as function of a reference frequency (which can be the speed of G_1 or the speed of G_2 as well) defining uniquely the excitations (both parametric and external) of the overall system. As a matter of fact, setting a certain speed for G_1 means also setting the speed of G_2 since the speed of the gears are linked by the gear ratio. As consequence, the mesh stiffness and mesh force, representing the parametric and the force excitations respectively, directly depend on the speed of the two gears. Finally, the reference frequency describes uniquely the operational conditions of the overall system. In the next section an example of forced response is computed on a dedicated test case and a comparison with Direct Time Integration (DTI) method is made to validate the MMTS methodology, developed into this paper.

7. Forced Response Computed on Test-Cases

In this Section, a study of the forced response of a test case is presented. The aim is to show when MMTS is applied for such applications and why it is convenient to use it. Such a problem can be studied, on the other hand, by Direct Time Integration (DTI) of the equations of motion but this is a very time-consuming method which makes difficult a detailed study of the forced response over a wide range of operational frequency. Anyway, here DTI is used to validate the methodology that is developed in this paper. More in detail, given a certain system model, characterized by given mass, stiffness and damping matrices, two parallel studies are developed on the system, applying MMTS and

DTI respectively. The validation of the MMTS methodology is made by comparing the peak-to-peak (*P2P*) measures of the multi-harmonic response developed in the time domain, calculated with the two methods. Here, the *P2P* result is plotted against a reference frequency, which is decided at the beginning of the calculation and defines uniquely all the excitations of the system (both parametric and external excitation). The reference frequency chosen for the *P2P* plots is the speed of Gear-1, Ω_1. As a matter of fact, the speed of Gear-2, Ω_2, is directly connected to Ω_1 through the gear ratio η. Then, all the parametric and external excitations are directly defined. Through the test-case analysis, the dynamic coupling phenomenon is investigated, by remarking its causes and consequences. It is demonstrated that the dynamic coupling, caused by the presence of a time-variant mesh stiffness, leads to a nodal coupling of certain nodal diameters of the meshing gears. To clearly note the phenomenon, a specific test case is built. In more detail, the example which has been used several times in this paper is considered. That is the case of a couple of gears (G_1 and G_2) having respectively $Z_1 = 10$ teeth and $Z_2 = 20$ teeth. The system model of each gear, as it was introduced at the beginning of the paper (Section 2) is constituted by two nodes per sector of the gear (the number of sectors is equal to the number of teeth). As consequence, each gear has number of dof equal to twice the number of its teeth. Being the dimension of the gear model twice the number of the sectors, two modal families of natural frequencies can be derived by performing modal analysis of both the gears considering them as stand-alone components. In Figure 9a,b the frequency vs. nodal diameter diagram of the two gears (considered as stand-alone components) is reported, given the mechanical characteristics of the gears shown in table Tables 1 and 2.

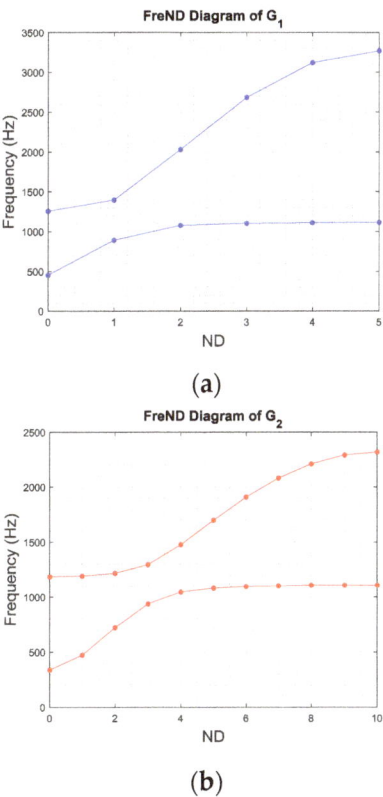

Figure 9. (**a**) *Frequency vs. ND diagram of Gear-1.* (**b**) *Frequency vs. ND diagram of Gear-2.*

Table 1. Mechanical characteristics of G1 model.

GEAR-1		
Z1	10	Teeth
ND	[0,...,5]	
Mechanical characteristics		
mb	0.2	(kg)
mc	1	(kg)
ka	10^7	(N/m)
kb	10^7	(N/m)
kc	10^8	(N/m)
damping ratio	0.005	(-)

Table 2. Mechanical characteristics of G2 model.

GEAR-2		
Z2	20	Teeth
ND	[0,...,10]	
Mechanical characteristics		
mb	0.2	(kg)
mc	2	(kg)
ka	10^7	(N/m)
kb	10^7	(N/m)
kc	10^8	(N/m)
damping ratio	0.005	(-)

The gears are coupled by means of a mesh stiffness of the same type described in Section 3. So, it assumes for the n_{th} pair of meshing teeth a constant value equal to k_t when the pair is in contact, it assumes a null value when it is not. In this test-case k_t is equal to 10^6 N/m. The mesh stiffness causes a modal coupling between some modes characterized by specific nodal diameters of the two gears respectively. It is good to remember that the mesh stiffness does not have a remarkable influence on the natural frequencies, which remains practically equal to those of the two gears considered as stand-alone components, even though it may have an important influence on the dynamic response of the overall system. In such a case the numbers of teeth Z_1 and Z_2 has a strong relation with each other. This condition emphasizes the nodal coupling between specific nodal diameters of the gears. It is worth to remind that such a system represents an unusual system because, in practice, gear systems are never designed with such numbers of teeth to avoid the same couple of teeth meshes too often. Nevertheless, this choice, which does not affect the validity of the methodology, aims to boost the effect of the analyzed phenomenon of the dynamic coupling so to better understand it. In Figure 10 an example of nodal coupling is reported for the system under analysis. That is the case of a nodal coupling between the nodal diameter ND-5 of G_1 and the nodal diameter ND-10 of G_2. By recalling the vector of the physical coordinates of the system defined in Section 2 (Equation (1)), the mode shape shown in Figure 10 contains both the nodal diameters. It means that an excitation of ND-5 of G_1 affects the vibration of G_2 which will vibrate at the same natural frequency with a ND-10 shape. It is important to remark that the ND-5 of G_1 (considered as stand-alone component) has a natural frequency $\omega_5 = 1118$ Hz (see Figure 9a). When the G_1 is coupled to G_2 by means of the mesh stiffness, the ND-5 of G_1 still has natural frequency ω_5, but the mode shape of the coupled system associated to that natural frequency shows an ND-5 mode shape coupled to an ND-10 of G_2. In other words, it is numerically demonstrated that a mesh stiffness with such a value of k_t causes the coupling between the nodal diameters of the gears without changing remarkably the natural frequencies with respect to those of the gears (considered as stand-alone components). The choice to keep a value of k_t that does not cause a remarkable change in the natural frequencies is a reasonable assumption that verifies what is experimentally found in the industrial applications. As a matter of fact, real test cases are characterized by mode shapes showing modal couplings between nodal diameters at given natural frequencies, which are practically the same of those of the stand-alone gears. Thus, the test case under exam aims at simulating a real coupled system where the dynamic characteristics of the mode shapes and natural frequencies remain practically unchanged.

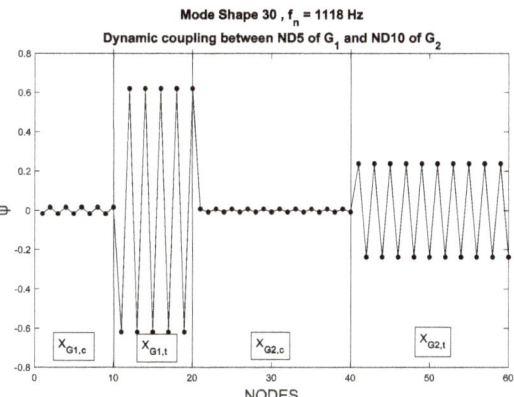

Figure 10. Mode shape of the system. *Nodal Coupling* between ND-5 of G_1 and ND-10 of G_2.

As it was described in Section 5, the external force acting on the system is a mesh force which travels from one tooth to another one with the mesh stiffness. In other words, a force of value F_m acts on the teeth nodes (with same value but with opposite direction) of a specific n_{th} teeth pair when it is in contact. So, as for the mesh stiffness, the Fourier series of the mesh force is studied, as described in Section 5. In Figure 11a,b the harmonic content (or Engine Orders, *EO*) of the forces, acting on the two gears respectively, is shown in terms of amplitudes of the various *EO*.

As anticipated in Section 6.2 (Equation (73)), a harmonic index *EO* of the travelling force excites mode shapes characterized by a specific ND. Since the ND under analysis for G_1 is 5, the *EO* excitation that have been selected are: 5, 15, 25, 35. MMTS allows the computation of the modal response of the mode shape of interest due to the selected *EO*. The forced response in the physical coordinates is easily derived through Direct Modal Transformation (DMT). Here, the forced response (expressed as the *P2P* measure of the multi-harmonic response developed in the time domain) of the teeth of the two gears is computed, in a given operational speed range (the reference speed is the speed of G_1) where the excitation of the mode shape in Figure 8 occurs. What is expected is to see a resonance of the G_1 due to the excitation of the ND-5 by some *EO* of the mesh force and an "induced" resonance of the G_2 due to the action of the latter *EO* exciting the G_1. The fact that the second resonance is induced by the first one is demonstrated by the fact that no excitation of the ND of G_2 should be present for that operational frequency conditions. As a matter of fact, by looking at the Campbell diagrams of the gears, you can note that for G_1 (Figure 12a) the involved *EO* crosses the natural frequency line in that operational speed range, while for G_2 (Figure 12b) no crossing of the natural frequency lines occurs by the involved *EO*. Thus, the conclusion is that the second resonance on G_2 is caused by the first one on G_1.

The *P2P* measure of the multi-harmonic response of the two gears computed using MMTS is reported in Figure 13. Here, also the *P2P* measure computed by means of DTI is shown in order to make a comparison between the two results. It is worthy to note that there is a big difference in terms of computational time for the construction of the forced response using the two methodologies (MMTS and DTI). In more detail, a DTI study can take some hours, and the computational time can increase considerably as the number of dof of the system increases as well as the resolution of the operational speed range and the integration time interval increase. As consequence, the computational efforts can be practically unsustainable for systems with a large number of dof and very low damping ratios whereby DTI must integrate for larger time interval in order to reach a steady-state response, where the transient part is completely extinguished (i.e., a real test-case of gear system where damping ratio is lower than 0.1%). On the other hand, MMTS, operating in the frequency domain, results in a very fast calculations of the forced response which can take few minutes and then it allows to select the *EO* of interest which have an influence in a given operational speed range, neglecting the minor effects of other *EO* and so reducing considerably the computational efforts. In both the analysis, two resonance peaks are visible in a given operational speed range (speed of G_1 from 210 Hz to 240 Hz). The blue curves are the resonances of G_1 computed respectively with MMTS and DTI. The same for the red curves. The resonance of G_2 is induced by the resonance of G_1, as anticipated before. By looking at the figures Figure 14a,b showing respectively the FFT of the time domain responses, computed through DTI, of the gears (respectively shown in the figures Figure 15a,b), it is easy to note that the main harmonic component of the multi-harmonic response in both cases is exactly the natural frequency of the mode shape shown in Figure 10 which couples ND5 of G_1 to ND10 of G_2. This represents an additional proof that the resonance of G_2 is directly induced by the excitation of that single mode shape by the mesh force on the G_1. This is a clear example of dynamic coupling between two meshing gears and MMTS allowed to forecast the resonance of G_2 induced by the excitation of the ND of G_1 and this could not be possible if you have considered the gears as stand-alone components. In that case, no interaction between the ND of the gears can be studied.

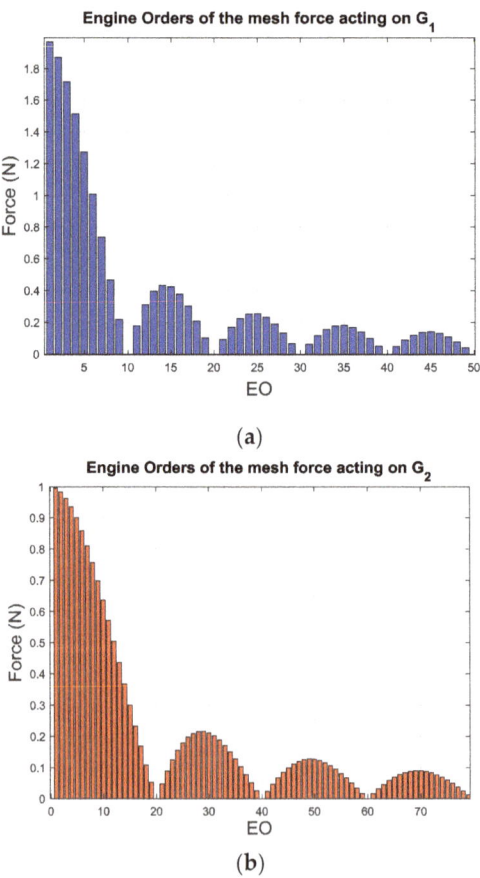

Figure 11. (a) Engine Orders of mesh force acting on G_1. (b) Engine Orders of mesh force acting on G_2.

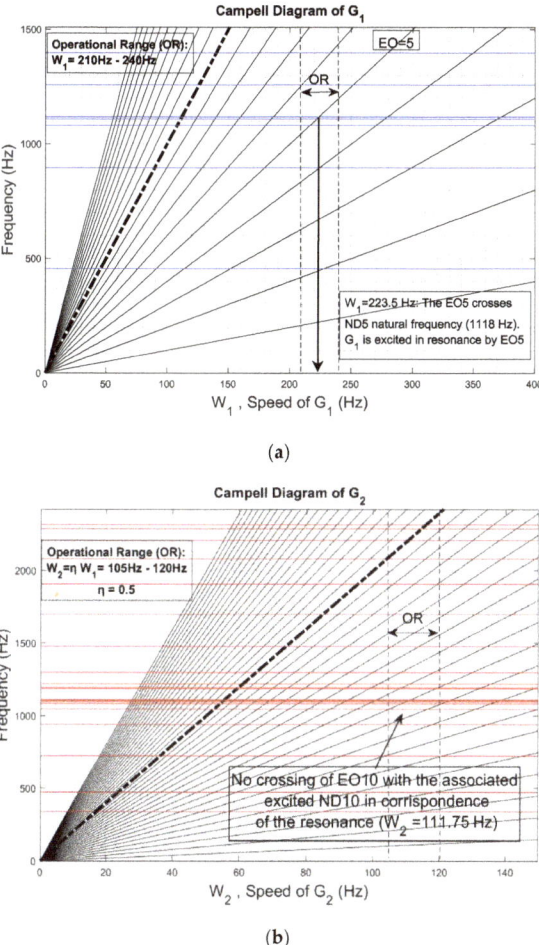

Figure 12. (**a**) Campbell diagram of G_1. (**b**) Campbell diagram of G_2.

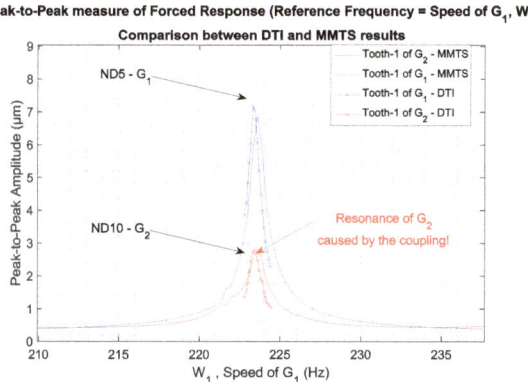

Figure 13. Peak-to-Peak measure of the multi-harmonic response of the gears. Comparison between DTI and MMTS.

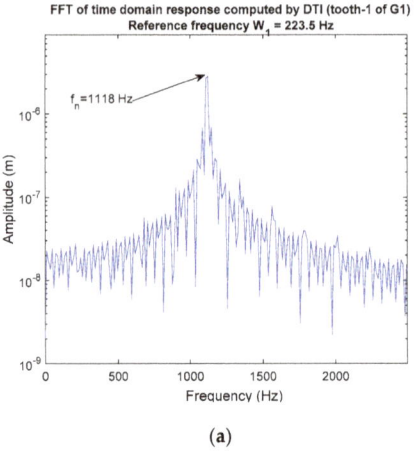

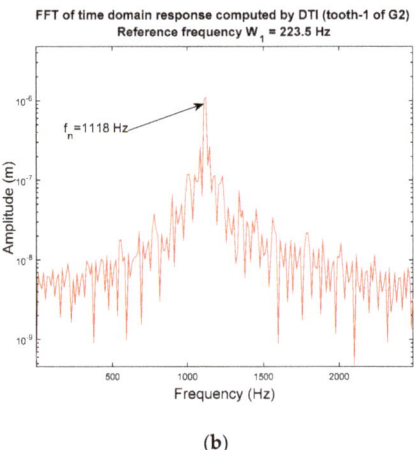

Figure 14. (a) FFT of the response of G_1 (Figure 14a). (b) FFT of the response of G_2 (Figure 14b).

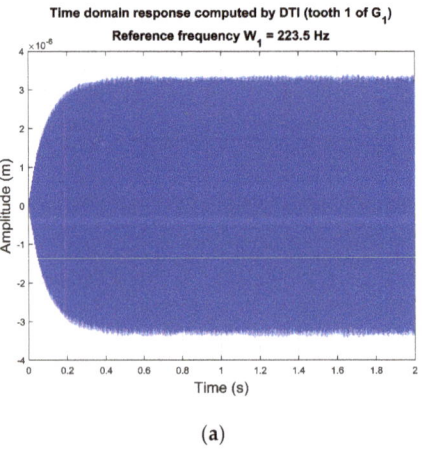

Figure 15. *Cont.*

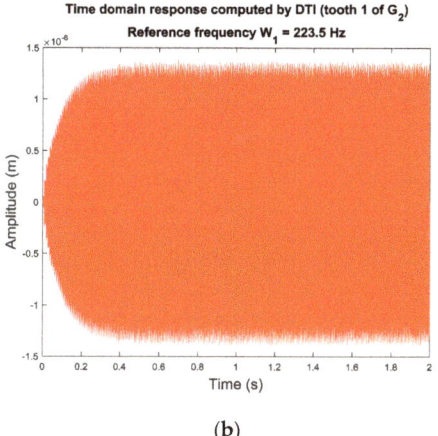

(b)

Figure 15. (a) Time domain response of G_1 by DTI. (b) Time domain response of G_2 by DTI.

8. Conclusions

The objective of this paper is to investigate the mutual interactions (dynamic coupling) which affect the response of a couple of meshing gears, by developing a methodology able to compute easily, with limited computational efforts, the forced response of the gears, without loosing the generality and complexity of the system. As a matter of fact, here the gears are considered as compliant bodies. As opposed to other methodologies which were developed in the past, whereby the assumption of the gears as rigid bodies needed to be supported by the introduction of the transmission error to simulate the compliance of the gears. Here, the challenge is to couple two compliant gears (whose dynamic characteristics are automatically included) and to investigate how the dynamics of a gear interacts with the other one when phenomena of mesh stiffness fluctuations occur. In addition to that, the methodology provides guidelines for an analytical solution to the problem, allowing the researcher to compute the forced response of a complex system undergoing both a parametric and external excitation. The choice of MMTS for the mathematical solution of the linear time-variant problem is addressed to its capability to provide an analytical solution to the problem, by strongly simplifying it. This is a great advantage for applications like gear coupling where the simplicity of the methodology (MMTS) compensates for the complexity of the system and allows for the analysis of the behavior of the gears in a considerably wide range of operation.

Author Contributions: Methodology, E.P., C.M.F and S.Z.; software, E.P.; validation, E.P., C.M.F and S.Z.; writing—original draft preparation, E.P.; writing—review and editing, C.M.F. and S.Z.; supervision, C.M.F.

Funding: This research was funded by GE Avio.

Acknowledgments: Authors would like to express their gratitude to Marco Moletta, Paolo Calza and Luca Ronchiato for technical assistance and sharing ideas to solve problem of engineering relevance and GE Avio for allowing the publication of this paper.

Conflicts of Interest: The authors declare no conflict of interest.

Appendix A Forced Response Computation

Complete p_{th} equation of motion in modal coordinates including all the harmonic sets of both the parametric and the excitation force, expressed in Fourier Series:

$$\begin{aligned}
\ddot{u}_{p1} + \omega_p^2 u_{p1} = & -2\left[i\omega_p A'_p e^{i\omega_p t} + CC\right] \\
& - \sum_{r=1}^{N} \sum_{s_1=1}^{\infty} \left\{ \left[D_{1pr}^{s_1} e^{i s_1 \Omega_1 t} + \overline{D_{1pr}^{s_1}} e^{-i s_1 \Omega_1 t} \right] \left[A_r e^{i\omega_r t} + CC \right] \right\} \\
& - \sum_{r=1}^{N} \sum_{s_2=1}^{\infty} \left\{ \left[D_{2pr}^{s_2} e^{i s_2 \eta \Omega_1 t} + \overline{D_{2pr}^{s_2}} e^{-i s_2 \eta \Omega_1 t} \right] \left[A_r e^{i\omega_r t} + CC \right] \right\} \\
& - \sum_{r=1}^{N} \sum_{s_3=1}^{\infty} \left\{ \left[D_{3pr}^{s_3} e^{i s_3 \frac{\Omega_1}{\eta_T 1} t} + \overline{D_{3pr}^{s_3}} e^{-i s_3 \frac{\Omega_1}{\eta_T 1} t} \right] \left[A_r e^{i\omega_r t} + CC \right] \right\} \\
& - c_p \left[i\omega_p A_p e^{i\omega_p t} + CC \right] + P_{10,p} + \sum_{k_1=1}^{\infty} \left[P_{1p}^{k_1} e^{i k_1 \Omega_1 t} + \overline{P_{1p}^{k_1}} e^{-i k_1 \Omega_1 t} \right] \\
& + P_{20,p} + \sum_{k_2=1}^{\infty} \left[P_{2p}^{k_2} e^{i k_2 \eta \Omega_1 t} + \overline{P_{2p}^{k_2}} e^{-i k_2 \eta \Omega_1 t} \right].
\end{aligned} \quad (A1)$$

Further manipulations of Equation (A1) allows to write the right-hand side (RHS) of the equation in a clearer form, by grouping the CC terms (Equation (A2)):

$$\begin{aligned}
RHS = & -2i\omega_p A'_p e^{i\omega_p t} - \sum_{r=1}^{N} \sum_{s_1=1}^{\infty} \left[D_{1pr}^{s_1} A_r e^{i(s_1\Omega_1 + \omega_r)t} + D_{1pr}^{s_1} \overline{A_r} e^{i(s_1\Omega_1 - \omega_r)t} \right] \\
& - \sum_{r=1}^{N} \sum_{s_2=1}^{\infty} \left[D_{2pr}^{s_2} A_r e^{i(s_2 \eta \Omega_1 + \omega_r)t} + D_{2pr}^{s_2} \overline{A_r} e^{i(s_2 \eta \Omega_1 - \omega_r)t} \right] \\
& - \sum_{r=1}^{N} \sum_{s_3=1}^{\infty} \left[D_{3pr}^{s_3} A_r e^{i(s_3 \frac{\Omega_1}{\eta_T 1} + \omega_r)t} + D_{3pr}^{s_3} \overline{A_r} e^{i(s_3 \frac{\Omega_1}{\eta_T 1} - \omega_r)t} \right] \\
& -ic_p \omega_p A_p e^{i\omega_p t} + \frac{1}{2} P_{1,0p} + \sum_{k_1=1}^{\infty} \left[P_{1p}^{k_1} e^{i k_1 \Omega_1 t} \right] + \frac{1}{2} P_{2,0p} + \sum_{k_2=1}^{\infty} \left[P_{2p}^{k_2} e^{i k_2 \eta \Omega_1 t} \right] + CC.
\end{aligned} \quad (A2)$$

Appendix B Elimination of *secular terms*

Here, the possible *secular terms* are analyzed. Two types of resonant terms can be distinguished inside the equation Equation (54) which can give *secular terms*: the first type gives "exact" *secular terms* and they are reported in Equations (A3) and (A4); the second type can give *secular terms* as the excitation frequency $k_1 \Omega_1$ approaches to ω_p, and they are reported in Equations (A5)–(A7). These terms are called *nearly secular terms*.

$$-2i\omega_p A'_p e^{i\omega_p t}; \quad (A3)$$

$$-ic_p \omega_p A_p e^{i\omega_p t}; \quad (A4)$$

$$D_{1pr}^{s_1} \overline{A_r} e^{i(s_1 \Omega_1 - \omega_r)t}; \quad (A5)$$

$$D_{2pr}^{s_2} \overline{A_r} e^{i(s_2 \eta \Omega_1 - \omega_r)t}; \quad (A6)$$

$$D_{3pr}^{s_3} \overline{A_r} e^{i(s_3 \frac{\Omega_1}{\eta_T 1} - \omega_r)t}. \quad (A7)$$

Nearly secular terms become "exact" *secular terms* in specific conditions. Below, each *nearly secular term* is analyzed to find the critical conditions which cause *secular terms*.

$$k_1 \Omega_1 = \omega_p + \varepsilon \sigma. \quad (A8)$$

- $D_{1pr}^{s_1} \overline{A_r} e^{i(s_1 \Omega_1 - \omega_r)t}$:

It produces *secular terms* for $r = p$ and $s_1 \Omega_1$ approaching to $2\omega_p$. The condition which verifies this case is $s_1 = 2k_1$. As consequence, by substituting the latter relation, it follows:

$$s_1 \Omega_1 = \frac{s_1}{k_1} \omega_p + \varepsilon \frac{s_1}{k_1} \sigma = 2\omega_p + 2\sigma\varepsilon. \quad (A9)$$

- $D_{2pr}^{s_2} \overline{A_r} e^{i(s_2 \eta \Omega_1 - \omega_r)t}$:

It produces *secular terms* for $r = p$ and $s_2 \eta \Omega_1$ approaching to $2\omega_p$. The condition which verifies this case is $s_2 = 2 \eta k_1$. As consequence, it follows:

$$s_2\eta\,\Omega_1 = \frac{s_2\,\eta}{k_1}\omega_p + \varepsilon\frac{s_2\,\eta}{k_1}\sigma = 2\omega_p + 2\sigma\varepsilon\,. \tag{A10}$$

- $D_{3pr}^{s_3}\,\overline{A_r}\,e^{i(s_3\frac{\Omega_1}{n_{T_1}} - \omega_r)t}$:

It produces secular terms for $r = p$ and $s_3\frac{\Omega_1}{n_{T_1}}$ approaching to $2\omega_p$. The condition which verifies this case is $s_3 = 2\,n_{T_1}k_1$. As consequence it follows:

$$\frac{s_3\Omega_1}{n_{T_1}} = \frac{s_3\,\omega_p}{k_1\,n_{T_1}} + \varepsilon\frac{s_3\,\sigma}{k_1\,n_{T_1}} = 2\omega_p + 2\sigma\varepsilon\,. \tag{A11}$$

Now it is possible to eliminate *secular terms* by applying the critical conditions analyzed before, summing up the resonant terms and forcing them to zero. It follows:

$$-2i\omega_p A'_p e^{i\omega_p t} - D_{1pp}^{s_1}\,\overline{A_p}\,e^{i(s_1\Omega_1 - \omega_p)t} - D_{2pp}^{s_2}\,\overline{A_p}\,e^{i(s_2\,\eta\,\Omega_1 - \omega_p)t} - \\ D_{3pp}^{s_3}\,\overline{A_p}\,e^{i(\frac{s_3}{n_{T_1}}\Omega_1 - \omega_p)t} - ic_p\omega_p A_p e^{i\omega_p t} + P_{1p}^{k_1}\,e^{i k_1\Omega_1 t} = 0. \tag{A12}$$

Substituting the equations Equations (A9)–(A11) into Equation (A12):

$$-2i\omega_p A'_p e^{i\omega_p t} - D_{1pp}^{(2\,k_1)}\,\overline{A_p}\,e^{i(2\sigma\varepsilon)t}\,e^{i\omega_p t} - D_{2pp}^{(2\eta k_1)}\,\overline{A_p}\,e^{i(2\sigma\varepsilon)t}\,e^{i\omega_p t} - \\ D_{3pp}^{(2\,n_{T_1}k_1)}\,\overline{A_p}\,e^{i(2\sigma\varepsilon)t}\,e^{i\omega_p t} - ic_p\omega_p A_p e^{i\omega_p t} + P_{1p}^{(k_1)}\,e^{i\sigma\varepsilon t}\,e^{i\omega_p t} = 0. \tag{A13}$$

It is possible to eliminate the common term $e^{i\omega_p t}$ into Equation (A13) and write the equation, considering that εt is exactly equal to τ:

$$-2i\omega_p A'_p - D_{1pp}^{(2k_1)}\,\overline{A_p}\,e^{i2\sigma\tau} - D_{2pp}^{(2\eta k_1)}\,\overline{A_p}\,e^{i2\sigma\tau} - D_{3pp}^{(2n_{T_1}k_1)}\,\overline{A_p}\,e^{i2\sigma\tau} - ic_p\omega_p A_p \\ + P_{1p}^{k_1}\,e^{i\sigma\tau} = 0. \tag{A14}$$

References

1. Chen, Z.; Shao, Y. Mesh stiffness calculation of a spur gear pair with tooth profile modification and tooth root crack. *Mech. Mach. Theory* **2013**, *62*, 63–74. [CrossRef]
2. Walha, L.; Fakhfakh, T.; Haddar, M. Nonlinear dynamics of a two-stage gear system with mesh stiffness fluctuation, bearing flexibility and backlash. *Mech. Mach. Theory* **2009**, *44*, 1058–1069. [CrossRef]
3. Kahraman, R.S. Interactions between time-varying mesh stiffness and clearance non-linearities in a geared system. *J. Sound Vib.* **1991**, *146*, 135–156. [CrossRef]
4. Theodossiades, S.; Natsiavas, S. Non-linear dynamics of gear-pair systems with periodic stiffness and backlash. *J. Sound Vib.* **2000**, *229*, 287–310. [CrossRef]
5. Kahraman, R.S. Non-linear dynamics of a spur gear pair. *J. Sound Vib.* **1990**, *142*, 49–75. [CrossRef]
6. Gregory, R.W.; Harris, S.L.; Munro, R.G. Dynamic behavior of spur gears. *Proc. Int. Mech. Eng.* **1963**, *178-1*, 207–218. [CrossRef]
7. Harris, S.L. Dynamic loads on the teeth of spur gears. *Proc. Inst. Mech. Eng.* **1958**, *172*, 87–112. [CrossRef]
8. Parker, R.G.; Vijayakar, S.M.; Imajo, T. Nonlinear dynamic response of a spur gear pair: modeling and experimental comparisons. *J. Sound Vib.* **2000**, *237*, 435–455. [CrossRef]
9. Liu, G.; Parker, R.G. Nonlinear dynamics of idler gear systems. *Nonlinear Dyn.* **2008**, *53*, 345–367. [CrossRef]
10. Lin, J.; Parker, R.G. Mesh stiffness variation instabilities in two-stage gear systems. *J. Vib. Acoust.* **2002**, *124*, 68–76. [CrossRef]
11. August, R.; Kasuba, R. Torsional vibrations and dynamic loads in a basic planetary gear system. *J. Vib. Acoust. Stress Reliab. Des.* **1986**, *108*, 348–353. [CrossRef]
12. Parker, R.G.; Agashe, V.; Vijayakar, S.M. Dynamic response of a planetary gear system using a finite element/contact mechanics model. *J. Mech. Des.* **2000**, *122*, 305–311. [CrossRef]

13. Velex, P.; Flamand, L. Dynamic response of planetary trains to mesh parametric excitations. *J. Mech. Des.* **1996**, *118*, 7–14. [CrossRef]
14. Maria, F.C.; Stefano, Z. Passive control of vibration of thin-walled gears: advanced modelling of ring dampers. *Nonlinear Dyn.* **2014**. [CrossRef]

© 2019 by the authors. Licensee MDPI, Basel, Switzerland. This article is an open access article distributed under the terms and conditions of the Creative Commons Attribution (CC BY) license (http://creativecommons.org/licenses/by/4.0/).

Article

Study on the Dynamic Characteristics of a Hydraulic Continuous Variable Compression Ratio System

Jiadui Chen *, Bo Wang, Dan Liu and Kai Yang

Key Laboratory of Advanced Manufacturing Technology, Ministry of Education, Guizhou University, Guiyang 550025, China; bwang-gzu@hotmail.com (B.W.); dliu@gzu.edu.cn (D.L.); kyang3@gzu.edu.cn (K.Y.)
* Correspondence: jdchen1@gzu.edu.cn

Received: 11 September 2019; Accepted: 21 October 2019; Published: 23 October 2019

Abstract: Variable compression ratio (VCR) technology has long been recognized as a method for improving the engine performance, efficiency, and fuel economy of automobiles, with reduced emissions. In this paper, a novel hydraulic continuous VCR system based on the principle of an adjustable hydraulic volume is introduced. The continuous variable compression ratio of the VCR system is realized by the hydraulic system controlling the rotation of the eccentric pin to change the positions of the top dead center (TDC) and the bottom dead center (BDC). The construction of the mathematical model and simulation model of the VCR system is also presented in this paper. The piston motion characteristics, flow characteristics, and pressure characteristics of the hydraulic system of the VCR system at different engine speeds and adjustment quantities are studied by simulation in this paper. The simulation results show that the VCR system has a fast response and good dynamic characteristics, and can achieve continuous adjustment of the compression ratio.

Keywords: variable compression ratio; adjustable hydraulic volume; mathematical model; dynamic characteristics; simulation

1. Introduction

Due to the increasingly severe energy crisis and environmental pollution [1], increasingly stringent emission and fuel economy standards have been proposed by many governments [2], which make improving the efficiency of engines an important topic. In China, the limit of fuel consumption will be 5 L per 100 km up to 2020 [3]. In the United States and Japan, passenger vehicles will have to achieve 23 km/L on average from 2025 [4]. Additionally, in Europe, the CO_2 emission limit will be 60 g per kilometer from 2025 [5]. The major challenges in the automotive industry are improving the engine efficiency by reducing the fuel consumption and reducing the engine emissions to meet the emission standards. Researchers have found that variable compression ratio (VCR) technology can provide further degrees of freedom to optimize the engine performance for various operating conditions. At low power levels, the engine operates at a higher compression ratio to capture the benefits of a higher thermal efficiency, while at high power levels, the engine operates at a lower compression ratio to prevent knocking. By continuously changing the compression ratio, an engine enables the optimum combustion efficiency to be obtained at all engine speed and load conditions, resulting in a better engine performance, lower fuel consumption, and lower exhaust emissions [6–8].

Owing to the advantages of VCR, researchers and manufacturers have devoted much effort to studying VCR technology. According to a survey, thousands of patents all over the world have been published, and more than 120 different kinds of VCR mechanisms have appeared, since 2000, and the numbers have kept increasing rapidly from 2013 up to the present date [3,8]. Forced Evolution Virus (FEV) has developed a VCR engine in which the crankshaft bearing is carried in an eccentrically mounted carrier that can rotate to raise or lower the top dead center (TDC) positions of the pistons in the cylinder. The compression ratio of FEV's VCR engine is adjustable by varying the rotation of the eccentric carrier [9,10]. Nissan uses a multi-link system to achieve VCR by inserting a control linkage system between the connecting rod and the crankshaft, and connecting this to an actuator shaft [11,12]. The Saab VCR engine achieves VCR by dynamically modifying the cylinder head position [13]. Ford has patented a means of varying the combustion chamber volume by using a secondary piston or valve [14]. Gomecsys has proposed a VCR engine in which moveable crankpins form an eccentric sleeve around the conventional crankpins and are driven by a large gear [15]. Honda has patented a VCR engine in which the deck height of the piston is varied [16]. Hiyoshi et al. have proposed a VCR engine in which the compression ratio could vary by adjusting the position of the added triangle that totally changes the piston movement process [17]. Kadota et al. have invented a VCR engine with a dual-piston system in which the compression ratio is varied by moving the TDC position with a hydraulic system and spring [18]. Kleeberg et al. have realized VCR by moving the TDC of the piston by adjusting the phase of the eccentric carrier inserted between the small-end of the connecting rod and the piston with a hydraulic system placed on both sides of the connecting rod [19].

The idea of varying the compression ratio CR is almost as old as the combustion engine itself. Over the years, various means have been proposed to realize VCR. However, there are many problems in existing VCR solutions. For example, most of the VCR solutions require comprehensive changes to existing engine architectures, few systems can realize the VCR continuously, and most of the VCR systems are too complex [4,13,20]. Hence, a novel hydraulic continuous VCR system based on the principle of an adjustable hydraulic volume is proposed in this paper. The hydraulic continuous VCR system can change the top and bottom dead center positions by the cubic capacity of the volume regulator, thereby allowing the compression ratio to be continuously changed. Due to the dynamic characteristics of the hydraulic system being the key factors affecting the adjustment performance of the hydraulic continuous VCR system, they are studied by establishing a mathematical model and simulation model.

2. Structure and Working Principle of the VCR System

2.1. Structure of the VCR System

Figure 1 shows the structure of the connecting rod. The connecting rod is composed of eccentric piston pin suspension, a piston pin, a connecting rod shank, a connecting rod cap, a left link, a left cylinder, a right link, a right cylinder, a left globe valve, a right globe valve, and so on. The connecting rod small end is equipped with an eccentric sleeve which houses the piston pin. The cylindrical surface of the eccentric sleeve is matched with the hole of the connecting rod small end and thus the eccentric sleeve can rotate in the hole of the connecting rod small end. The eccentric piston pin suspension is in splined connection with eccentric sleeve. The eccentric piston pin suspension is connected to the left link of left cylinder and the right link of right cylinder respectively by means of articulated connection. Therefore, the eccentric sleeve would be rotated by changing the positions of the left cylinder and the right cylinder. Thereby the effective connecting rod length and thus the compression ratio can be varied by rotating the eccentric sleeve. Figure 2 is a schematic diagram of the hydraulic continuous VCR system. Figure 2 shows that the hydraulic system of the continuous VCR system is composed of a low-pressure oil source, high-pressure oil source, check valve, solenoid valve, left globe valve, right globe valve, left cylinder, right cylinder, volume regulator solenoid valve, volume regulator, and so on.

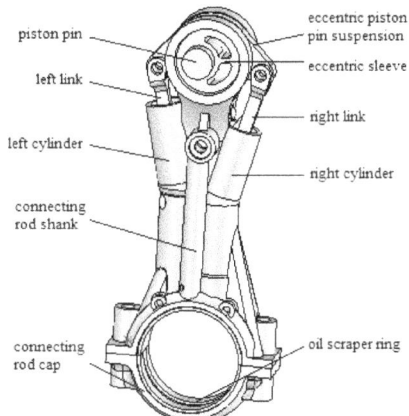

Figure 1. Structure of the connecting rod.

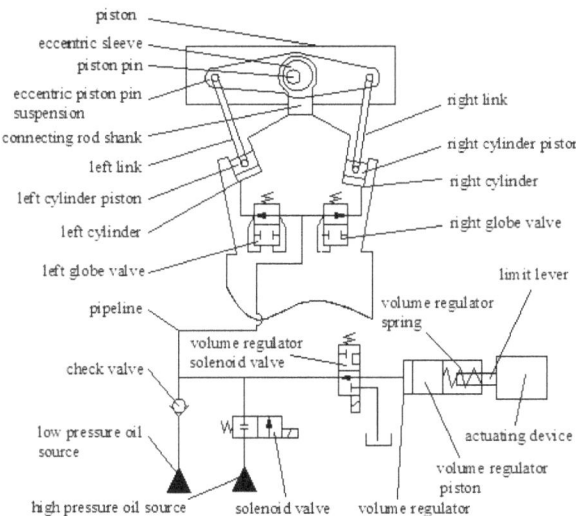

Figure 2. Schematic diagram of the hydraulic continuous variable compression ratio (VCR) system.

2.2. Working Principle of the VCR System

In the hydraulic continuous VCR system, compression ratio adjustment is realized by varying the position of the piston pin installed in the rotating eccentric sleeve that is controlled by the left cylinder and the right cylinder. As a result, the compression ratio adjustment completely relies on forces and moments related to movement of the piston for actuation and no external energy is necessary. The working principle of the hydraulic continuous VCR system is as shown below.

When it is necessary to adjust the compression ratio of the cylinder, the limit lever should be adjusted to a reasonable position by the actuating device, according to the speed and load of the engine. When the cylinder of the engine is running to the expansion stroke, the high-pressure oil circuit is disconnected by closing the solenoid valve of the high-pressure oil source. The system is supplied by the low-pressure oil source. At the same time, the solenoid valve of the volume regulator is turned on. The oil in the oil circuit connecting the volume regulator, the left globe valve, and the right globe valve flows into the volume regulator through the solenoid valve of the volume regulator, and the pressure of the connecting oil circuit then decreases. When the pressure of the connecting oil circuit is lower than the closing pressure of the right globe valve, the right globe valve is turned on. At the same time, due to the action of the high gas pressure on the surface of the engine piston, the oil pressure in the left cylinder is still very high. As a result, the left globe valve that is controlled by the pressure of the oil circuit is also closed. As the cylinder runs to the exhaust stroke, the gas pressure on the piston surface of the cylinder gradually decreases. When the engine runs to the later stage of the exhaust stroke and the early stage of the inlet stroke of the cylinder, the oil pressure of the left cylinder will be lower than that set by the left globe valve, and the left globe valve will be turned on. At this stage, the inertia force of the engine piston, piston pin, etc., which can move relative to the engine-connecting rod, will point in the upward direction shown in the figure. Under the combined action of the inertia force and the force of the oil pressure on the left piston, the center of the piston pin moves upward, which drives the eccentric piston pin suspension to rotate clockwise, as shown in Figure 3a. The clockwise rotation of eccentric piston pin suspension drives the left piston to move upward in the left cylinder through the left link, and the volume of the left cylinder gradually increases. On the contrary, the right piston moves downward in the right cylinder under the action of clockwise rotation of the eccentric piston pin suspension, so that the volume of the right cylinder will gradually decrease. Since the diameter of the left cylinder is larger than that of the right cylinder and the change of volume of the left cylinder is also greater than that of the right cylinder, all the oil flowing out of the right cylinder flows into the left cylinder. The same occurs for the left cylinder, so all the oil flowing out of the volume regulator flows into the left cylinder until the piston of the volume regulator returns to the front of the volume regulator by the spring force. While the pressure of the system is less than that of the low-pressure oil source, the low-pressure oil source supplies oil to the system through the check valve, and makes the system pressure the same as its oil pressure. As the right piston moves to the bottom of the right cylinder, the left piston reaches its highest position and stops moving. At the same time, the low-pressure oil source stops supplying oil to the system and the check valve is closed. A closing system with the largest volume is formed by the left cylinder, the right cylinder, the volume regulator, and their connecting circuits. The same maximum volume condition will be reached each time the compression ratio is adjusted.

When the cylinder runs to the later stage of the inlet stroke and the early stage of the compression stroke, the inertia force direction of the piston, piston pin, and other parts whose movements are relative to the engine piston becomes downward. Under the combined action of the inertia force and the force of oil pressure on the left piston and the right piston, the center of the piston pin moves upward, which drives the eccentric piston pin suspension to rotate counterclockwise, as shown in Figure 3b. The counterclockwise rotation of eccentric piston pin suspension drives the left piston to move downward in the left cylinder through the left link, and the volume of the left cylinder will gradually decrease. On the contrary, the right piston moves upward in the right cylinder under the action of the counterclockwise rotation of the eccentric piston pin suspension, so that the volume of the right cylinder will gradually increase. However, the volume variation of the left cylinder is larger than that of the right cylinder; that is to say, only a part of the oil flowing out of the left cylinder can flow into the right cylinder. Therefore, the oil pressure of the system increases gradually. When the system pressure is greater than the pre-tightening force of the volume regulator spring, the piston of the volume regulator is pushed into the regulating rod by the oil. The piston of the volume regulator stops moving after it makes contact with the limit lever, and the movement of the left piston and the right piston stops at the same time. The solenoid valve of the volume regulator is turned off. Then, medium in the left cylinder and the right cylinder becomes incompressible so that the eccentric piston pin suspension no longer rotates, as shown in Figure 3c. The oil in the volume regulator flows back to the tank through the solenoid valve of the volume regulator and the piston of the volume regulator returns to the front of the volume regulator by the spring force.

From above, the stroke of the volume regulator piston can be controlled by adjusting the position of the limit lever properly, according to the required compression ratio. This means that the oil volume flowing into the volume regulator cylinder can be controlled. The volume of oil is equal to the sum of the volumes of all hydraulic components in a closed hydraulic system. Therefore, when the cubic capacity of the volume regulator is determined, the volume of the left cylinder and the volume of the right cylinder are also determined because of their linkage and constant flow difference. Accordingly, the stop positions of the left piston and the right piston are determined, and the position of the eccentric piston pin suspension is also determined. Furthermore, the distance between the piston pin and the hole of the connecting rod end is determined too. The effective length of the connecting rod can be adjusted in this way, so the compression ratio of the engine is adjusted. As the cylinder compression stroke continues, the pressure of gas acting on the piston surface continues to rise, and the left piston will continue to move downward, which makes the system pressure rise rapidly. The left globe valve closes when the system pressure is greater than the set pressure of the left globe valve and this is maintained until the next compression ratio adjustment. The right globe valve is the same as the left globe valve. At the same time, the solenoid valve of the high-pressure oil source is turned on, which connects the high-pressure oil source and the connecting oil circuit. The supply of the high-pressure oil source ensures that the left globe valve and the right globe valve are always closed.

If the compression ratio needs to be adjusted again, the above process only needs to be repeated.

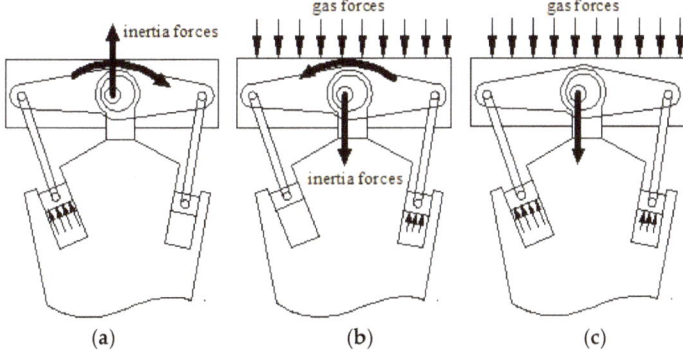

Figure 3. Schematic diagram of the adjustment process. (**a**) First stage of adjustment; (**b**) second stage of adjustment; (**c**) adjustment completion.

3. Mathematical Modeling

According to the working principle of VCR, the eccentric piston pin suspension rotates around the center of the eccentric sleeve. As shown in Figure 4, the eccentric piston pin suspension, the left cylinder, and the right cylinder form two crank slider mechanisms by using the engine-connecting rod as the reference for motion. L0 is the eccentric size, L1 is the left arm length of eccentric piston pin suspension, L2 is the left link length, L3 is the right arm length of eccentric piston pin suspension, L4 is the right link length, θ is the angular displacement of eccentric piston pin suspension, α is the angular displacement of the left link, β is the angular displacement of the right link, δ is the installation angle of the left cylinder, φ is the installation angle of the right cylinder, F is the force of the engine piston acting on the eccentric piston pin suspension, F_{L1} is the force of the left link acting on the eccentric piston pin suspension, F_{L2} is the force of the left link acting on the piston of left cylinder, F_{LP} is the force of oil pressure on the piston of the left cylinder, F_{R1} is the force of the right link acting on the eccentric piston pin suspension, F_{R2} is the force of the right link acting on the piston of the right cylinder, and F_{RP} is the force of oil pressure on the piston of the right cylinder.

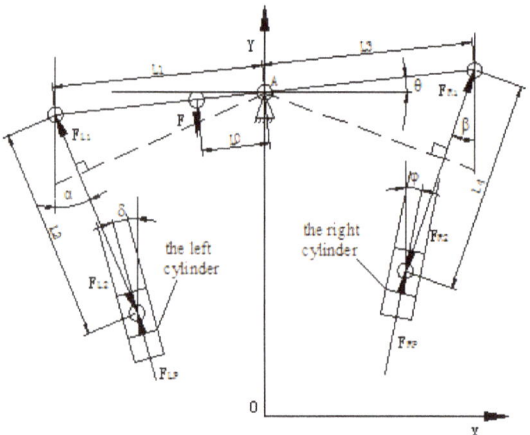

Figure 4. Schematic diagram of force.

According to the working principle of the VCR system and Figure 4, the system equations are derived one by one, as follows.

The piston displacement of the left cylinder s_{dL} can be expressed as

$$s_{dL} = \sqrt{(L1\cos\theta - L2\sin\alpha)^2 + (L1\sin\theta + L2\cos\alpha)^2} \tag{1}$$

The piston displacement of the right cylinder s_{dR} is expressed as

$$s_{dR} = \sqrt{(L3\cos\theta - L4\sin\beta)^2 + (L3\sin\theta + L4\cos\beta)^2} \tag{2}$$

The equilibrium equation of eccentric piston pin suspension can be expressed as

$$F \times L0\cos\theta + F_{R1} \times L3\cos(\beta + \theta) - F_{L1} \times L2\cos(\alpha - \theta) = J\frac{d^2\theta}{dt^2} \tag{3}$$

where J is the moment of inertia of eccentric piston pin suspension.

The equilibrium equation of the left cylinder piston is expressed as

$$F_{L2}\sin(\alpha - \delta) - F_{LP} - m_{LP}\frac{d^2 s_{dL}}{dt^2} - c\frac{ds_{dL}}{dt} = 0 \tag{4}$$

where c is the damping coefficient, m_{LP} is the mass of the left cylinder piston, δ is the installation angle of the left cylinder, $F_{L2} = -F_{L1}$, $F_{LP} = \frac{\pi d_L^2}{4} p_L$, d_L is the diameter of the left cylinder piston, and p_L is the operating pressure of the left cylinder.

The equilibrium equation of the right cylinder piston is expressed as

$$F_{R2}\sin(\beta - \varphi) - F_{RP} - m_{RP}\frac{d^2 s_{dR}}{dt^2} - c\frac{ds_{dR}}{dt} = 0 \tag{5}$$

where c is the damping coefficient, m_{RP} is the mass of the right cylinder piston, φ is the installation angle of the left cylinder, $F_{R2} = -F_{R1}$, $F_{R2} = \frac{\pi d_R^2}{4} p_R$, d_R is the diameter of the right cylinder piston, and p_R is the operating pressure of the left cylinder.

The flow continuity equation in the left cylinder is as follows:

$$Q_L = \begin{cases} \frac{\pi d_L^2}{4}\frac{ds_{dL}}{dt} - \beta_e\left(V_{L0} - \frac{\pi d_L^2}{4}S_{dL}\right)\frac{dp_L}{dt} & \text{the left globe valve opened} \\ 0 & \text{the left globe valve closed} \end{cases} \quad (6)$$

where β_e is the elastic coefficient of the hydraulic medium and V_{L0} is the initial volume of the left cylinder.

The flow continuity equation in the right cylinder is as follows:

$$Q_R = \begin{cases} \frac{\pi d_R^2}{4}\frac{ds_{dR}}{dt} + \beta_e\left(V_{R0} + \frac{\pi d_R^2}{4}S_{dR}\right)\frac{dp_R}{dt} & \text{the right globe valve opened} \\ 0 & \text{the right globe valve closed} \end{cases} \quad (7)$$

where V_{R0} is the initial volume of the right cylinder.

The flow continuity equation in the volume regulator is as follows:

$$Q_R = \begin{cases} \frac{\pi d_V^2}{4}\frac{ds_{dV}}{dt} + \beta_e\left(V_{V0} + \frac{\pi d_V^2}{4}S_{dV}\right)\frac{dp_V}{dt} & \text{the solenoid valve opened} \\ 0 & \text{the solenoid valve closed} \end{cases} \quad (8)$$

where s_{dV} is the piston displacement of the volume regulator, V_{V0} is the initial volume of the volume regulator, d_V is the diameter of the volume regulator piston, and p_V is the operating pressure in the volume regulator.

The equilibrium equation of the volume regulator piston is as follows:

$$\frac{\pi d_V^2}{4}p_V - m_V\frac{d^2 s_{dV}}{dt^2} - c\frac{ds_{dV}}{dt} - F_{V0} - k_{V0}s_{dV} = 0 \quad (9)$$

where F_{V0} is pre-tightening force of volume regulator spring, and k_{V0} is the stiffness of volume regulator spring.

The flow continuity equation of the check valve is as follows:

$$Q_C = \begin{cases} A_{cd}\sqrt{\frac{2\times(0.25-p_L)}{\rho}}, & p_L < 0.25 \text{ and the left globe valve opened} \\ 0, & p_L > 0.25 \text{ or the left globe valve closed} \end{cases} \quad (10)$$

where A_{cd} is the effective flow area of the check valve and ρ is the density of the hydraulic medium.

System flow balance equation:

According to the working principle of VCR, the system flow balance equation can be divided into two cases. The first case is when the resultant force is upwards, and the flow of the left cylinder is equal to the total of the flow of the right cylinder and the flow of check valve:

$$Q_L = Q_R + Q_C \quad (11)$$

Another case is when the resultant force is downwards, and the flow of the left cylinder is equal to the total of the flow of the right cylinder and the flow of the volume regulator:

$$Q_L = Q_R + Q_V \quad (12)$$

4. Simulation Model and Simulation Results

4.1. Simulation Model

Advanced Modeling Environment for performing Simulation of engineering system (AMESim) was selected as the simulation software. The simulation model was established by the AMESim software according to the working principle and construction of VCR. Figure 5 is the simulation model of VCR. The compressibility of hydraulic oil and pressure loss existing in the physical system are considered in the simulation model. The planar mechanical library of AMESim was used for the modeling of the mechanical structure of VCR. Additionally, the hydraulic library of AMESim was used for the modeling of the hydraulic system of VCR. In the simulation model, the variable hydraulic orifices are used to simulate the hydraulic valve, including the solenoid valve, left globe valve, right globe valve, and volume regulator globe valve, in the physical model. The switch of all variable hydraulic orifices is controlled by the signal of the sensors. The low-pressure oil source is simulated by a constant-pressure oil source. The high-pressure oil source is simulated by a subsystem that consists of a fixed displacement hydraulic pump, pressure relief valve, and so on. An electromotor is used to simulate the rotation of the engine driving the rotation of the crank axle. The motion between the cylinder and piston of the engine is simulated by a driven prismatic pair. Furthermore, the force of the engine piston is applied by the piecewise linear signal source. The parameter setting of the simulation model should refer to the physical system, and the simulation parameters are shown in Table 1.

Table 1. Simulation parameters.

Item	Parameter
Diameter of left cylinder piston (mm)	15.4
Diameter of right cylinder piston (mm)	13.8
Diameter of volume regulator piston (mm)	10
Diameter of volume regulator rod (mm)	5
Diameter of pipeline (mm)	6
Stroke of left cylinder (mm)	16
Stroke of right cylinder (mm)	16
Initial displacement of left cylinder piston (mm)	11.7
Initial displacement of right cylinder piston (mm)	4.3
Initial displacement of volume regulator piston (mm)	0
Initial pressure of right cylinder (bar)	10
Initial pressure of left cylinder (bar)	10
Initial pressure of volume regulator (bar)	0
Initial pressure of pipeline (bar)	2.5
Density of hydraulic medium (kg/mm^3)	850×10^{-9}
Effective flow area of check valve (mm^2)	10
Maximum flow coefficient	0.7
Pre-tightening force of volume regulator spring (N)	35
Stiffness of volume regulator spring (N/mm)	1000
Hydraulic oil elastic modulus (bar/mm^2)	7
Mass of crank axle (kg)	2
Mass of engine connecting rod (kg)	0.4
Mass of eccentric piston pin suspension (kg)	0.07
Length of left arm of eccentric piston pin suspension (mm)	26
Length of right arm of eccentric piston pin suspension (mm)	26
Length of pipeline (m)	0.3
Eccentric size (mm)	4
Pressure of low pressure oil source (bar)	2.5

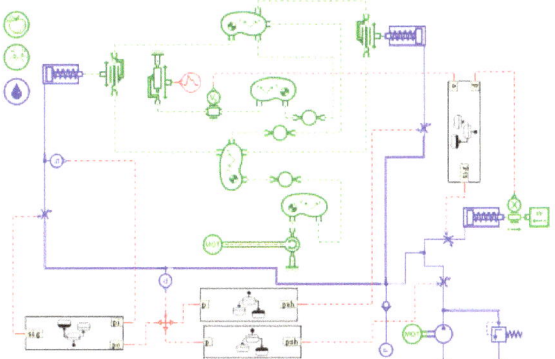

Figure 5. Advanced Modeling Environment for performing Simulation of engineering system AMESim simulation model of variable compression ratio (VCR) system.

4.2. Simulation Results

In order to study the dynamic characteristics of the hydraulic system of VCR, the adjustment process of different engine speeds and different regulating quantities was simulated. According to the working principle and construction of VCR, the initial displacement of the left cylinder piston, the initial displacement of the right cylinder piston, and the initial displacement of the volume regulator piston were set to 11.7 mm, 4.3 mm, and 0 mm, respectively. The simulation results are shown below.

Figure 6 presents the simulation results of left cylinder piston displacement and right cylinder piston displacement at different engine speeds with the adjustment of 3 mm. From Figure 6, we know that the displacement of the left cylinder piston increases gradually to the stroke of the left cylinder, and then decreases gradually, in contrast to the left piston, where the displacement of the right piston first decreases to 0 and then increases gradually. It can also be seen that the greater the engine speed, the greater the crank angle required to complete the adjustment process. There is a fluctuation of piston displacement before the piston position is stable, and the fluctuation increases with the engine speed. There are the same adjustment effects of piston displacement at 800 r/min, 1200 r/min, and 1500 r/min. Once the piston position is determined, the compression ratio is also determined according to the structure of the VCR system. Therefore, 800 r/min, 1200 r/min, and 1500 r/min can have the same compression ratio at the same adjustment quantity. However, the piston displacement at 2000 r/min is greater than that at other engine speeds. In other words, when the engine speed is higher than a certain speed, i.e., 1500 r/min, the adjustment effect of the VCR system will be affected by the engine speed and it cannot obtain the same compression ratio as the lower engine speed. In order to ensure effective adjustment of the compression ratio, the VCR system must work under a certain engine speed, i.e., 1500 r/min. From Figure 6, we can also see that the adjustment of the piston position at the different engine speeds is completed before 180°, which enables the VCR system to achieve adjustment of the compression ratio in one engine duty cycle.

Figure 7 presents the simulation results of the left cylinder flow and right cylinder flow at the adjustment of 3 mm under different engine speeds. Figure 7 shows that both the inflow and outflow of the left cylinder are larger than those of the right cylinder at each engine speed, with the same adjustment. The figure also shows that the inflow of the left cylinder is greater than the outflow of the left cylinder at each engine speed, with the same adjustment, and the outflow of the right cylinder is smaller than the inflow of the right cylinder at each engine speed, with the same adjustment. This is due to the diameter of the left cylinder being larger than that of the right cylinder and the linkage between the two cylinders. It can be also seen that the flow increases with the engine speed at the same adjustment quantity. The higher the engine speed, the less time it takes for the engine to rotate the same angular displacement. However, the flowing volume caused by the same adjustment quantity is the same. Therefore, the higher the engine speed, the higher the flow.

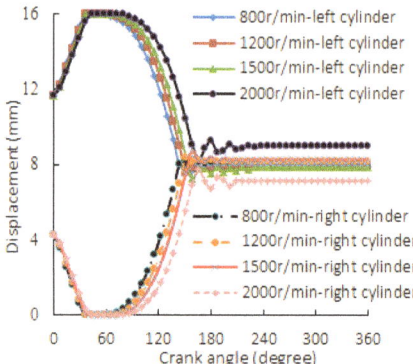

Figure 6. Displacement of the cylinder piston at different engine speeds with the same adjustment quantity.

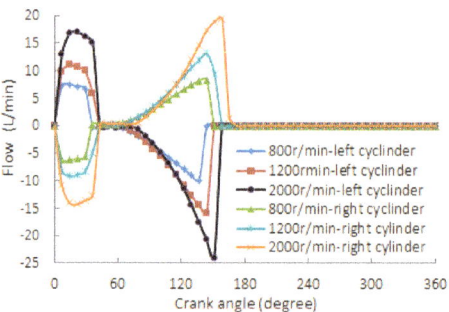

Figure 7. Flow of cylinders at different engine speeds with the same adjustment quantity.

The simulation results in Figure 8 show that the pressure drops from the initial pressure of the left cylinder and right cylinder to the same pressure as the system, and then gradually increases. When the system pressure reaches the closing pressure of the variable hydraulic orifices, the variable hydraulic orifices close. The pressure of the left cylinder and the pressure of the right cylinder rise sharply with fluctuation after the variable hydraulic orifices close. The simulation results indicate that the pressure fluctuation at a high engine speed is more severe than that at a low engine speed. We can also see that the cylinder pressure increases with the engine speed, and the maximum pressure of the left cylinder reaches 1800 bar at 2000 r/min, while the maximum pressure of the right cylinder is close to 2000 bar. It is necessary to take measures to reduce the pressure in the left cylinder and right cylinder to ensure the working reliability of the system.

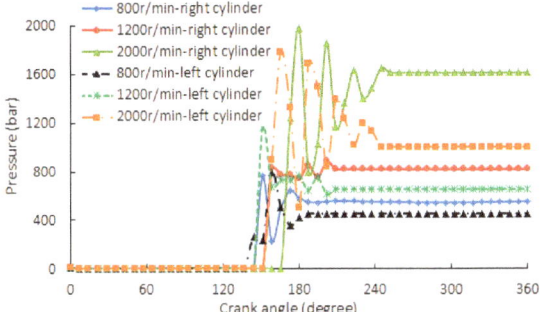

Figure 8. Pressure of the left cylinder at different engine speeds with the same adjustment quantity.

Figure 9 presents the simulation results of left cylinder piston displacement and right cylinder piston displacement at 800 r/min, with different adjustment quantities. From the figure, we can see that the displacement of the left cylinder piston increases gradually to the stroke of the left cylinder, and then decreases gradually, and in contrast to the left cylinder piston, the displacement of the right cylinder piston first decreases to 0 and then increases gradually. All piston displacement curves are almost similar before the piston stops moving. The displacement of the left cylinder piston decreases with the increase of the displacement adjustment of the volume regulator piston. However, the displacement of the right cylinder piston increases with the increase of the displacement adjustment of the volume regulator piston. According to the structure of the VCR system, if the positions of the left cylinder piston and right cylinder piston are changed, the compression ratio of the engine will be changed. The adjustment of the piston position at the different adjustment quantities is completed before 180°. Therefore, the VCR system can achieve adjustment of the compression ratio in one engine duty cycle, which greatly ensures the reliable operation of the engine. In sum, the VCR system can change the compression ratio of the engine by adjusting the maximum displacement volume regulator piston.

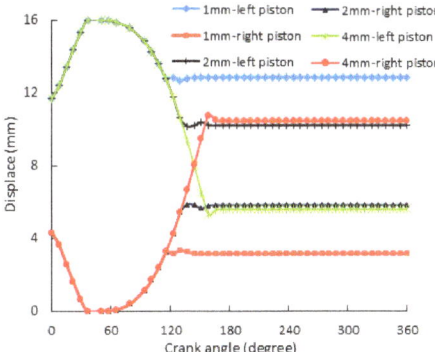

Figure 9. Displacement of the cylinder piston at 800 r/min with different adjustment quantities.

The simulation results in Figures 10 and 11 show the flow changes with the change of displacement adjustment of the volume regulator piston at 800 r/min. There are the same flows for the left cylinder, the right cylinder, and the check valve at different adjustment quantities before a 60° crank angle. However, the flow of the volume regulator is zero during this period. The flow of the left cylinder is equal to the sum of the flow of the right cylinder and the flow of the check valve before a 60° crank angle. As the adjustment quantity is increased, both the duration angle and the flow of the medium are increased when the crank angle is more than 60°. During the period, the flow of the check valve is zero and the flow of the left cylinder is equal to the sum of the flow of the right cylinder and the flow of the volume regulator. More hydraulic medium flows out of the left cylinder with the increase of adjustment quantity, which increases the flow of each cylinder and prolongs the flow time of the medium.

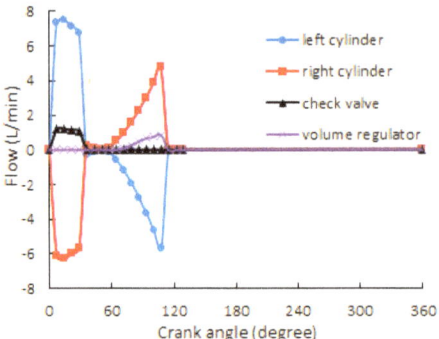

Figure 10. Flow of 1 mm adjustment at 800 r/min.

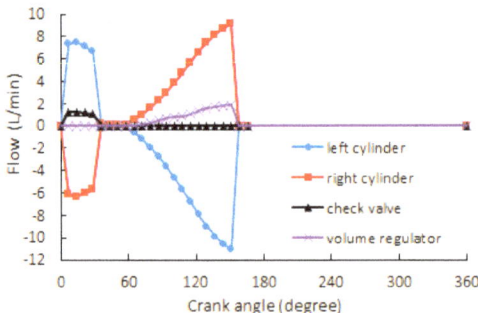

Figure 11. Flow of 4 mm adjustment at 800 r/min.

Figure 12 shows the simulation results of the pressure in the left cylinder and the right cylinder with different displacement adjustment of the volume regulator piston at 800 r/min. The pressure drops from the initial pressure of the left cylinder and right cylinder to the same pressure as the system. When the piston of the left cylinder moves downward, the system pressure rises gradually. The pressure fluctuation in the left cylinder is strengthened as the adjustment increases after the variable hydraulic orifices close. The pressure fluctuation of the right cylinder is opposite to that of the left cylinder. There is no regularity in the stable pressure of the left cylinder and the right cylinder. The stable pressure of 1 mm adjustment is smaller than that of 2 mm adjustment, but larger than that of 4 mm adjustment.

In summary, this novel hydraulic continuous VCR system can continuously adjust the compression ratio of the engine in one duty cycle of the engine, as long as the volume regulator is properly adjusted according to the engine speed.

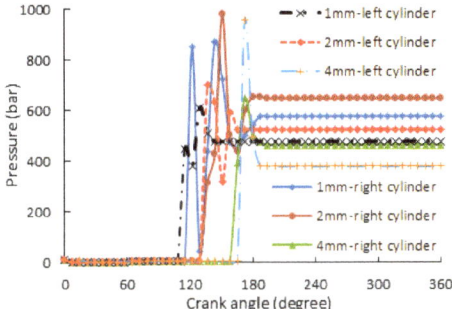

Figure 12. Pressure of the cylinders at 800 r/min with different adjustment quantities.

5. Conclusions

Various approaches of VCR technology have been realized. However, up to now, no VCR engines have been found in series production. One reason for this might be the fact that most of the VCR solutions require comprehensive changes to existing engine architectures. A novel hydraulic continuous VCR system based on the principle of an adjustable hydraulic volume is introduced in this paper. The mathematical model of the novel VCR system, with an engine-connecting rod as the reference material, is established, which clearly states the working principle of the VCR system from the theory.

An AMESim simulation model of the novel VCR system was produced in this study. The dynamic characteristics of the hydraulic system of the VCR system were comprehensively studied using the simulation model. The simulation results show that the hydraulic system of the VCR system has good adjusting dynamic characteristics. The left cylinder piston and the right cylinder piston have their respective identical and defined positions at different engine speeds with the same adjustment quantity under the low engine speed. The final position of the left cylinder piston decreases as the adjustment increases, but the right cylinder piston reverses. Both the flow of the left cylinder and the flow of the right cylinder increase with the engine speed and adjustment quantity. The pressure fluctuation and stable pressure of the cylinders also increase with the engine speed after the variable hydraulic orifices close. However, there is no regularity in the stable pressure of the left cylinder and the right cylinder.

The simulation results also show that the novel hydraulic continuous VCR system can achieve continuous adjustment of the compression ratio of an engine.

Author Contributions: Conceptualization and methodology, J.C. and B.W.; investigation, J.C., B.W., D.L., and K.Y.; resources, J.C. and B.W.; writing—original draft preparation, J.C. and B.W.; writing—review and editing, J.C., B.W., D.L., and K.Y.; Funding acquisition, J.C.

Funding: This research was funded by the Foundation of Guizhou Educational Committee (KY [2017]106) and Guizhou University talent fund (no. 2015-50).

Acknowledgments: We gratefully acknowledge the support of Wang (Ziqin Wang), who helped guide this research.

Conflicts of Interest: The authors declare no conflicts of interest.

References

1. Dhananjay, K.S.; Avinash, K.A. Combustion characteristics of a variable compression ratio laser-plasma ignited compressed natural gas engine. *Fuel* **2018**, *214*, 322–329.
2. Feng, D.Q.; Wei, H.Q.; Pan, M.Z. Comparative study on combined effects of cooled EGR with intake boosting and variable compression ratios on combustion and emissions improvement in a SI engine. *Appl. Therm. Eng.* **2018**, *131*, 192–200. [CrossRef]
3. Yang, S.; Lin, J.S. A theoretical study of the mechanism with variable compression ratio and expansion ratio. *Mech. Based Des. Struct. Mach.* **2018**, *46*, 267–284. [CrossRef]

4. Hoeltgebaum, T.; Simoni, R.; Martins, D. Reconfigurability of engines: A kinematic approach to variable compression ratio engines. *Mech. Mach. Theory* **2016**, *96*, 308–322. [CrossRef]
5. Westerloh, M.; Twenhövel, S.; Koehler, J.; Schumacher, W. Worldwide Electrical Energy Consumption of Various HVAC Systems in BEVs and Their Thermal Management and Assessment. *SAE Tech. Pap. Ser.* **2018**. [CrossRef]
6. Caio, H.R.; Janito, V.F. Kinematics of a variable stroke and compression ratio mechanism of an internal combustion engine. *J. Braz. Soc. Mech. Sci. Eng.* **2018**, *40*, 476–489.
7. Jiang, S.; Smith, M.H. Geometric Parameter Design of a Multiple-Link Mechanism for Advantageous Compression Ratio and Displacement Characteristics. *SAE Tech. Pap. Ser.* **2014**, *1*. [CrossRef]
8. Karsten, W.; Frank, G.; Jakob, A.; Mario, M.; Vitor, C.; Thompson, L. Experimental investigation of a variable compression ratio system applied to a gasoline passenger car engine. *Energy Convers. Manag.* **2019**, *183*, 753–763.
9. Asthana, S.; Bansal, S.; Jaggi, S.; Kumar, N. A Comparative Study of Recent Advancements in the Field of Variable Compression Ratio Engine Technology. *SAE Tech. Pap. Ser.* **2016**, *1*. [CrossRef]
10. Mane, P.; Pendovski, D.; Sonnen, S.; Uhlmann, A.; Henaux, D.; Blum, R.; Sharma, V. Coupled Dynamic Simulation of Two Stage Variable Compression Ratio (VCR) Connecting Rod Using Virtual Dynamics. *SAE Int. J. Adv. Curr. Prac. Mobil.* **2019**, *1*. [CrossRef]
11. Shelby, M.H.; Leone, T.G.; Byrd, K.D.; Wong, F.K. Fuel Economy Potential of Variable Compression Ratio for Light Duty Vehicles. *SAE Int. J. Engines* **2017**, *10*, 817–831. [CrossRef]
12. Kojima, S.; Kiga, S.; Moteki, K.; Takahashi, E.; Matsuoka, K. Development of a New 2L Gasoline VC-Turbo Engine with the World's First Variable Compression Ratio Technology. *SAE Tech. Pap. Ser.* **2018**. [CrossRef]
13. Romero, C.A.; Castañeda, E.D.J.H. Developing Small Variable Compression Ratio Engines for Teaching Purposes in an Undergraduate Program. *SAE Tech. Pap. Ser.* **2019**. [CrossRef]
14. Shaik, A.; Moorthi, N.S.V.; Rudramoorthy, R. Variable compression ratio engine: A future power plant for automobiles—An overview. *Proc. Inst. Mech. Eng. Part D J. Automob. Eng.* **2007**, *221*, 1159–1168. [CrossRef]
15. Wittek, K.; Geiger, F.; Andert, J.; Martins, M.; Oliveira, M. An Overview of VCR Technology and Its Effects on a Turbocharged DI Engine Fueled with Ethanol and Gasoline. *SAE Tech. Pap. Ser.* **2017**. [CrossRef]
16. Shi, H.; Al Mudraa, S.; Johansson, B. Variable Compression Ratio (VCR) Piston—Design Study. *SAE Tech. Pap. Ser.* **2019**. [CrossRef]
17. Hiyoshi, R.; Aoyama, S.; Takemura, S.; Ushijima, K.; Sugiyama, T. A Study of a Multiple-link Variable Compression Ratio System for Improving Engine Performance. *SAE Tech. Pap. Ser.* **2006**, *1*. [CrossRef]
18. Kadota, M.; Ishikawa, S.; Yamamoto, K.; Kato, M.; Kawajiri, S. Advanced Control System of Variable Compression Ratio (VCR) Engine with Dual Piston Mechanism. *SAE Int. J. Engines* **2009**, *2*, 1009–1018. [CrossRef]
19. Kleeberg, H.; Tomazic, D.; Dohmen, J.; Wittek, K.; Balazs, A. Increasing Efficiency in Gasoline Powertrains with a Two-Stage Variable Compression Ratio (VCR) System. *SAE Tech. Pap. Ser.* **2013**, *1*. [CrossRef]
20. Wolfgang, S.; Sorger, H.; Loesch, S.; Unzeitig, W.; Huettner, T.; Fuerhapter, A. The 2-Step VCR Conrod System—Modular System for High Efficiency and Reduced CO_2. *SAE Tech. Pap. Ser.* **2017**, *1*. [CrossRef]

© 2019 by the authors. Licensee MDPI, Basel, Switzerland. This article is an open access article distributed under the terms and conditions of the Creative Commons Attribution (CC BY) license (http://creativecommons.org/licenses/by/4.0/).

Article

Efficient Driving Plan and Validation of Aircraft NLG Emergency Extension System via Mixture of Reliability Models and Test Bench

Zhengzheng Zhu [1], Yunwen Feng [1], Cheng Lu [1] and Chengwei Fei [2,*]

1. School of Aeronautics, Northwestern Polytechnical University, Xi'an 710072, China
2. Department of Aeronautics and Astronautics, Fudan University, Shanghai 200433, China
* Correspondence: cwfei@fudan.edu.cn

Received: 23 July 2019; Accepted: 19 August 2019; Published: 1 September 2019

Abstract: The emergence extension system (a mechanical system) of nose landing gear (NLG) seriously influences the reliability, safety and airworthiness of civil aircrafts. To efficiently realize the NLG emergence extension, a promising driving plan of emergence extension is proposed in respect of the reliability sensitivity analyses with a mixture of models. The working principle, fault tree analysis and four reliability models are firstly discussed for NLG emergence extension. In respect of the mixture of models, the reliability sensitivity analyses of emergence extension are then performed under different flight speeds (270 Kts, 250 Kts, 220 Kts, and 180 Kts). We find dimpling torque and aerodynamic torques of forward and after doors are the top three failure factors and the start reliability is the most in emergence extension failures. Regarding the results, feasible driving plans of NLG emergence extension are developed by adjusting the aerodynamic torque of NLG forward door, and are validated by the aerodynamic torque experiment of forward door with regard to strut rotational angle under the flight speed 270 Kts. It is indicated that (1) the adverse torque generated by the new driving mechanism obviously reduces by about 24.8% from 1462.8 N·m to 1099.6 N·m, and the transmission ratio of aerodynamic torque (force) is greatly improved when the NLG strut is lowered near to 100°; (2) under different flight speeds (180 Kts, 220 Kts, 250 Kts, and 270 Kts), the new driving mechanism realizes the lower tasks of emergence extension which cannot be completed by the initial driving mechanism; and (3) the lowering time of the new driving mechanism shortens with the increasing flight speed. The proposed new driving mechanism is verified to be reliable for emergence extension of aircraft NLG besides normal extension and to be a promising feasible driving plan with high lowering reliability. The efforts of the paper provide an efficient driving mechanism for the design of NLG in civil and military aircrafts.

Keywords: landing gear; emergency extension; reliability sensitivity analysis; driving mechanism; mixture of models

1. Introduction

Landing gear is one of the key systems in aircrafts, which are used in take-off, landing and ground operation [1–4]. The landing gear of civil aircrafts includes both normal retractable system and emergency extension system. To improve the safety of aircraft landing, both Federal Aviation Regulation Part 25 (FAR 25) [5] and China Civil Aviation Regulations Part 25 (CCAR–25–R4) [6] stipulated that aircrafts must have an emergency measure to lower landing gear when its function fails normally. During the past decade, approximately 68 flight accidents happened for the related civil aircrafts such as Boeing, Airbus and other aircrafts, owing to the failure of landing gear system [7,8]. Obviously, the reliability of the retraction system seriously influences the safety of landing.

However, landing gear retraction is a sophisticated system, which further comprises many sub-systems like the mechanical, hydraulic, and control devices [9–11]. Hence, it is difficult to simulate the real conditions of emergency extension in engineering. Currently, the analysis of retraction system involves three categories of methods. (i) Theoretical analysis method performs the dynamic response analysis of emergency extension with a mathematical model. Owing to the complexity of both equations and ensuing calculations, the corresponding mathematical models are often simplified by compromising the accuracy of calculations. (ii) Test method acquires the accurate results approximate to the real conditions. However, this method is excessively time-consuming and costly. (iii) Simulation method accurately analyzes the integrated system of landing gear emergency extension through reasonable equivalence. The simulation analysis is generally facilitated with a co-simulation model of all the mechanical, hydraulic and control systems.

In fact, scholars have carried out a lot of works in this area. For instance, Chang et al. [12] analyzed the reliability of the steering mechanism of nose landing gear (NLG) by combining the dynamic simulation model and first-order second-moment method with an artificial neural network. Choi et al. [13] investigated the operational dynamic behaviors of a T-50 landing gear system using ADAMS and discussed the effects of temperature, aerodynamics and maneuver load on normal/emergency operation of landing gears and doors. Yin et al. [14] conducted the fault analysis of the emergency lowering of an aircraft NLG by co-simulation method and then obtained the dynamic response characteristics of the landing gear retraction system. Meanwhile, Yin et al. [15] established the dynamics model of landing gear with the limit state equation to analyze the influence of key parameters on the reliability of the retraction system. McClain et al. [16] introduced the improved landing gear system with the efforts involving the design, test and integration of components and the failure investigation. Öström et al. [17] discussed the co-simulation of two models including commercial-off-the-shelf software and low-cost flight simulation model for landing gear. Lin et al. [18] proposed the deployment and locking theory of mechanism and analyzed the reliability of a landing gear system from qualitative and quantitative perspectives, by constituting a fault tree. Zhang et al. [19] examined the dynamic performance of a landing gear system using ADAMS, during retraction (extension) operations under various flight speeds and hydraulic fluid temperatures. Chen et al. [20] presented an efficient prognostic tool based on stochastic filtering-based method, to predict the failure time of the landing gear retraction system. However, these mentioned efforts hardly involve the efficient reliability model and the sensitivity of different torques to the emergency extension of landing gear, which is significant for precisely designing the sophisticated emergency extension system of landing gear. The main reasons are that in engineering, the reliability of emergency extension is synthetically affected by start failure, continuous movement failure, movement precision failure, and static strength failure. In this case, it is urgent to develop an integrated reliability approach by considering many reliability models (i.e., starting reliability model, continuous movement reliability model, movement precision reliability model, and static strength reliability), for the comprehensive reliability sensitivity analyses of the emergency extension system.

To address this issue, this paper will systematically analyze the emergency extension fault of the NLG under the actual loading conditions. Firstly, the working principle of detailed emergency extension and failure modes are introduced. Secondly, the safety boundary equation and variables distribution of the reliability model are determined to establish the reliability models comprising starting reliability model, continuous movement reliability model, movement precision reliability model, and static strength reliability. In respect of the starting reliability model, the calculation approach of failure probability is introduced in detail under different flight speeds. Next, the sensitivity analysis of different torques to the emergency extension of landing gear is conducted to find the torques with high influence level. Finally, a promising design plan is explored to solve the problem of emergency extension fault and improve the reliability of NLG emergence extension for aircrafts, in respect of experimental validation.

The rest of the paper is organized as follows. In Section 2, the reliability models of NLG emergence extension are discussed including working principle, fault tree analysis, reliability approach, and mixture of models. The reliability sensitivity analyses of emergence extension are performed to find the failure probabilistic of emergence extension and the effect levels of bottom events in the fault tree model in Section 3. Section 4 proposes and validates efficient driving plans (driving mechanisms) by the experiment of NLG emergence extension. Conclusions and findings are summarized in Section 5.

2. Reliability Modeling of NLG EmergencyExtension

For civil aircrafts, the retraction (extension) mechanism seriously influences the safety and airworthiness of flights and the reliability of the landing gear system, which was strictly required in the relevant clauses in FAR 25 and FAR 25.729 (c) [5,6]. Therefore, it is urgent to provide the emergency measures to ensure the reliable lowering and lock of the NLG. The experiments on the effects of aerodynamic force on NLG emergency extension were conducted under four flight speeds (i.e., 270 Kts, 250 Kts, 220 Kts, and 180 Kts) during the flight-test program of an aircraft, in which Kts is the unit of flight speed. The test results show that the NLG could not lower to the designed angle degree and did not possess the reliable lowing function required. To meet the requirements of airworthiness clauses, it is necessary to address the emergency extension fault of civil aircrafts.

2.1. Working Principle of Emergency Extension

Landing gear generally consists of a retractable actuator, lock mechanism, shock absorber strut, cabin door driving mechanism, and so forth. The doors and retraction mechanism of landing gear are mechanically linked. In the case of an emergency, the emergency extension crank is rotated by the emergency extension cable mechanism to overcome the spring force of the lock and to realize the unlocking. Subsequently, under gravity and other external forces, the shock absorber strut of the landing gear is lowered to a specific position to lock the lower-lock, so that the emergency extension process is completed. In line with the function and geometrical analyses of emergency extension, the normal operation of the NLG mechanism is correlated with landing gear gravity, door gravity, aerodynamic force, damping force, unlocking force, and locking spring force. Figure 1 shows the structure diagraph of NLG emergency extension.

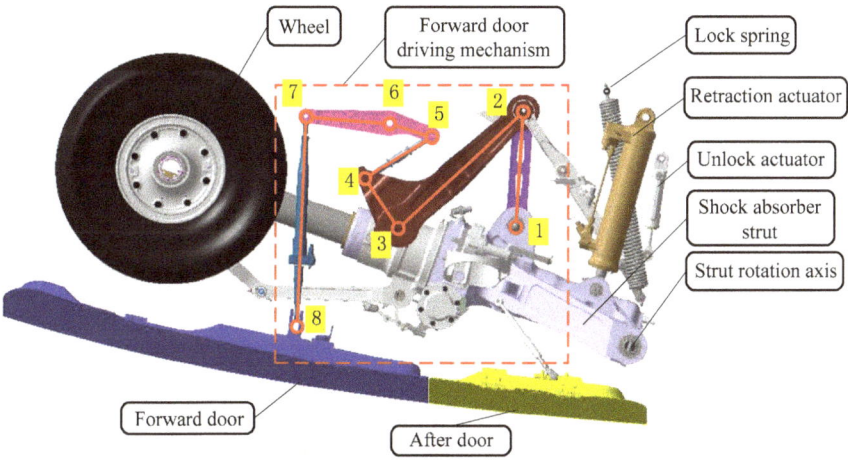

Note: points (1,2, ... , 8) indicate the connecting points between structures

Figure 1. Structure diagraph of emergency extension.

In respect to Figure 1, the process of emergence extension is described as follows. The landing gear is lowered and rotated around the strut rotation axis. When the landing gear is lowered, the shock absorber strut rotates around the strut rotation axis. The rocker arm 2–3 is driven by the connecting rod 1–2. The rocker arms 2–3 and 3–4 are rigid structures and move together. Then the connecting rod 5–6 isdriven by the connecting rod 4–5. The rocker arm 6–7 moves with the rocker 5–6, and the rocking arm 7–8 is carried by the connecting rod 6–7. In this time, the NLG is lowered and the forward door of the landing gear is opened. Obviously, the parts 1–2–3–4–5–6–7–8 together constitute the driving mechanism of the forward door. For the complex driving mechanism analysis of the forward door, the driving mechanism was reduced as a planar four-link mechanism [21] and spatial four-link mechanism [22] for movement reliability analyses with acceptable accuracy. The two simplifications models will be applied to the reliability of movement precision in the next work.

2.2. Fault Tree Analysis of Emergency Extension

During the process of emergency extension, it must be ensured that the NLG always rotates around the strut rotation axis. The torques of all components are induced by external loads and impact on the emergency extension. These torques are listed in Table 1.

Table 1. Explanations of torques.

Symbol	Meaning
M_1	Torque on strut rotation axis produced by the gravity of NLG
M_2	Torque on strut rotation axis produced by the gravity of door
M_3	Torque on strut rotation axis produced by the aerodynamic force of strut and wheel
M_4	Torque on strut rotation axis produced by the spring force of locking mechanism
M_5	Torque on strut rotation axis produced by damping force
M_6	Torque on strut rotation axis produced by friction
M_7	Torque on strut rotation axis produced by after-door aerodynamic force
M_8	Torque on strut rotation axis produced by forward-door aerodynamic force
M_9	Torque on strut rotation axis produced by the unlocking force of upper-lock
M_T	Total torque

Table 1 shows the analysis of the torque produced by the forces acting on the strut during emergency extension. The active torques produced by landing gear mass force play a leading role in the early stages of emergency extension. In the later period of emergency extension, the aerodynamic resistance torque of landing gear plays a leading role. Owing to the linkage design of the landing gear door and landing gear, the aerodynamic force of forward door and after door produces a torque during the whole process of emergency extension. The spring torque of the lock mechanism is inevitable. The spring force should ensure the safety of the lock mechanism. The retractable actuator cylinder is fixed on the airframe, because the piston rod of the retractable actuator cylinder is connected and moves with the shock absorber strut in the process of lowering and rotation. Meanwhile, the hydraulic oil rapidly flows in the actuator cylinder and return pipeline, to generate the damping torque of emergency extension, which is determined by piston speed, return pipeline length, and so on. From the above analysis, we can summarize the main reasons inducing the failure of emergency extension T as follows.

1. Excessive resistance torque on strut rotation axis (denoted by X_1). The aerodynamic force of the landing gear door is too large for the driving mechanism to overcome the passive torque by the active torque of shock absorber strut, so that the landing gear cannot be lowered and locked.
2. Imprecise extension of landing gear (denoted by X_2). The landing gear cannot be lowered to a predetermined position owing to the motion accuracy of the mechanism.
3. Unlocked upper-lock. The upper-lock is unlocked (denoted by X_3).
4. The lower-lock is not locked (denoted by X_4).
5. Static strength failure of landing gear emergency extension (denoted by X_5).

In respect of the five fault reasons, the fault tree of emergency extension is established as shown in Figure 2, in which the five bottom events indicate five failure reasons in the fault tree. Obviously, the failure of one bottom event will lead to the failure of emergency extension.

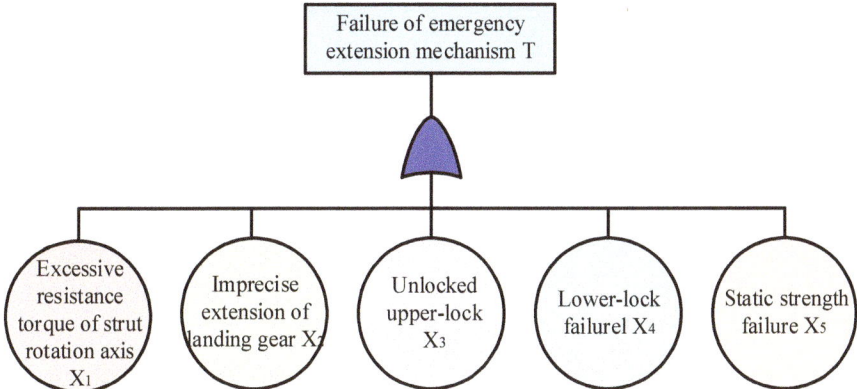

Figure 2. Fault tree of emergency extension.

2.3. Reliability Method and Mixture of Models

The reliability models of the emergency extension mechanism include the mechanism starting model, mechanism continuous movement model, mechanism movement precision model, and static strength model [23]. These reliability models can be established based on the fault tree analysis of the emergency extension mechanism and the mechanism reliability analysis theory [24]. For reliability analysis, the four reliability models actually have similar procedures. In this study, the starting reliability model is considered as a typical example to explain the reliability method. To start the mechanism from stationary to motion state, it is ensured that the driving torque M_d is larger than the resistance torque M_r, i.e.,

$$M = M_d - M_r > 0 \tag{1}$$

The starting failure probability P_f is defined as the probability that driving torque is less than resistance torque, i.e.,

$$P_f = \{M_d - M_r < 0\} \tag{2}$$

With regard to the distribution characteristics of driving torque and resistance torque, the starting reliability index of the mechanism is expressed as

$$\beta = \frac{\overline{M}_d - \overline{M}_r}{\sqrt{C_d^2 \overline{M}_d^2 + C_r^2 \overline{M}_r^2}} \tag{3}$$

where C_d is the variable coefficient of driving torque; C_r is the variable coefficient of resistance torque; and $\overline{M}_d$ and $\overline{M}_r$ denote the mean values of driving torque M_d and resistance torque M_r.

In terms of the standard normal cumulative distribution function $\Phi(\bullet)$ [25,26], the starting reliability R is

$$R = \Phi(\beta) \tag{4}$$

The starting failure probability P_f is

$$P_f = 1 - R = \Phi(-\beta) \tag{5}$$

In this case, the first-order second-moment method is used to calculate the starting reliability. The rest of the reliability models are not repeatedly respected. The reliability model, safety boundary equation, reliability index, and parameter distributions of the bottom event [23,27–29] are shown in Table 2.

Table 2. Reliability model and distribution of bottom events in fault tree.

No.	Bottom Events	Reliability Models	Safe Boundary Equation	Reliability Index	Distribution
1	Excess passive torque X_1	Starting reliability M	$M = M_d - M_r$	$\beta = \dfrac{\overline{M}_d - \overline{M}_r}{\sqrt{C_d^2 \overline{M}_d^2 + C_r^2 \overline{M}_r^2}}$	Normal
		Continuous movement reliability M_ω	$M_\omega = \omega - \omega^*$	$\beta = \dfrac{\overline{\omega} - \overline{\omega}^*}{\sqrt{C_\omega^2 \overline{\omega}^2 + C_{\omega*}^2 \overline{\omega}^{*2}}}$	Normal
2	Imprecise lowing X_2	Movement precision Reliability θ_0	$\theta_0 = \theta - \theta^*$	$\beta_\theta = \dfrac{\mu_\theta - \mu_{\theta*}}{\sqrt{V_{\theta*}^2 + V_\theta^2}}$	Normal
3	Unlocked upper-lock X_3				
4	Failed locked lower-lock X_4				
5	Static strength failure X_5	Static strength reliability	$M_s = S - L$	$\beta_M = \dfrac{\mu_S - \mu_L}{\sqrt{C_S^2 \mu_S^2 + C_L^2 \mu_L^2}}$	Normal

Note: $\overline{M}_d$—the mean of driving torque M_d; $\overline{M}_r$—the mean of resistance torque M_r; C_d—the variable coefficient of driving torque; C_r—the variable coefficient of resistance torque; $\overline{\omega}$—the mean of angular velocity ω; C_ω—the variable coefficient of angular velocity; $\overrightarrow{\omega^*}$—the mean of allowed angular velocity ω^*; $C_{\omega*}^2$—the variable coefficient of velocity or angular velocity; μ_θ—the mean of θ; $\mu_{\theta*}$—the mean of θ^*; V_θ—the standard deviation of θ; $V_{\theta*}$—the standard deviation of θ^*; μ_S—the mean of mechanism static strength; μ_L—the mean of mechanism load; C_S—the variable coefficient of S; C_L—the variable coefficient of L.

In terms of the above fault tree analysis, the relationship between bottom events and top events is "or" for emergency extension. All the bottom events are minimal cut sets. The occurrence probability P_T of the top event is equal to the sum of the probability of occurrence for all the minimal cut-sets. In this case, the failure probability of emergency extension is

$$P_T = \sum_{i=1}^{5} P(X_i) \tag{6}$$

where X_i ($i = 1, 2, \ldots, 5$) is ith bottom event in the fault tree; and $P(.)$ indicates the failure probability.

First, the failure probability of each bottom event with the reliability model is obtained by the related parameters, such as geometric sizes, material performance parameters, lowering speed, and so forth. With respect to Equation (6), the failure probability of the top event is obtained in the fault tree of emergency extension.

As for the bottom event X_1 (excess passive torque), the reliability models involve starting reliability M and continuous movement reliability M_ω. In the unlocking stage of emergency extension for the strut rotation angle in [0°, 2.66°], the starting reliability model is applied to compute the failure probability of the starting driving mechanism, while after unlocking (i.e., the strut rotation angle of landing gear excess 2.66°), the continuous movement reliability model is employed to evaluate the failure probability of lowering the driving mechanism. In this paper, the maximum failure probability in the whole motion process is regarded as the failure probability of the bottom event X_1.

3. Reliability Sensitivity Analysis of Emergence Extension

In this section, the reliability sensitivity analyses of the NLG emergence extension system comprising reliability analysis and sensitivity analysis are investigated to provide a reference for the development and validation of a feasible plan for reducing emergency extension failure and faults.

3.1. Reliability Analysis

3.1.1. Starting Reliability Analysis of Emergency Extension

As illustrated in Table 1, the amplitudes and directions of all the forces acting on the NLG and the corresponding torque on shock absorber strut rotation axis vary during the process of emergency extension. Therefore, it is necessary to carry out force balance analysis and reliability analysis in the whole process of emergency extension. To ensure that the NLG can be lowered in an emergency, the relationship between active torques and passive torques is investigated. The above nine forces act on strut rotation axis in the form of torque. The total torque is expressed as

$$M_T = \sum_{i=1}^{9} M_i = M_1 + M_2 + M_3 + M_4 + M_5 + M_6 + M_7 + M_8 + M_9 \tag{7}$$

Based on the starting reliability, the emergency extension is successful as $M_T > 0$, or the emergency extension fails. Obviously, the nine torques significantly influence the normal start-up of emergency extension. When $M_T < 0$, the emergency extension cannot be realized. It is necessary to further analyze the relationship between torque and strut angle. Regarding the flight speed 270 Kts, the changing trend of each torque and total torque with strut rotation angle during emergency extension are shown in Figure 3.

As shown in Figure 3, (1) all the torques vary with the strut rotation angle during the emergency extension. Merely, the torque on strut rotation axis M_9 stops to act on the emergence extension system after about 2.66 degrees (°) of landing gear angle, because the point at 2.66° of landing gear angle indicates that the upper-lock is unlocked to finish its mission, and its unlocking force on the emergence extension system does not exist after 2.66° of landing gear angle; (2) the torques M_1 and M_9 are conductive to the emergence extension in the whole process, and the torque M_3 lowers the emergence extension system in [0~2.66°] and stop work after 2.66°; (3) the torques M_5 and M_6 provide a resistance against emergence extension, in which M_5 is much larger than M_6, which is very small in the whole lowering process; (4) the torques M_2, M_4, M_7, and M_8 have both a positive and negative effect on emergence extension, in which M_2, M_7 and M_8 are firstly positive torques at 21.8°, 69.2° and 35.1°, respectively, and then negative while M_4 is firstly adverse and then beneficial for emergence extension; (5) the torque of M_T reaches the minimum at 2.66° (unlocking completion stage) and 100.1° (ready locking stage), which indicate the two critical points inducing the failure of the emergence extension system.

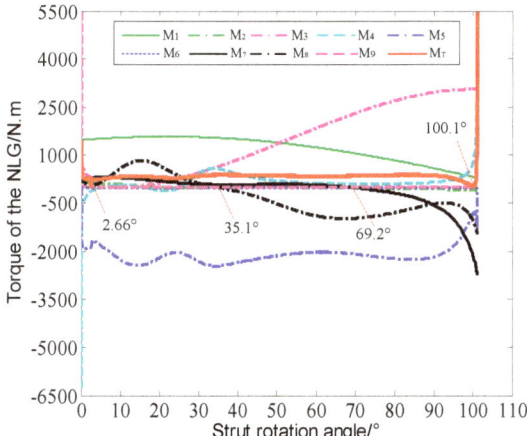

Figure 3. Change of NLG torques with strut rotation angle at 270 Kts.

By importing the nine torques and variable coefficients C_i ($i = 1, 2, \ldots, 9$) with strut rotation angle into Equation (3), the starting reliability index β can be computed by

$$\beta = \frac{\sum_{i=1}^{9} M_i}{\sqrt{\sum_{i=1}^{9} C_i^2 M_i^2}} \tag{8}$$

where i is the ith torque.

With respect to Equation (8), the failure probability can be obtained by the standard normal distribution function. Regarding the relevant engineering tests and empirical data [30–33], the variable coefficients of nine torques are listed in Table 3.

Table 3. Variable coefficients of nine torque.

Torques	M_1	M_2	M_3	M_4	M_5	M_6	M_7	M_8	M_9
Variable coefficient C_i	0.03	0.03	0.08	0.03	0.03	0.03	0.08	0.08	0.03

3.1.2. Reliability Analysis of Emergency Extension Based on Fault Tree

With regard to the reliability model, the failure probability P_{Xi} of ith bottom event X_i can be calculated quantitatively when the NLG emergency extension cannot be lowered at different flight speeds. According to engineering experiments, we find that the maximum failure of bottom event X_1 occurs at [0°, 2.66°]. Therefore, we only consider the starting failure probability of the emergence extension mechanism as the failure probability of bottom event X_1 in this paper. The results on the failure probability P_{Xi} under different flight speeds (270 Kts, 250 Kts, 220 Kts and 180 Kts) are shown in Table 4.

Table 4. Failure probability of bottom events in fault tree.

Failure Probability	Flight Speed			
	270 Kts	250 Kts	220 Kts	180 Kts
P_{X1}	1.2728×10^{-5}	2.2401×10^{-5}	4.0683×10^{-5}	6.1819×10^{-5}
P_{X2}		5.7905×10^{-8}		
P_{X3}		5.7026×10^{-8}		
P_{X4}		5.7026×10^{-8}		
P_{X5}		4.2655×10^{-6}		
P_{T1}	1.7165×10^{-5}	2.6838×10^{-5}	4.5120×10^{-5}	6.6256×10^{-5}

As shown in Table 4, the failure probability of the top event (emergence extension) increases with the decreasing flight speed. When the flight speed is at 180 Kts, the failure probability increases to the maximum 6.6256×10^{-5}. The main reason is that the strut aerodynamic torque of landing gear and the aerodynamic torques of forward and after doors reduce with the decline of the flight speed. In this condition, the total torque of NLG and the angular acceleration of the rotating axis drop. Therefore, when the emergency extension of NLG is put down, a large angular velocity is obtained for the rotating axis due to large flight speeds, besides smaller failure probability for emergency extension. In other words, low flight speed and angular velocity obtained against the rotating shaft will lead to larger failure probability of emergency extension.

From the above analysis, it is seen that the starting reliability is the main factor influencing the failure probability of emergency extension. Therefore, the sensitivities of the nine torques on starting reliability were discussed to determine the specific influence of torques on the start reliability of emergence extension system.

3.2. Sensitivity Analysis

To identify the main influencing factors on emergency extension fault, the sensitivity analysis is implemented under the flight speed of 270 Kts in this section.

3.2.1. Sensitivity Analysis of Torques

Sensitivity is defined by the ratio of change (or gradient) of failure probability to the change of random variables. According to the reliability index of the NLG emergency extension mechanism in Equation (3) and partial derivative, the sensitivity of the ith torque is

$$\frac{\partial \beta}{\partial M_i} = \frac{\sum_{i=1}^{9} C_i^2 M_i^2 - C_i^2 M_i \cdot \sum_{i=1}^{9} M_i}{\left(\sqrt{\sum_{i=1}^{9} C_i^2 M_i^2}\right)^3} \tag{9}$$

in which $\frac{\partial \beta}{\partial M_i}$ reflects the influence levels of the ith torque on the failure probability of emergency extension. The larger the variation of $\frac{\partial \beta}{\partial M_i}$ is, the larger the influence of the corresponding torque on reliability index and failure probability is.

By substituting the torque and the corresponding variable coefficient into Equation (9), the sensitivities of each torque to start reliability are obtained as shown in Figure 4. When the landing gear angle is 2.66°, the corresponding unlocking is completed, and then M_9 disappears. Therefore, the sensitivity of M_9 only exists in $[0°, 2.66°]$.

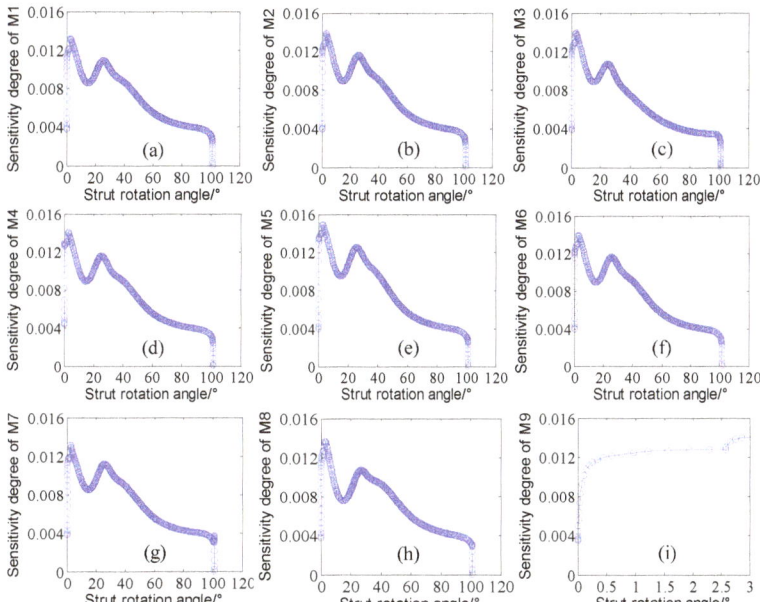

Figure 4. Sensitivities degree of nine torques to strut rotation angle: (**a**) Sensitivity degree of M_1; (**b**) Sensitivity degree of M_2; (**c**) Sensitivity degree of M_3; (**d**) Sensitivity degree of M_4; (**e**) Sensitivity degree of M_5; (**f**) Sensitivity degree of M_6; (**g**) Sensitivity degree of M_7; (**h**) Sensitivity degree of M_8; (**i**) Sensitivity degree of M_9.

Appl. Sci. 2019, 9, 3578

As displayed in Figure 4, the sensitivities of nine torques have different change trends irregularly with strut rotation angle. Therefore, it is difficult to specifically identify which torques are the main factors influencing the failure probability of emergency extension with the variation of strut rotation angle.

3.2.2. Effect of Main Torque on Failure Probability of Emergency Extension

To determine the significant torques on the failure probability of emergency extension, ranking method is adopted to vividly order by the sensitivities of torques to the failure probability of emergency extension with the changes of strut rotation angle. The ranking is shown in Figure 5.

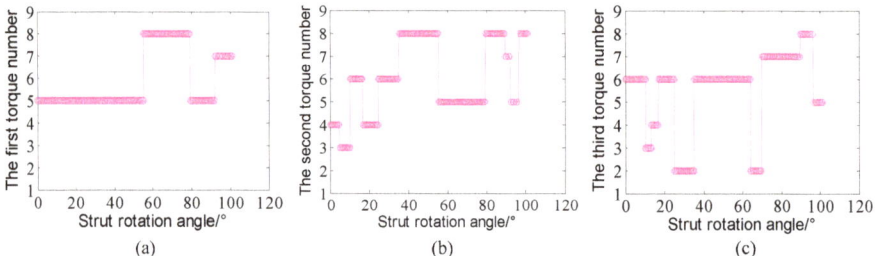

Figure 5. Influences of the top three torques on the failure probability of emergency extension under strut rotation angles: (**a**) the first torque number; (**b**) the second torque number; (**c**) the second torque number.

As revealed in Figure 5a, both the damping torque M_5 and the aerodynamic torques (M_7 and M_8) of after and forward door are the three largest influencing the failure probability of emergency extension. When the strut rotation angle changes in [0°, 55.1759°], M_5 has the largest factor affecting the failure probability of emergency extension. M_8 will take the place of M_5 and becomes the most influential torque when the strut rotation angle varies in [55.2889°, 79.2622°]. With the further increase of the strut rotation angle, M_5 and M_7 will become the most influential torques as the strut rotation angle in [79.5208°, 92.0710°] and [92.5061°, 101.1959°], respectively. Similarly, we can see the variations of the second and third torques with the strut rotation angle, as shown in Figure 5b,c. The ranking of torques is useful to guide the design of a feasible plan in emergence extension reliability improvement.

4. Proposal and Validation of New Driving Mechanism Plan

The foregoing sensitivity analyses shows that the damping torque and the aerodynamic torques of forward and after doors are the three main factors affecting the emergency extension failure. To address the engineering problem of emergence extension faults, two feasibility plans are proposed at the flight speed of 270 Kts in this paper.

Firstly, the NLG relies on kinetic energy to realize the emergency extension by reducing the damping coefficient of the actuating cylinder. Through the transient simulations of the angle variation of forward door with time under different damping coefficients of the actuating cylinder, the simulation results are drawn in Figure 6.

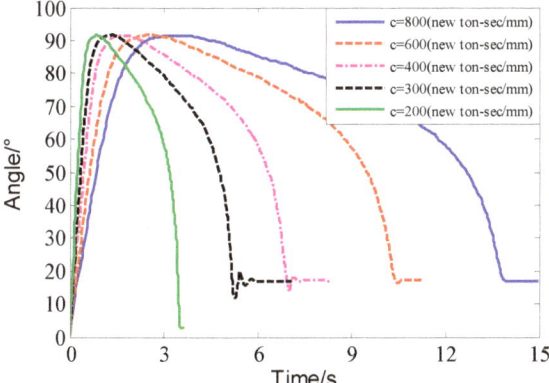

Figure 6. Angle change of forward door under different damping coefficients.

As demonstrated in Figure 6, (1) with the reduction of the damping coefficient, the lower NLG strut is faster, but the position at which the NLG strut finally stop remains unchanged; (2) the strut oscillates near the equilibrium position and has larger shock for smaller damping. When the damping coefficient drops below 200, the strut can be lowered and locked. In this case, however, landing gear have a larger impact on the aircraft since the lower strut only costs 3.75 s.

We verified the above simulation findings by ground tests. The test results show that it is unacceptable to break through the dead point with the assistance of the kinetic energy, by reducing the damping coefficient for the NLG. It is urgent to seek analternative plan.

Because the driving mechanism of the forward door is highly sensitive to aerodynamic force, considering the location selection and assembly requirements of the driving point of the landing gear strut, a new driving mechanism was proposed by selecting the actuating cylinder lug of landing gear as the driving point. The design principles include the below steps.

(1) The point of intersection between the main structure and mechanism of NLG keeps invariability.
(2) The main force-transferring path of the forward door structure is unchanged.
(3) The mechanism has strong capacity forresisting forward door load.

The new driving mechanism of forward door is design as depicted in Figure 7.

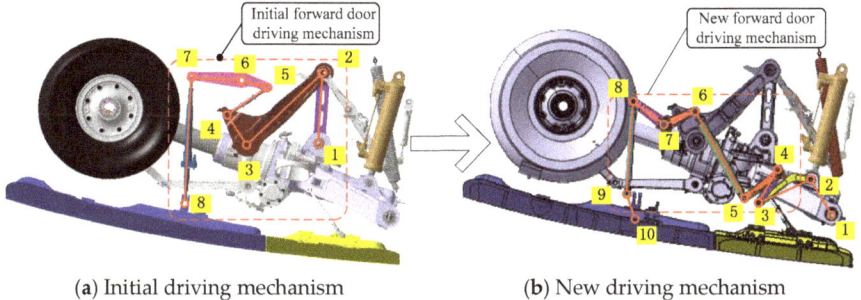

(a) Initial driving mechanism   (b) New driving mechanism

Figure 7. New driving mechanism and force transmission path of forward door.

As shown in Figure 8a, since the working principle of the emergency extension of the initial driving mechanism has been introduced in Section 2.1, we do not repeat it here. As shown in Figure 8b, the new driving mechanism of the forward door is linked to the shock absorber strut by two rocker arms and pull rods, in which the rotating axis of the rocker arm is fixed on the side plate of the landing gear door. Therefore, the switch of forward door can be driven by the shock absorber strut. When the landing gear is lowered, the shock absorber strut rotates around the strut rotation axis 1. The rocker arm 3–4 can be driven by the connecting rod 2–3. The rocker arms 3–4 and 4–5 are rigid and can move together. The rocker arm 6–7 is driven by the connecting rod 5–6. The rocker arm 7–8 moves with the rocker 6–7, and the rocking arm 9–10 is started by the effect of the connecting rod 8–9. Through the above transmission of torques from point 1 to point 10 under the driving force provided by the hydraulic system, the lowering function of NLG and the opening and closing of the landing gear forward door are realized. Obviously, the driving mechanism of the forward door consists of nine flexible manipulators (1–2, 2–3, 3–4, 4–5, 5–6, 6–7, 7–8, 8–9 and 9–10) and 10 driving pivots (1, 2, ... , 10). The working principle of the emergency extension of the initial and new driving mechanism is vividly shown in Figure 8.

To validate the effectiveness and feasibility of the new design plan with the failure reason studied in Section 3, the NLG system designed by the new plan of emergence extension is tested by the iron bird test rig. The sketch map of the iron bird test system is illustrated in Figure 9 where Figure 9a is the geometric model of the iron bird test rig, and Figure 9b indicates the test system of emergence extension.

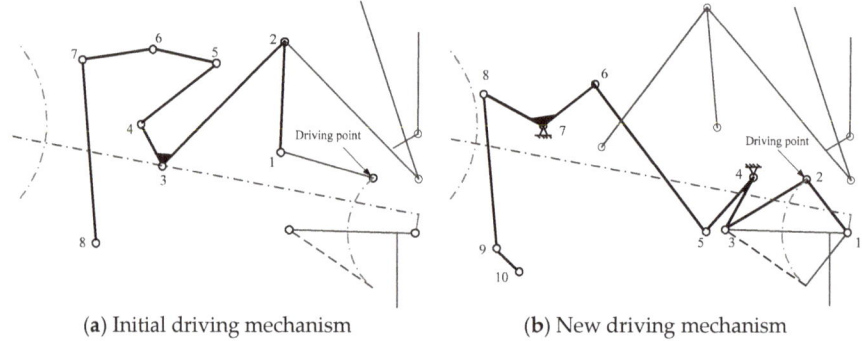

(a) Initial driving mechanism (b) New driving mechanism

Figure 8. Working principle of initial and new driving mechanism.

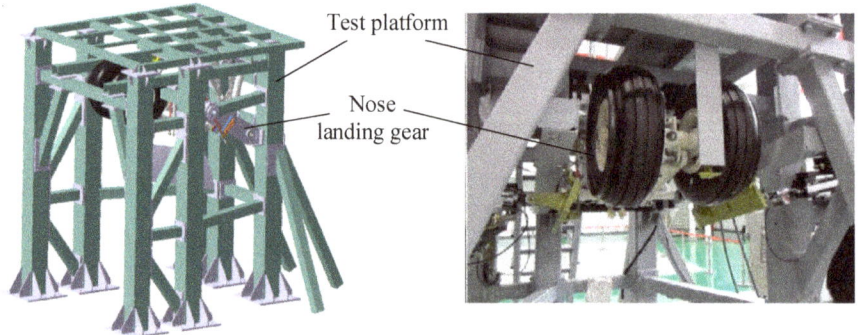

(a) Geometric model of iron bird test rig (b) Test equipment of emergency extension

Figure 9. Sketch map of iron bird test system.

Through the experiment of the aerodynamic torque of forward door for the new emergence extension system with regard to strut rotational angle under the flight speed 270 Kts, the test results of new and initial driving mechanisms including the change curve and transmission efficiency of aerodynamic torque with strut rotation angle are drawn in Figure 10. In Figure 10, the transmission ratio ζ [34] is calculated by

$$\zeta = \frac{T_2}{T_1} = \frac{d\gamma}{d\alpha} \tag{10}$$

where T_1 is the aerodynamic torque of the landing gear forward door; T_2 is the required lowering torque to overcome T_1; α indicates the strut rotation angle of NLG; and γ is the rotation angle of forward door a certain state.

On the premise of limited design space and guaranteeing transmission efficiency of the mechanism, the distribution of the transmission ratio directly affects the design size, weight, and transmission life of driving mechanism. Therefore, the transmission ratio is regarded to reflect the transmission performance of the driving mechanism.

As shown in Figure 10a, the adverse torque generated by the new driving mechanism is obviously smaller than that generated by the initial driving mechanism during the emergency extension. Compared with the initial driving mechanism, the adverse torque changes from −1462.8 N·m to −1099.6 N·m and thus reduces by about 24.8%. The reduction of the negative aerodynamic torque is promising to more easily open the forward door and improve the reliability of emergence extension, which enhances the landing safety of aircrafts.

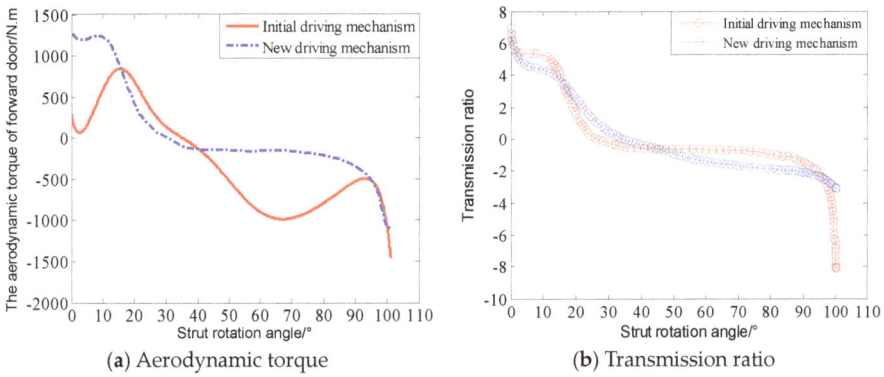

Figure 10. Test results of initial and new driving mechanisms at 270 Kts.

As indicated in Figure 10b, relative to the initial driving mechanism of the forward door, the transmission ratio of aerodynamic torque (force) for the new driving mechanism is greatly improved when the NLG strut is lowered near to 100°, because the transmission ratio of the new driving mechanism markedly decreases from −8.1 to −2.9 with respect of transmission ratio calculation. Therefore, the new driving mechanism can efficiently transmit the torques. The feasibility of the new design plan for emergence extension is verified again.

As seen from the above analysis, the emergency extension fault of NLG is mainly induced by unreasonable mechanism design and weak ability of withstanding aerodynamic loads. The weak ability may lead to abortively lock the landing gear for the forward door, resulting from the aerodynamic resistance of the forward door, which can be offset by the gravity and aerodynamic forces. The results show that the new mechanism has high transmission efficiency and transmission ratio although it is more complex than the initial driving mechanism. Moreover, the new driving mechanism can also slow down the closing speed of the forward door at the final stage in the lowering process of the landing gear. To support the rapidity of the new driving mechanism in opening the forward door, we tested the lowering time of the emergence extension and normal extension of aircraft NLG based on the new and initial forward door driving mechanisms, under different flight speeds simulated on the ground. The test results are listed in Table 5.

Table 5. The lowering time of the initial and new driving mechanism.

Driving Mechanisms	Flight Speed /Kts	Lowering Time of NLG/s	
		Emergency Extension	Normal Extension
Initial driving mechanism	270	Failed lowering and lock	10.69
	250	Failed lowering and lock	10.81
	220	Failed lowering and lock	11.02
	180	Failed lowering and lock	11.38
New driving mechanism	270	14.66	11.09
	250	15.24	11.25
	220	15.95	11.28
	180	17.17	11.61

As shown in Table 5, (1) for the normal extension, the lowering time of the new driving mechanism is about 11 s under different flight speeds (180 Kts, 220 Kts, 250 Kts and 270 Kts), which is slightly longer than that of the initial driving mechanism. The reason is that the new driving mechanism with nine arms has a more complex transmission path of force and torque, than the initial driving mechanism with seven arms. However, the complexity only has a big impact on the transmission time due to small increments in lowering time; (2) for emergence extension, the new driving mechanism realized the lower tasks under different flight speeds of 180 Kts, 220 Kts, 250 Kts and 270 Kts, at which the initial driving mechanism cannot be completed. Specifically, the emergence extension of the new driving mechanism costs 17.17 s, 15.95 s, 15.24 s and 14.66 s against 180 Kts, 220 Kts, 250 Kts and 270 Kts, respectively. In addition, it is indicated that the lowering time of the new driving mechanism shortens with the increasing flight speed.

Summarily, the proposed new driving mechanism can reliably lower the emergence extension system of aircraft NLG, besides the normal extension system. The proposed new driving mechanism is illustrated to be a promising feasible driving plan with high reliability.

5. Conclusions

The purpose of this paper is to propose an efficient driving plan (i.e., new driving mechanism) of emergence extension for aircraft nose landing gear (NLG) through reliability sensitivity analyses with a mixture of models, to address the emergence extension fault of the landing gear for the initial driving mechanism. Through the reliability sensitivity analyses of the initial driving mechanism and the validation of the proposed new driving mechanism, some major conclusions and findings are summarized as follows:

1. Through the reliability analysis of emergence extension, it is illustrated that the start reliability has the most failure probability in five reliability modes (starting reliability, continuous movement reliability, movement precision reliability, and static strength reliability), indicating that the event X_1 (excessive resistance torque on strut rotation axis) seriously influences the reliability of NLG emergence extension.
2. From the sensitivity analysis of NLG emergence extension, the effect levels of nine torques (M_1, M_2, ... , M_9, explained in Table 2) on the failure probability of NLG emergence extension are determined, and the sensitivity degrees of the torques M_5, M_8 and M_7 are the top three by the order by sensitivity degree for nine torques. The conclusion is promising guidance for the driving plan design of NLG emergence extension.
3. Two driving plans of NLG emergence extension are designed in this paper. One is to adjust the damping coefficients of the actuating cylinder, and the other is to the aerodynamics of forward door. Through the comparison and validation of the two plans, the second driving plan is acceptable, because this plan can reduce the adverse torque of emergence extension by about 24.8%, increase the transmission ratio of the driving mechanism, and address the emergence extension fault problem and reliably realize the emergence extension of aircraft NLG besides the

normal extension system. The proposed new driving mechanism is illustrated to be a promising feasible driving plan with high reliability. The developed driving mechanism is promising in the application of civil and military aircrafts, which supports the safety and airworthiness in flight.

Author Contributions: Conceptualization, C.-W.F.; Data curation, C.L.; Formal analysis, Z.-Z.Z.; Funding acquisition, Y.-W.F.; Investigation, C.L.; Methodology, Z.-Z.Z., C.L. and C.-W.F.; Project administration, Y.-W.F.; Resources, C.-W.F.; Software, Z.-Z.Z. and C.L.; Supervision, Y.-W.F. and C.-W.F.; Writing—original draft, Z.-Z.Z.; Writing—review & editing, C.-W.F.

Funding: This research was funded by [National Natural Science Foundation of China] grant number [51875465, 51605016 and 51975124], [Special Science and Technology Program of Civil Aircraft of China] grant number [MJZ-Y-2018-92], and [Research Start-up Funding of Fudan University] grant number [FDU38341]. The APC was funded by [FDU38341].

Conflicts of Interest: The authors declare that there is no conflict of interests regarding the publication of this article.

References

1. Krüger, W.R.; Morandini, M. Recent developments at the number simulation of landing gear dynamics. *CEAS. Aeronaut. J.* **2011**, *1*, 55–68. [CrossRef]
2. Infante, V.; Fernandes, L.; Freitas, M.; Baptista, R. Failure analysis of a nose landing gear fork. *Eng. Fail. Anal.* **2017**, *82*, 554–565. [CrossRef]
3. Platz, R.; Gotz, B.; Melz, T. Approach to evaluate and to compare basic structural design concepts of landing gears in early stage of development under uncertainty. *Model. Validation. Uncertain. Quan.* **2016**, *3*, 167–175.
4. Thoai, N.; Alexandra, S.; Paul, E. Method for analyzing nose landing gear during landing using structural finite element analysis. *J. Aircraft.* **2012**, *49*, 275–280.
5. FAR-25. *Federal Aviation Regulations Part 25: Transport Category Airplanes*; Federal Aviation Administration: Washington, DC, USA, 2011.
6. CCAR-25-R4. *China Civil Aviation Regulations Part 25: Transport Aircraft Airworthiness Standards*; Civil Aviation Administration of China: Peking, China, 2011.
7. Civil Aviation Safety. World Civil Aviation Accident date-base [EB/OL]. Available online: http//www.air-safety.net (accessed on 10 August 2019).
8. Flight Safety Foundation. Aviation Safe Network [EB/OL]. Available online: aviation-safety.net (accessed on 10 August 2019).
9. Rahmani, M.; Behdinan, K. On the effectiveness of shimmy dampers in stabilizing nose landing gears. *Aerosp. Sci. Technol.* **2019**, *91*, 272–286. [CrossRef]
10. Knowles, J.; Krauskopf, B.; Lowenberg, M. Numerical continuation analysis of a three-dimensional aircraft main landing gear mechanism. *Nonlinear Dyn.* **2013**, *71*, 331–352. [CrossRef]
11. Rankin, J.; Krauskopf, B.; Lowenberg, M.; Coetzee, E. Operational parameter study of aircraft dynamics on the ground. *J. Comput. Nonlinear Dyn.* **2010**, *5*, 021007. [CrossRef]
12. Chang, Q.C.; Xue, C.J. Reliability analysis and experimental verification of landing gear steering mechanism considering environmental temperature. *J. Aircraft.* **2018**, *53*, 1154–1164. [CrossRef]
13. Choi, S.; Kwon, H.B.; Chung, S.J.; Jung, C.R.; Sung, D.Y. An operational analysis and dynamic behavior for a landing gear system using ADAMS. *J. Korean. Soc. Aeronaut. Space Sci.* **2003**, *31*, 110–117.
14. Yin, Y.; Nie, H.; Wei, X.H.; Chen, H.; Zhang, M. Fault analysis and solution of an airplane nose landing gear's emergency lowering. *J. Aircraft.* **2016**, *53*, 1022–1032. [CrossRef]
15. Yin, Y.; Hong, N.; Huajin, N.; Ming, Z. Reliability analysis of landing gear retraction system influenced by multifactors. *J. Aircraft.* **2016**, *55*, 713–724. [CrossRef]
16. McClain, J.G.; Vogel, M.; Pryor, D.R.; Heyns, H.E. The United States Air Force's landing gear systems center of excellence—A unique capability. In Proceedings of the 2007 US Air Force T&E Days, Destin, FL, USA, 13–15 February 2007.
17. Öström, J.; Lähteenmäki, J.; Viitanen, T. F18 hornet landing simulations using ADAMS and Simulink co-Simulation. In Proceedings of the AIAA Modeling and Simulation Technologies Conference and Exhibit, Honolulu, HI, USA, 18–21 August 2008; pp. 2008–6850.

18. Lin, Q.; Nie, H.; Ren, J.; Chen, J.B. Investigation on design and reliability analysis of a new deployable and lockable mechanism. *Acta Astronaut.* **2012**, *73*, 183–192. [CrossRef]
19. Zhang, H.; Ning, J.; Schmelzer, O. Integrated landing gear system retraction/extension analysis using ADAMS. In Proceedings of the 2000 International ADAMS User Conference, Orlando, FL, USA, 19–21 June 2000.
20. Chen, J.; Ma, C.B.; Song, D. Multiple failure prognosis of landing gear retraction/extension system based on H filtering. *P. I. Mech. Eng. G J. Aer.* **2015**, *229*, 1543–1555. [CrossRef]
21. Ting, K.L.; Zhu, J.M.; Watkins, D. The effects of joint clearance on position and orientation deviation of linkages and manipulator. *Mech. Mach. Theory.* **2000**, *35*, 391–401. [CrossRef]
22. Jhuang, C.S.; Kao, Y.Y.; Chen, D.Z. Design of one DOF closed-loop statically balanced planar linkage with link-collinear spring arrangement. *Mech. Mach. Theory.* **2018**, *130*, 301–312. [CrossRef]
23. Song, L.K.; Fei, C.W.; Bai, G.C.; Yu, L.C. Dynamic neural network method-based improved PSO and BR algorithms for transient probabilistic analysis of flexible mechanism. *Adv. Eng. Inform.* **2017**, *33*, 144–153. [CrossRef]
24. Lu, C.; Feng, Y.W.; Liem, R.P.; Fei, C.W. Improved kriging with extremum response surface method for structural dynamic reliability and sensitivity analyses. *Aerospace Sci. Technol.* **2018**, *76*, 164–175. [CrossRef]
25. Zhang, C.Y.; Wei, J.S.; Jing, H.Z.; Fei, C.W. Reliability analysis of blisk low fatigue life with generalized regression extreme neural network method. *Materials* **2019**, *12*, 1545. [CrossRef] [PubMed]
26. Lu, C.; Feng, Y.W.; Fei, C.W.; Feng, X.X.; Choy, Y.S. Weighted regression-based extremum response surface method for structural dynamic fuzzy reliability analysis. *Energies.* **2019**, *12*, 1588. [CrossRef]
27. Liu, J.Y.; Song, B.F.; Zhang, Y.G. Competing failure model for mechanical system with multiple functional failures. *Adv. Mech. Eng.* **2018**, *10*, 1–16. [CrossRef]
28. Zhang, J.W.; He, S.H.; Wang, D.H.; Liu, Y.P.; Yao, W.B.; Liu, X.B. A new reliability analysis model of the chegongzhuang heat-supplying tunnel structure considering the coupling of pipeline thrust and thermal effect. *Materials* **2018**, *12*, 236. [CrossRef] [PubMed]
29. Zhao, Y.G.; Ono, T. A general procedure for first/second-order reliability method. *Str. Saf.* **1999**, *21*, 95–112. [CrossRef]
30. He, X.; Oyadiji, O. Application of coefficient of variation in reliability-based mechanical design and manufacture. *J. Mater. Process. Technol.* **2001**, *119*, 374–378. [CrossRef]
31. Michaael, D.; Desmond, F.; Ktrang, N. Assessment of the reliability of calculations of the coefficient of variation for normal and polymegethous human corneal endothelium. *Optometry. Vision. Sci.* **1993**, *70*, 759–770.
32. He, X.F.; Zhai, B.; Dong, Y.M.; Liu, W.T. Safe-life analysis accounting for the loading spectra variability. *Eng. Fail. Anal.* **2010**, *17*, 1213–1220. [CrossRef]
33. Zhang, J.H. *Guidelines for Structural Strength Reliability Design of Missiles and Launch Vehicles (Metal Structural Part)*; China Astronautic publishing house: Beijing, China, 1994.
34. Yao, M.Y.; Qin, D.T.; Zhou, X.Y.; Zhan, S.; Zeng, Y.P. Integrated optimal control of transmission ratio and power split ratio for a CVT-based plug-in hybrid electric vehicle. *Mech. Mach. Theory* **2019**, *136*, 52–71. [CrossRef]

© 2019 by the authors. Licensee MDPI, Basel, Switzerland. This article is an open access article distributed under the terms and conditions of the Creative Commons Attribution (CC BY) license (http://creativecommons.org/licenses/by/4.0/).

MDPI
St. Alban-Anlage 66
4052 Basel
Switzerland
Tel. +41 61 683 77 34
Fax +41 61 302 89 18
www.mdpi.com

Applied Sciences Editorial Office
E-mail: applsci@mdpi.com
www.mdpi.com/journal/applsci

www.ingramcontent.com/pod-product-compliance
Lightning Source LLC
LaVergne TN
LVHW071943080526
838202LV00064B/6664